THE GENETIC TESTING OF CHILDREN

THE GENETIC TESTING OF CHILDREN

Angus Clarke
Institute of Medical Genetics,
University of Wales College of Medicine,
Cardiff, UK

βIOS
SCIENTIFIC
PUBLISHERS

Oxford • Washington DC

A CIP catalogue record for this book is available from the British Library.

ISBN 1 85996 052 9

BIOS Scientific Publishers Ltd
9 Newtec Place, Magdalen Road, Oxford OX4 1RE, UK
Tel. +44 (0) 1865 726286. Fax. +44 (0) 1865 246823
WorldWide Web home page: http://www.bios.co.uk

Typeset by Saxon Graphics Ltd, Derby, UK

Contents

PART TWO: CARRIER TESTING IN CHILDHOOD

PART THREE: PREDICTIVE GENETIC TESTING

Contributors

Priscilla Alderson works in the Social Science Research Unit at the Institute of Education, University of London. She has researched parents' and children's consent to surgery and women's views about breast cancer treatment and research. She is currently the UK partner in a European Commission-funded programme on 'Prenatal Screening: Past, Present and Future 1996 – 1999'. Part of this programme involves interviews with adults with genetic and other congenital conditions, such as cystic fibrosis, sickle cell disease, thalassaemia, spina bifida and Down's syndrome.

Chris Barnes trained in general and psychiatric nursing and then studied social administration, gaining her BA in 1985. She then gained experience in nursing research, looking at ways in which nurse education could help to prevent or manage violent behaviour in clinical settings. She has worked for the past ten years in genetic counselling, and has a particular interest in familial chromosome rearrangements. She completed the Manchester University MSc course in Genetic Counselling in 1996, undertaking the research described in this volume as a part of that course.

Julia Binedell received a Master's degree in Research Psychology from the University of Cape Town, South Africa. Her research interests include community mental health care and applications of health psychology. She is currently a trainee on the South Wales doctoral course in Clinical Psychology.

Don Bradley trained originally in pharmacy and medicinal chemistry before turning to biochemistry and setting up the newborn screening programme for Wales. His research interests lie in expanding the concept of newborn screening, the effect of screening on families and the nature of informed consent to the screening process.

Lucy Brindle studied Genetics and Psychology at the University of Newcastle-upon-Tyne. Her postgraduate research has examined families' experiences of living with an adult-onset genetic disorder of the kidneys. She analyses the issues surrounding pre-natal, childhood and presymptomatic diagnosis of genetic disorders from a constructionist perspective. Her research interests include public understanding of genetics; family communication; 'risk culture'; discourse analysis and the genetic counselling process.

Peter Campion is Professor of Primary Care Medicine, University of Hull, and a GP principal in Hull. His main research interests are the doctor–patient interaction, and the nature of 'functional' illnesses. His involvement in the genetics project reported in Chapter 17 was through the analysis of the doctor–patient interactions which were recorded in the genetic clinics.

Alison Chapple is Research Associate (Medical Sociology) at the National Primary Care Research and Development Centre, University of Manchester. Her research interests include genetic counselling, women's reproductive health, patients' use of cardiac services and primary health care.

Angus Clarke is Reader in Clinical Genetics at the Department of Medical Genetics, University of Wales College of Medicine, Cardiff. He works as a consultant clinical geneticist and teaches undergraduate medical students. His research interests include the social and ethical issues raised by advances in genetics, as well as specific genetic disorders including Rett syndrome, ectodermal dysplasia and some neuromuscular disorders. He was chairman of the Clinical Genetics Society's Working Party on the genetic testing of children.

Cynthia B. Cohen is a Senior Research Fellow at the Kennedy Institute of Ethics at Georgetown University. She has served as Executive Director of the National Advisory Board on Ethics in Reproduction, Associate for Ethical Studies at the Hastings Center and Chair of the Philosophy Department at the University of Denver. She has published numerous articles and books on ethical and legal questions that arise at the beginning and end of life, including *New Ways of Making Babies* and *Casebook on the Termination of Treatment and Care of the Dying.* Dr Cohen is a graduate of Barnard College, received her MA and PhD from Columbia University and her law degree from the University of Michigan.

Shirley Dalby is the Development Director of the Neurological Alliance, an umbrella group of voluntary organizations concerned with neurological conditions. She has previously been the Director of the Huntington's Disease Association and of Ataxia and she initiated the formation of the Genetic Interest Group (GIG). She is a co-opted GIG Trustee and her work for them has included chairing the GIG Working Party on the genetic testing of children.

John Gillott has been Policy Officer at the Genetic Interest Group (GIG) since 1994. He was a member of the GIG Working Party that drew up the GIG response to the Report of the Clinical Genetics Society's Working Party on the genetic testing of children.

Rogeer Hoedemaekers was a language teacher at a large comprehensive school in The Netherlands. He then studied moral theology and became a researcher at the Department of Ethics, Philosophy and History of Medicine at the School of Medical Sciences of the Catholic University, Nijmegen, The Netherlands and Visiting Fellow at the Centre for Professional Ethics at the University of Central Lancashire, Preston, UK. He is currently completing his PhD thesis on normative determinants of genetic screening and testing.

Outi Järvinen graduated from the Faculty of Medicine, University of Helsinki in 1990. She then worked in the Department of Paediatric Neurology at the Children's Castle Hospital in Helsinki. Since 1996 she has worked as a researcher in the Department of Medical Genetics at the Family Federation of Finland. She is preparing her doctoral thesis on genetic testing in childhood.

Anita Jolly is a junior hospital doctor. She undertook the research on which her contribution is based while she was a medical student studying for an intercalated degree in Genetics.

Helena Kääriäinen is the Chief Physician at the Department of Medical Genetics at the Family Federation of Finland. She is also a Docent (associate professor) in Clinical Genetics at the University of Helsinki. In her work and research she is involved with genetic counselling, the delineation of rare hereditary diseases, gene mapping and ethical aspects of modern genetics.

Seymour Kessler originally trained as a geneticist at Colombia University in New York City (PhD 1965). Dr Kessler later joined the faculty in the Department of Psychiatry at Stanford University and conducted research in behaviour and psychiatric genetics. In 1977 he completed a second PhD in Social and Clinical Psychology at the Wright Institute in Berkeley, CA and became the Director of the Genetic Counseling Program at the University of California in Berkeley. He is currently an Associate Clinical Professor in the Department of Pediatrics, University of California, San Francisco and maintains a private practice in clinical psychology, specializing in health issues.

Zarrina Kurtz is a freelance consultant in public health and health policy; formerly medical adviser to the Inner London Education Authority and Senior Lecturer in Paediatric Epidemiology at the Institute of Child Health, London. She carries out research and service development nationally, focused on the health of young people and with particular interests in chronic conditions and mental health, the health of minority ethnic groups and in children's rights.

Christine Lavery was appointed Director of the Society for Mucopolysaccharide Disease in 1993 following a period of 8 years' employment as National Development Officer for Specific Disease Groups at Contact A Family (CAF). She had the idea and then researched and edited the first editorial of the very successful publication, *The CAF Directory of Specific Diseases and Rare Syndromes in Children with their Family Support Networks*. Christine has pioneered a European database to identify the epidemiology and incidence of mucopolysaccharide diseases. Information has been collected so far on more than 2000 affected MPS children from all over Europe. Christine has undertaken research to establish the attitudes of parents to genetic testing for the healthy (unaffected) siblings of their affected children. In 1996, Christine was appointed to the UK Government's Department of Health Advisory Committee on Genetic Testing.

Theresa M. Marteau is Professor of Health Psychology and Director of the Psychology and Genetics Research Group at the United Medical and Dental Schools of Guy's and St Thomas', London. Over the past 15 years she has been conducting research into the psychological and social implications of genetic risk information presented to pregnant women about their fetuses, as well as to children and adults about risks to their own health. Through observational and experimental studies the work of her group has drawn attention to the problems that health professionals have in communicating such information and the consequences of this for patients.

Carl May is Senior Research Fellow (Medical Sociology) in the Department of General Practice at the University of Manchester. He has researched and published widely in the field of professional patient interaction and was the principal investigator on the MRC-funded project, 'The process of genetic counselling', reported in this volume.

Sheila A. M. McLean is a Professor and is the first holder of the International Bar Association Chair of Law and Ethics in Medicine, and Director of the Institute of Law and Ethics in Medicine, both at Glasgow University. She serves on a number of governmental and non-governmental committees and has been commissioned by the British government to review the consent provisions of the Human Fertilization and Embryology Act 1990. She has published extensively in the area of medical law and ethics.

Susan Michie is a Senior Research Fellow in Health Psychology. She is a qualified Clinical Psychologist who has worked for several years with children and families. Her doctoral thesis was on the cognitive development of children. She currently works in the Psychology and Genetics Research Group at Guy's Hospital (United Medical and Dental School) in London, funded by the Wellcome Trust. One of the studies in this programme of research is of the psychological impact of predictive genetic testing in adults and children.

Dietmar Mieth is Professor of Theological and Social Ethics at the Catholic Theological Faculty of the Eberhard-Karls University of Tubingen, Germany. He is Director of the Moral Theology Section of the international theological journal *Concilium* and Director of the Centre for Ethics in the Sciences and Humanities at the University of Tubingen. He was previously the Professor of Moral Theology in Freiburg, Switzerland. His main interests are the foundations of ethics, social ethics, bioethics and narrative ethics. He has given expert evidence to a number of German and European parliamentary hearings.

Evelyn Parsons is a Medical Sociologist who, for the past 8 years, has been Director of Social Science in the School of Nursing Studies, University of Wales College of Medicine, Cardiff. Her PhD in Psychosocial Genetics explored the implications for women of living with Duchenne's muscular dystrophy (DMD) in the family. In addition to her post in Nursing Studies, she is a Senior Research Fellow in the Department of Medical Genetics, where she is responsible for evaluating the all-Wales newborn screening programmes for DMD and cystic fibrosis. She is also involved in an MRC-funded study looking at the psychosocial implications of presymptomatic testing of women at risk of familial breast cancer.

Andrea Farkas Patenaude is a Clinical Psychologist. She is Director of Psycho-Oncology Research in the Department of Pediatric Oncology at the Dana–Farber Cancer Institute in Boston, MA and an Assistant Professor of Psychology in the Department of Psychiatry at Harvard Medical School. She is also Co-chair of the Advisory Council on Genetic Issues of the American Psychological Association. Dr Patenaude serves as the psychologist on several projects evaluating the emotional impact of genetic testing for the BRCA1, BRCA2 and p53 genes on members of high-risk families. She also for many years directed the psychology programme for paediatric oncology and bone marrow transplantation patients at the Dana–Farber Cancer Institute and the Children's Hospital, Boston.

Martin Richards is Professor of Family Research and Director of the Centre for Family Research at the University of Cambridge. His research concerns families and children. For the past decade he, together with colleagues at the Centre, have been working on a programme of research on the psychosocial implications of the new human genetics. His particular interests are in the ways in which family members cope with the inherited risk of cancer and lay-knowledge of inheritance. Together with Theresa Marteau, he edited *The Troubled Helix: social and psychological implications of the new human genetics* (Cambridge University Press, 1996).

Heather Skirton is Senior Genetic Nurse Specialist and Resource Manager in the Taunton and Somerset NHS Trust. She worked as a nurse and midwife for 19 years before becoming a Genetic Nurse Specialist in 1989. After completing an MSc at Exeter University, she obtained a Research and Development Studentship to study for a PhD investigating the effectiveness of genetic counselling from the client's perspective. She is currently the Chairman of the Association of Genetic Nurses and Counsellors and convenor of the Association's Working Parties on genetic nurse practice and education in the United Kingdom. She has published work on genetic nursing practice, counselling aspects of genetic services and the care of families with Huntington's disease.

Dorothy C. Wertz is Senior Scientist in the Division of Social Science, Ethics and Law at the Shriver Center for Mental Retardation in Waltham, MA. She received her PhD from Harvard in the study of religion and taught sociology and anthropology for 18 years. Since 1981, she has researched social and ethical aspects of human genetics, most notably through international surveys. She has recently completed (with John Fletcher and Kare Berg) *Guidelines on Ethical Issues in Medical Genetics and the Provision of Genetics Services* for the World Health Organisation. Her books include: *Lying In: A History of Childbirth in America*, (1989; with Richard W. Wertz) and *Ethics and Human Genetics: A Cross-Cultural Perspective* (with John C. Fletcher).

Acknowledgements

The production of this volume represents the collective achievement of the contributors, the publishers, the sponsors and the organizers of the meeting from which it was derived. Accordingly, I would like to thank all those who contributed to it in any form. This includes: Jonathan Ray, Fran Kingston and colleagues at BIOS; Veronica English, Fleur Fisher, Patricia Fraser, Gillian Romano-Critchley and Ann Somerville at the Ethics Department of the British Medical Association; the Genetic Interest Group itself as well as its individual contributors; Ruth Chadwick and the whole EUROSCREEN project, based at the Centre for Applied Ethics in the University of Central Lancashire and funded through the European Commission. I also had the support of the other members of the EUROSCREEN subgroup on the Genetic Testing of Children. I would also like to acknowledge the support of my family and their tolerance of my awkward pattern of working.

Preface

This book examines the ethical and social issues raised by the genetic testing of children. It consists of essays written from a number of different perspectives and is truly multidisciplinary. As such, it does not set out one position but many; it does not answer questions so much as raise them. I hope that everyone interested in these issues will learn through being challenged to look at them in unfamiliar ways. This may help us all to find a way towards our common goal – promoting the welfare of children and the families from which they come.

This volume arose from a meeting, held in London in June 1996 and organised by EUROSCREEN, the British Medical Association and the Genetics Interest Group. The spoken papers at that meeting have been revised and, after much discussion, they emerge here as chapters. The discussions at the meeting have not been included directly in the text, but have been important in the development of the chapters – authors and editor all owe a debt to the participants at the meeting.

Discussions about childhood genetic testing raise many of the difficult issues that surround genetic testing in general, including the issues of confidentiality and autonomy, genetic privacy, potential abuses of genetic information, and the appropriateness of predictive genetic testing. These discussions also touch on many issues that relate to the place of children within society. These topics all feature in this volume, and I hope that these chapters will be of interest to many professionals and families who are interested in these broader issues and not just to those involved in the genetic testing of children. The debates on this topic have extensive implications – wherever genetics impacts upon society and social forces impact upon children.

Angus Clarke

Introduction

Angus Clarke

This volume examines a range of issues raised by the genetic testing of children, especially in those circumstances where it is not to the direct benefit of the child. This introduction must therefore outline the general social and ethical issues relating to genetics, but should also consider the place of children in the family and in society. This will help to put subsequent chapters in context, as each of them addresses the way in which decisions are made about the genetic testing of children.

1. Social and ethical issues around genetics

The widespread application of genetic technologies raises many ethical and social issues, and this is particularly true for the application of genetics to medical practice. The use of molecular genetic methods to identify microbial infections or genetic changes confined to tumours is not so different from the conventional investigations carried out in any pathology laboratory. Analysis of the genetic constitution of the individual, however, is rather different. Such analyses can be carried out at any stage of life and not just as diagnostic tests of those already manifesting an illness. *Pre-natal diagnostic testing* raises questions about the termination of pregnancies. *Predictive or presymptomatic testing* of apparently healthy individuals will identify those who may be at risk of developing disease in the future. To what extent is it helpful to have such information when there may be little that can be done to prevent or treat the predicted illness? It is knowledge without power. Furthermore, individuals share genes with their relatives, so the result of a test of the genetic constitution of one individual may have profound implications for other members of their family.

The historical legacy of *eugenics* – in which the rights and welfare of individuals have been subordinated or sacrificed to the supposed interests of collective entities such as society, the state, the race or the species – ensures that contemporary clinical geneticists feel the need to distance themselves from these past abuses (Muller-Hill, 1988). There also exist less explicit forms of eugenics in which the forces of the 'free' market, aligned with social convention, may achieve the goals of the eugenics movement more effectively than it managed to do for itself (Duster, 1990); the response of professionals to these more subtle forms of eugenics is diverse.

The Genetic Testing of Children, A.J. Clarke (ed.).
© 1998 BIOS Scientific Publishers Ltd, Oxford.

The increasing emphasis on *respect for autonomy* as the predominant principle of medical ethics is in tune with the individualistic mood of the late twentieth century. It is now difficult for clinicians to justify their actions on the grounds of beneficence, which is understood as a form of paternalism; instead, patients make the decisions about their own health care on the basis of the full information provided by their physician. *Informed consent* is required before a genetic test can be performed, and the tested individual will expect to retain full control over access to the test result by third parties – whether they be other members of the family or institutions such as insurance companies or employers. Respect for *confidentiality* is one aspect of this respect for autonomy, and has been challenged by some interested parties keen to promote the social applications of genetic information (Boddington, 1994).

The focus upon genetic explanations of difference – between individuals or between social groups and populations – could detract attention from other factors important in contributing to the biological and social differences between individuals and groups. This process of *geneticization* leads us to look to the level of the genes to explain and to manage 'social difficulties' such as disability, criminality, alcoholism, homosexuality ... even if social, political and environmental factors may be just as important and rather more accessible to intervention. Furthermore, many people would not accept homosexuality or (certain types of) disability as 'social difficulties'; that need to be treated or prevented; there is a serious danger that medical scientists may accept as objective fact particular value-laden representations of human variation as medical conditions or biological disorders.

This leads directly to another set of issues surrounding genetics – the question of *respect for persons*. Respect for people with disease or *disability* has always been limited by social intolerance, discrimination and stigmatization but could be further compromised by the modern consumer's quest for the perfect body and the perfect image. Pre-natal screening programmes are already seen by some as a technical solution to the challenge of disability. Future advances in genetics are likely to reinforce the tendency for the parents of children with Down's syndrome (for example) to be seen as responsible or even guilty for giving birth to an affected child. This clearly provides a challenge to any notion of the intrinsic value of the individual – whether religious or humanistic in inspiration.

There is also the potential for (crude) genetics to undermine respect for persons by promoting *reductionism* – the explanation of complex human activities in terms of molecular biology, genetics and neuroscience (Lewontin, 1991; Ramsey, 1994). Such reductionism is inevitably flawed because complex phenomena, such as human social behaviours, can only be explained in terms appropriate to the level of complexity involved and could not even in principle be accounted for by explanations framed in the 'ultimate' reductionist terms of subatomic physics.

An analogy may clarify this point. While we have no doubt that computers obey the laws of the physical universe, no-one would try to account for, or to predict, a computer's solution to a particularly difficult calculation through considering the movement of individual electrons. Such an attempt would be ludicrous and is, in any case, impossible because of the uncertainty surrounding the properties (or at least our knowledge of the properties) of subatomic particles. Any accounting for the computer's calculation would need to be framed in abstract terms relating to the logical circuits involved, not in mechanistic terms as if we could ever have a perfect knowledge of the state of the physical world even within the word processor on which I am writing this sentence.

Reductionist explanations, assumptions and patterns of thought, however, have become part of the mindset of many of our contemporaries. This leads some philosophers, for example, to draw up criteria for deciding whether or not a life is 'worthwhile'; respect (and care) can then, it is argued, be withdrawn from certain individuals who fail to fulfill the relevant criteria (Harris, 1985). Accounts of a worthwhile life that reckon the value of the individual as largely equivalent to his or her contribution to society can easily lead on to one or other form of eugenics – crude or less crude.

It can be appreciated that the ethical and social issues discussed here – relating to pre-natal and predictive genetic testing, eugenics, autonomy, confidentiality, consent, geneticization, respect for persons, reductionism – are not distinct and separate but are closely and dynamically interrelated. For a more general review and a guide to further reading, see Harper and Clarke (1997).

2. Children and childhood

As well as the issues around genetics, there are numerous issues raised by the concept of childhood. Why, or perhaps when, should children be treated as any different from adults?

Every society provides a framework of care within which children are cared for, educated and allowed to develop towards adulthood. Some societies have rites of passage that celebrate the specific point at which a child becomes an adult; others – including our own – allow a child to accrue the competences and status of adulthood over a period of years marked by a series of less important milestones. While some societies diffuse responsibility for child care and education rather widely, others – including ours – focus responsibility for child care on the parents. This introduces particular difficulties if the parents are not both sharing a household with their child; a more diffuse parenthood might soften these 'single parent' and other problems.

The point at which children become adult in their capacity to take responsibility for health matters has been contested. In Britain, the decision of the House of Lords in the Gillick case has given legal minors the right to make decisions about such matters independently of their parents once they are sufficiently mature to understand the consequences of their decisions. This does not resolve the issues raised when parents wish to arrange genetic testing on young children who cannot even participate in a decision-making process. Do parents have a *right* to have such children tested? What if the relevant health professional fears that it may not be in the children's best interests to have genetic testing carried out when they are so young? Whose judgement should prevail? Is it not unhelpful either for professionals or for parents to approach such discussions in a confrontational manner, with minds already made up?

The role of the professional, the genetic counsellor, as a gatekeeper to genetic testing can be criticised on several counts. The days of such gatekeeper functions is limited because of the increasing availability of testing and other services on a global basis, across national boundaries, and often involving the Internet. Furthermore, the genetic counsellor could expose him- or herself to criticism as someone who wishes to impose unnecessary restrictions over the genetic testing process – and thereby enhance their professional power.

Espousing the interests of the child – protecting children from their misguided or abusing parents – is a very powerful position to adopt in a society, like ours, that has such a sentimental understanding of childhood. This manoeuvre has been very helpful in advancing the welfare of children over the last 150 years – but that does not mean that every contemporary appeal by professionals to their duty to protect children from their families will be legitimate. While the welfare of children can be protected by professionals acting on their behalf, such professionals as paediatricians, teachers and even geneticists can increase their social authority and professional power by staking a claim to regulate additional aspects of children's lives. Professionals can therefore be portrayed as having something to gain by pushing a particular policy about access to genetic (or other) testing.

This chapter is not the place to begin a detailed exposition of the rights of children contrasted with the competing rights of parents and others over them, nor do I have the competence to carry out such a task. Instead, I refer the interested reader to four books. Archard (1993) lays out a framework within which the rights of the child can be considered alongside the rights and duties of parents and society at large. Alderson (1993) explores the extent to which young children can contribute to important decisions being made about their health care. King (1997) dissects the claims made by different professional groups to speak on behalf of children, and Gittins (1998) examines the ways in which society – i.e. the world of adults – represent children and childhood. These four books do not address the specific issues raised in relation to the genetic testing of children, but they do challenge preconceptions about childhood and assist the reader to consider these issues afresh.

3. Issues for practitioners (genetic counsellors)

We now turn to the issues that a genetic counsellor may need to consider in the course of a consultation with a family. We must first attend to some definitions. We must distinguish between genetic counselling and genetic testing, and between the genetic testing of individuals and the genetic screening of a population.

Genetic counselling is '... a communication process dealing with the human problems associated with the occurrence, or the risk of occurrence, of a genetic disorder in a family...' (Fraser, 1974). It may or may not include a discussion about genetic testing. Genetic testing may be applied to an individual in relation to their specific personal and family context, or it may be part of a population-based programme of screening made available to a large group such as all newborn infants or pregnant women. Family-based genetic counselling is generally reactive, responding to pre-existing concerns in an informed family, while screening programmes are often proactive and can raise anxieties among large groups in the population – often with little knowledge of the disorder in question – and the genetic testing process will often not be very successful in allaying these concerns. Appreciating these distinctions is most important in understanding the context of any genetic counselling consultation.

3.1 Non-directiveness

It is easy to understand why contemporary clinical geneticists and genetic counsellors should be eager to distance themselves from eugenic abuses of human rights in the past. One way in which some achieve this is to reject the imposition of institutional measures of their effectiveness in terms of the numbers of births of affected infants or the number of terminations of pregnancy of affected fetuses. Another device for establishing distance between contemporary clinicians and the legacy of eugenics is to emphasise that all reproductive decisions about genetic disorders are made by the clients/patients/counsellees and are not imposed by the professionals. Adherence to this principle is termed 'non-directiveness', and this is one of the central tenets of genetic counselling. This term came to prominence at the same time as the non-directive style of counselling developed by Carl Rogers, and it doubtless received indirect support from the ethos of counselling in this wider sense. In addition, the professionalization of genetic counselling and the process of detachment from eugenics occurred during the same period as the principle of autonomy came to predominate over the other principles of medical ethics.

Unfortunately, this led to the social context in which individuals are expected to make reproductive decisions being ignored. Insufficient attention has been paid to social influences either at the macro level (e.g. social tolerance; the political willingness to support those with genetic disorders and other types of disability) or at the micro level (e.g. the way in which genetic information or testing is made available in routine clinical practice; the way in which family, friends and neighbours respond when a genetic condition is identified; personal experiences of stigmatization or discrimination). It can be argued that some claims to non-directiveness are disingenuous, given these social constraints on the freedom of individuals to make their own decisions. How 'free' can a decision about pre-natal screening for a genetic disorder be if the parents have no confidence in the willingness of society to support them and their child, if their child is affected?

'Non-directiveness' is a crucial term in discussions about the goals and the ethos of genetic counselling, and practitioners differ in their understandings of it. Does compliance with this principle make it improper for genetic counsellors ever to challenge a client's statement of intent, even if it is tentative or ill-considered?

For clients to make their own, autonomous decisions they will obviously require access to full information about the various options ahead. The concepts of autonomy, informed consent and non-directiveness are closely interwoven. Clients may also find it helpful for a genetic counsellor to explain the possible implications of a genetic test result before the test is carried out; rehearsal of the various scenarios that might arise in the future may help clients reach decisions that they are less likely to regret subsequently. When will such scenario-based decision counselling lead clients towards particular, counsellor-approved outcomes and when will it function, as intended, to improve the quality of the client's decisions without directing them to make a particular choice?

3.2 Genetic privacy – confidentiality

Another issue that is closely related to autonomy concerns the control of personal genetic information. Problems arise in relation to the control of such information if an individual does not want to pass information on to other members of his family that they could find helpful in making their health care decisions or reproductive plans. Respecting the confidentiality of the first individual's genetic information may entail denying his relatives the opportunity to make their own informed and autonomous decisions. How can such conflicts of principle be resolved? Do professionals ever have a duty to break a client's confidentiality in order to give information to a relative who may find it of value? – or will such policies so undermine clients confidence in genetic counsellors that it undermines their credibility and integrity as professionals? (Harper and Clarke, 1997).

4. Genetic testing in childhood

Several of these separate issues arise in the realm of genetic testing in childhood. Where there is clear medical benefit from genetic, or any other type, of diagnostic testing then it is ethically unexceptionable. Where genetic testing is predictive of a disorder likely to manifest in the course of childhood, few ethical difficulties are raised. But where the disorder in question is usually of late onset – not developing until well into adult life – there are ethical and practical problems raised by the testing of a child. If the child is old enough to request such testing in their own right – where they meet the criteria of 'Gillick competence' – then these issues evaporate, but that will clearly not be possible for young children.

There are five particular problems raised by predictive testing for an untreatable, late-onset disorder – or by carrier testing, of reproductive significance only – in a young child:

(i)   the 'right' of parents to arrange genetic testing on their child; (do such rights exist? or is it more helpful to consider the responsibility of parents and society to act in the best interests of the child?)
(ii) the judgement as to whether the child is sufficiently mature and of sufficient understanding to make their own decision, or to participate in a collective decision process;
(iii) the abrogation of the child's future capacity to make their own informed, autonomous decision as an adult;
(iv) the likely loss of confidentiality concerning the child's genetic status, which would be protected if they were not tested until older: and
(v)   the possibility of negative emotional consequences for the child resulting from discrimination, stigmatization or altered parental expectations associated with the genetic test results.

Given the uncertainty surrounding these questions in the context of genetic testing for any particular child, it can be appreciated that the professional has to balance many competing claims in discussing the possibility of genetic testing with the parents and perhaps also with the child. Observing such discussions allows us to examine the way in which the professional and the parents together focus on the issues relevant to their circumstances and work towards a resolution. The issues of non-directiveness, parental rights, child autonomy, child competence, genetic privacy and confidentiality, stigmatization and discrimination may all appear in just a single discussion about a single family.

At whatever age predictive or carrier genetic testing is carried out on a child, it is in adolescence and early adult life that they will have to come to terms with the facts and implications of their genetic situation. In many contexts, it seems to me that adolescents will adjust better to unwelcome circumstances if pains have been taken to give them as much explicit control as possible over their situation. My lay understanding of personal emotional development and my observations of adolescents in clinical medicine and genetics have provided me with a stock of supporting anecdotes. Together, these may not count as firm evidence, but they are the reason underlying my cautious approach to the genetic testing of children. Research is clearly needed to examine these assumptions and judgments.

5. The relevant genetic contexts

Genetic testing of children will raise issues and concerns when the child is unlikely to be affected by the disorder until well into adult life (predictive testing) or when the testing will determine the child's carrier status – when the child will not be affected him- or herself but his/her own future children may be affected.

It is worth distinguishing here between two large classes of genetic carrier state, both of which are only of relevance to the child's own future reproductive decisions and future children rather than his/her own health. If the child is a carrier of an autosomal recessive disorder, then his/her own future children would be at risk if their partner was also a carrier of the same condition. For a child to be affected, both parents of an affected child have to be carriers and to hand on their altered copy of the relevant gene. In contrast, if the child is a carrier for a sex-linked disease or for a balanced chromosomal translocation, then her future children would be at risk irrespective of the genetic constitution of her partner.

The concerns about the possible adverse effects of genetic testing in childhood are especially strong in relation to predictive testing for untreatable, late-onset disorders such as Huntington's disease (HD). This is reinforced by the low uptake rate of predictive testing for HD among adults at 50% risk. The issues are similar in relation to carrier testing but the concerns are then more restricted to the reproductive domain – they should not be regarded as trivial, however, because the expectations held by a child's parents in relation to their future pattern of relationships, marriage and career can have a profound influence upon the direction and quality of their future life – especially, perhaps, in the context of sex-linked disorders or chromosomal rearrangements (Parsons and Atkinson, 1992; 1993; Parsons and Clarke, 1993).

It is also clear that the healthy siblings of children affected by autosomal recessive diseases have many needs for information, counselling and support whether or not they turn out to be carriers (Fanos and Johnson, 1995a,b). Testing the unaffected sibs of children with a serious autosomal recessive disease therefore does not tidy up or resolve the family situation for either the parents or the children; one of the strongest reasons for a cautious approach towards the genetic carrier testing of children is the fear that the testing of the child may be regarded inappropriately as the end of the matter. As with any genetic testing, one must consider precisely what questions will be resolved by genetic testing and what issues will remain afterwards.

6. Discussions about testing children

Most genetic professionals in Europe and North America are unwilling to carry out predictive genetic tests for the late-onset, untreatable neurodegenerative disorders such as HD – for the reasons outlined above. A number of professional reports and lay groups agree with this approach (Clinical Genetics Society, 1994; Institute of Medicine, 1994; Wertz et al., 1994; Dalby 1995; American Society of Human Genetics/American College of Medical Genetics, 1995; British Paediatric Association, 1996), although there is discussion about the definition of a child – by age? by competence? – and about predictive testing for disorders that may manifest in childhood and for which useful surveillance or prevention may be available. These cautious policies have been developed, despite the lack of evidence that testing causes harm, because of professionals' unwillingness to introduce such tests in young children.

There have been accounts of the genetic testing of children for family cancer disorders, such as familial adenomatous polyposis coli (FAP), in which screening of the bowel for tumours by annual colonoscopy may be started from around the age of 10 years (Michie et al., 1996). The context of genetic testing for such a disorder is so different from that of testing for the late-onset, untreatable neurodegenerative disorders, however, that experience gained in relation to testing for FAP tells us little that is relevant to the HD context. Instead, the discussion with the family has to focus on when testing will be most helpful to all concerned, given the range of different possible test results for each of those at risk (Clarke, 1997).

Some predictive genetic testing of healthy children has been carried out in the USA despite policies advising against this, but little is known about the impact of these tests on the families – and nothing is known about the long-term consequences for the children (Wertz and Reilly, 1997). Where such testing has been performed, it will be important for the experiences of the families to be studied and reported.

In the different context of carrier testing, some families express a very strong desire to test a child for carrier status. There is a wide variation in professional and laboratory practices in relation to such testing (Balfour-Lynn et al., 1995; Wertz and Reilly, 1997). There may be an unstated family agenda underlying some requests, such as a desire to test paternity, but this is unusual and it is much more commonly a straightforward but strong 'desire to know'. The family context will often be of parents who have lost a brother, sister, son or daughter with a serious and most distressing disease; unresolved grief in the parents may – very understandably – intrude into the decisions being made about the child. In my experience, it is not helpful for professionals to refuse to perform carrier tests under these circumstances. It is very reasonable to explore the motivation and reasoning underlying the parents' request, to indicate that testing will not resolve all the issues even if the child is shown not to be a carrier and to raise for discussion the possible difficulties that may arise from testing in childhood; but if the parents continue to insist on the child being tested then it may well be appropriate to agree to this.

If professionals adopt a rigid, anti-testing position that is unwilling to listen to the parents and to acknowledge their reasoning and the strength of their feelings – whether spoken or unspoken – then antagonisms may be generated that are unnecessary and that will be unhelpful for all concerned. One approach that professionals can use in this context – a parental request to clarify the carrier status of their child – is, after a thorough explanation of the issues, to agree to test the child in principle but to introduce a delay of a few weeks, after which the family has to take the initiative to arrange the testing; the decision is put entirely in their hands. This may give the family a chance to reassess their decision without the fear that the test will be denied them, so that for the first time they become free to examine all the factors for and against testing and then to make up their own minds.

7. Lessons from screening children for susceptibility to disease

Experience with screening children for genetic susceptibility to common diseases is limited, but the experience that has been accumulated amounts to a cautionary tale. A report from Bergman and Stamm some 30 years ago introduced the term 'cardiac nondisease' to describe the emotional and physical effects of inappropriately labelling healthy children who had benign cardiac murmurs as being sick (Bergman and Stamm, 1967). In the newborn period, a Swedish programme of screening to identify infants affected by alpha-1-antitrypsin deficiency was stopped because of damaging social consequences. Affected children are more susceptible to lung disease in adult life and should be protected from exposure to cigarette smoke, but the fathers of the identified infants actually smoked more heavily than the fathers of control infants (Thelin et al., 1985). In the US, children found to have above average serum levels of cholesterol have sometimes been subjected to excessively restrictive diets, even leading to growth failure from malnutrition (Lifschitz and Moses, 1989; Newman et al, 1990), and sometimes their possible special health needs have been ignored (Bachman et al., 1993).

8. The 'adoption' model of genetic information

There are close parallels between handling genetic information about children and handling information about their adoption. Both categories of information are likely to be unwelcome but witholding the information from the child for ever is likely to cause further problems. The practical issue becomes how best to impart the relevant information to the child. Not to mention that there is important information until suddenly broaching the topic with an unsuspecting teenager is not likely to be the most successful approach – and many teenagers will not in fact be totally unsuspecting, even if they have never been told anything about their adoption or the family illness.

Most children ask questions about their origin and their family that allow some information to be imparted at an early age – even if it is just the fact that there is more information to be imparted later. It will usually be unhelpful to mislead a child even if the full facts are not disclosed at once. Gradually increasing quantities of information can be given – about the genetic disorder or about their adoption – as the child increases in maturity. In the case of adoption in Britain, the child is able to seek out his/her natural parents once he/she reaches 18 years of age. In the context of genetic information, it is possible for the adolescent or young adult to seek genetic testing once they have decided for themselves that they want this information. In both contexts, it is possible for the family to have explained that there is a potential 'genetic' problem for the child to consider once older – relating to a disease or to parentage – and full control over this process can be given explicitly to the adolescent as a way of helping him/her to come to terms with the possibly unwelcome information.

9. The contents of this volume

9.1 Setting the scene

In the first section, the issues that are raised by the genetic testing of children are explored from three different perspectives. McLean introduces these issues from a legal perspective, giving an overview of the social, ethical and legal issues raised by genetic testing in general and by childhood genetic testing in particular; she retains a very practical focus on issues that can arise in clinical practice. Alderson speaks from her background in the social sciences, with an interest in social processes on the grand scale and their effect on the individual. She considers both the relations between genetics and society and the involvement of children in this relationship. Mieth writes from the perspective of an ethicist about diagnostic testing in childhood, as well as the issues specifically raised by genetic testing and by the testing of children.

9.2 Carrier testing in childhood

The next section concerns testing for genetic carrier status – when the results will be relevant to the child principally in the future when they come to make their own decisions about relationships and reproduction. Lavery writes about the experiences of families with children affected by a group of severe inherited disorders, the mucopolysaccharidoses. These are recessive conditions, most being autosomal recessive and one being sex-linked in inheritance, so that the healthy brothers or sisters of the affected child may well be carriers. She argues that the whole experience of being the brother or sister of an affected child needs to be considered very carefully and sensitively and that a blanket policy not to make carrier testing available to them until they are 16 years can be unhelpful in dismissing their real needs for information and support.

Barnes reports a questionnaire survey of families with balanced chromosomal rearrangements where the healthy children could be carriers, who would than have a risk of bearing children affected by serious congenital anomalies or developmental problems or of pregnancies ending in miscarriage. She gathered information about the number of potential carrier children who had been tested; parental knowledge of the implications of carrier status, parental plans to discuss testing or test results with the children and parental reactions to the identification of their child as a carrier. Professionals may regard the diagnosis of the carrier state in a child as not being a major problem – especially in the context of a pre-natal diagnosis where testing is really aimed at identifying a fetus as 'affected' or 'unaffected', and with carrier status regarded as a subcategory of 'unaffected'. In contrast to this, families reported receiving the news that their child was a carrier as markedly distressing.

The study reported by Jolly, Parsons and Clarke addressed similar issues but was an interview-based study of fewer families. This study interviewed some of the parents and some of the 'children' in these families, now adolescent or adult, who had been tested years before. There were similar findings, with some parents having not raised the genetic issues with their children or having found it difficult to do so, but there was no evidence of serious family problems resulting from the carrier testing in childhood. This is a lengthy chapter containing many brief extracts from the interviews. We hope this allows the feelings of family members to come across and speak directly to the reader.

A report by Järvinen and Kääriäinen in Finland presents the preliminary findings of a retrospective, questionnaire-based study of parents and their (now adult) children from families in which carrier testing of children had been performed 8–12 years previously in relation to the sex-linked disorders haemophilia and the Xp21-related muscular dystrophies and chromosomal translocations.

While there was plenty of evidence in these three studies that carrier testing in childhood did not adequately resolve the relevant issues in practice, there was little evidence that testing in childhood actually produced additional problems. The final chapter in this section is by Gillott, who gives the view of the Genetic Interest Group (GIG) – an umbrella organization representing the views and interests of the lay societies supporting individuals and families affected by genetic diseases in Britain. The GIG assessment of the Clinical Genetics Society's (CGS) report on the genetic testing of children largely supported the CGS perspective on predictive testing for late-onset, untreatable disorders but was sharply critical of the CGS view on carrier testing.

9.3 Predictive genetic testing

This section of the volume opens with an account of how families affected by Huntington's disease have – or have not – talked about the issue of genetic risk with their (often adult) children. Skirton interviewed 15 members of 12 families with HD; most parents certainly intended to tell their children of their disease risk, despite the distress this was likely to cause, but this was often delayed because they found it difficult to choose 'the right moment'.

Kessler discusses the ways in which families function in relation to threats such as Huntington's disease. If one member shows signs of the disease, or is tested for HD so that their genetic status becomes known, then there are implications for

the functioning of the family as a whole. Family processes are discussed in relation to shame, guilt, preselection, predictive testing and the psychological needs
of children.

In her chapter, Binedell discusses the handling of requests from adolescents for
predictive testing for HD. If the age of legal majority is not chosen as a simple,
legal criterion then what criteria of competence would be appropriate to guide
professionals in how to respond to these requests? One suggestion is that the reasons behind the request should be thoroughly explored in case the request for predictive testing is symptomatic of a different question or concern arising in the
adolescent. It may often be that simple compliance with the request for testing
may fail to meet the individual's needs and may be unhelpful in the long term.
Similar issues also arise in requests for predictive testing from adults, of course,
but extra time may be required to tease apart these issues when the request for
testing comes from an adolescent.

The next two chapters consider predictive testing for a range of conditions, not
just for HD. Cohen argues that parental requests for their children to be tested
must be considered very carefully and should not be dismissed by health professionals – who respect parents as the best judges of their children's welfare in most
circumstances and must have good reasons for diverging from that practice here.
Society must also take steps, she argues, to ensure that those identified as destined
to develop untreatable diseases in the future do not suffer social discrimination
but are accepted as essential members of the human community and are nurtured
accordingly. There may be occasions, she urges, where the predictive genetic testing of children could be helpful.

The issues that are raised by the testing of children for inherited susceptibility
to specific cancers are considered by Patenaude. There are some circumstances
where this may be of practical benefit, or at least assist in making clinical decisions. There are also complex issues to consider in relation to family processes
and family functioning. Patenaude discusses the ways in which families attribute
responsibility for the occurrence of their cancers, and how this can influence their
readiness to accept or seek predictive testing for their children. She concludes
that further work is needed to define the circumstances under which such testing
is likely to be helpful and to guide professionals in their discussions with families
about the possibility of such testing.

9.4 Research perspectives

Richards addresses two issues relating to the genetic testing of children. He sifts
the limited evidence relating to the ability of children to understand the facts of
inheritance. This complements the work of Alderson, that examines the ability of
children to contribute to important decisions about their health care. Richards
also examines the attitudes of relatively well-informed members of the British
public towards parents' right to seek predictive genetic testing of their children.
He finds that a majority approve of parents having this right and thinks it likely
that they will eventually win the argument – in practice if not in principle.

Michie and Marteau review the psychological literature for evidence that will
inform decisions about predictive genetic testing of children. They find several

deficiencies in our knowledge and recommend research in three particular areas: (i) the attitudes of children, parents and health professionals; (ii) the criteria for the informed consent of children; (iii) the psychological consequences of predictive genetic testing in childhood. They conclude, contentiously, that there is sufficient information to proceed cautiously with testing, if requested, subject to two conditions: (i) that it is performed within the context of good counselling and informed decision-making; (ii) that it is performed as part of a research protocol. We are grateful to the authors and the *British Journal of Health Psychology* for permission to publish this paper as a chapter in this volume.

The third chapter in this section challenges the claims of experimental psychology to provide an objective account of reality, of the truth. Brindle argues from a social constructionist position that the usual notion of the subject – the self – is flawed; the subject is not a discrete, rational entity that exists and functions independently of the social context within which it is situated. It would be misleading to treat the results of the investigator's measurements as if what they measured were discrete, independent entities – this would be to reify the investigator's concepts instead of treating them as heuristic devices. Brindle argues that the formulation of the ethical issues surrounding the genetic testing of children needs itself to be challenged. It may be more fruitful to examine, for example, the accomplishment of competence by children within social interactions rather than the measurement of specific abilities within the child – as if such measures could define the child's competence – as if competence were a discrete and quantifiable entity.

The chapter by Chapple, May and Campion contains accounts of three discussions in a genetic counselling clinic about the genetic testing of children. This examination of the process of genetic counselling and the chapter's focus upon the words used in the clinic, upon the transcript, emphasises the importance of attending to the discursive nature of genetic counselling. These accounts show how difficult it can be for the genetics professional to comply with the expectations of the family seeking testing for their child while simultaneously observing (what he/she considers to be) the appropriate professional guidelines. The authors conclude that more work is required both to understand the reason for a family's request for genetic testing of their children and to understand the concerns of the professional and the constraints within which he/she is operating. We are grateful to the authors and the *European Journal of Genetics in Society* for permission to publish this paper as a chapter in this volume.

Finally in this methodological section, there is a report from two of my close colleagues that examines the newborn screening programme for Duchenne's muscular dystrophy (DMD) in Wales. Newborn screening for DMD is controversial because it is of no particular benefit to the affected child although it may be helpful to the family as a whole; this chapter is part of an ongoing social evaluation of its impact. Because of the small numbers of families likely to be identified, the initial research tools employed were qualitative – especially the thematic analysis of in-depth interviews of the parents of affected boys. Subsequently, some elements of a questionnaire-based research design have been introduced. It has proved very interesting that the information obtained through these two routes – interviews and questionnaires – has proved to be complementary rather than merely confirmatory. Neither approach alone would have identified the full range of responses to the unwelcome news that a family's infant son might be affected by this lethal muscle disorder.

9.5 The wider context

A second chapter by Alderson makes a forceful appeal for us all to consider the wider social issues relating both to genetics and to childhood and children. Our attitudes to these topics and how we view the practical and ethical issues around them may be heavily influenced by the ways in which we talk. Powerful social forces may thereby shape the way in which we think. An emphasis on individual autonomy, for example, may obscure the ways in which social factors, such as economic inequality, undermine the very autonomy that is ostensibly being upheld. Alderson helps us to move beyond abstract ethical postures and to examine the real world and its difficulties as they are experienced by our fellow human beings. To what extent have advances in genetics served to help the lot of mankind?

The next chapter, by Kurtz, examines the notion of paternalism. This has become almost a term of abuse in medical ethics and yet there are circumstances in which the interests of children must be protected by their parents or by society as a whole – when paternalism will be entirely appropriate. Different formulations of appropriate paternalism are discussed, and these have implications for the proper organization and delivery of health services.

Hoedemaekers undertakes a conceptual clarification of the concept of 'risk' in relation to predictive genetic testing. This traditional philosophical task is especially necessary in this case, where familiar words (e.g. 'risk') can be used in both familiar and unfamiliar ways. Hoedemaekers distinguishes crucially between 'calculated risk' as a precise and quantifiable value, 'fuzzy risk' as an imprecise, incalculable hazard or danger and 'uncertainty'. He also makes the distinction between risk as the chance of developing a genetic disease and risk as a possible adverse outcome of genetic testing. The personal significance of particular risks given to individuals is also highly variable – the same objective risk figure could be understood and interpreted in numerous different ways. Hoedemaekers draws attention to the importance of biographical factors to account for these differences.

Dalby then contributes a brief chapter outlining the issues raised by the possible commercialization of genetic testing in Britain from the perspective of the (British) Genetic Interest Group (GIG). She suggests that most families would prefer genetic testing to be made available through the National Health Service (NHS) but that they would be willing to use commercial testing facilities if the NHS channels were slow or otherwise difficult. Families might also prefer to use commercial facilities if they required strict privacy to ensure that insurance companies did not have access to their test results.

The final chapter in this section and in the volume, by Wertz, reports the attitudes of genetics professionals from many countries towards the issues raised by genetic testing in childhood. The attitudes of public, parents and primary health care professionals in the US are also examined. It appears that genetics service providers take a more restrictive approach to these issues than public, parents or primary care health professionals. There are also highly significant differences between genetics service providers from different countries – those from northern and western Europe and English-speaking parts of the globe were much more restrictive than those from all other continents – only a minority of the former would test a healthy young child for Huntington's disease at parental request while a substantial majority of the latter (service providers from other continents) would do so. Possible reasons for these differences are discussed.

9.6 Appendices

Appendix 1 is the report of the Clinical Genetics Society's Working Party on the Genetic Testing of Children (1994). This sets out the initial grounds for concern about childhood genetic testing and examines a number of issues not explored in detail elsewhere in this volume – including adoption and the legal context of childhood genetic testing in England. This report is reproduced by permission of the *Clinical Genetics Society* and the *Journal of Medical Genetics*.

Appendix 2 presents the previously unpublished results of an attitude survey of members of the European Society of Human Genetics. The results of this survey cannot be regarded as representative because the response rates were too low; they were also too low to permit valid comparisons between countries. There are interesting parallels, however, between the results of this attitude survey and those reported in the chapter by Wertz.

References

Alderson, P. (1993) *Children's Consent to Surgery*. Open University Press, Buckingham.

American Society of Human Genetics/American College of Medical Genetics Report (1995) Points to consider: ethical, legal and psychological implications of genetic testing in children and adolescents. *Am. J. Hum. Genet.* 57: 1233–1241.

Archard, D. (1993) *Children. Rights and Childhood*. Routledge, London.

Bachman R.P., Schoen, E.J., Stembridge, A., Jurecki, E.R. and Imagire, R.S. (1993) Compliance with childhood cholesterol screening among members of a prepaid health plan. *Am. J. Dis. Child.* 147: 382–385.

Balfour-Lynn, I., Madge, S. and Dinwiddie, R. (1995) Testing carrier status in siblings of patients with cystic fibrosis. *Arch. Dis. Child.* 72: 167–168.

Bergman, A.B. and Stamm, S.J. (1967) The morbidity of cardiac nondisease in schoolchildren. *New. Eng. J. Med.* 276: 1008–1013.

Boddington P. (1994) Confidentiality in genetic counselling. In: *Genetic Counselling: Practice and Principles* (ed A. Clarke). Routledge, London.

British Paediatric Association Ethics Advisory Committee (1996) Testing Children for Late Onset of Genetic Disorders. British Paediatric Association, London.

Clarke, A. (1997) Parents' responses to predictive genetic testing in their children. *J. Med. Genet.* 34: 174.

Clinical Genetics Society (1994) Report of the Working Party on the Genetic Testing of Children. *J. Med. Genet.* 31: 785–797.

Dalby, S. (1995) Genetics Interest Group response to the UK Clinical Genetics Society report 'The genetic testing of children'. *J. Med. Genet.* 32: 490–491.

Duster, T. (1990) *Backdoor to Eugenics*. Routledge, London.

Fanos, J.H. and Johnson, J.P. (1995a) Perception of carrier status by cystic fibrosis siblings. *Am. J. Hum. Genet.* 57: 431–438.

Fanos, J.H. and Johnson, J.P. (1995b) Barriers to carrier testing for adult cystic fibrosis sibs: the importance of not knowing. *Am. J. Med. Genet.* 59: 85–91.

Fraser, F.C. (1974) Genetic counselling. *Am. J. Hum. Genet.* 26: 636–659.

Gittins, D. (1998) *The Child in Question*. Macmillan, London.

Harper, P.S. and Clarke, A.J. (1997) *Genetics, Society and Clinical Practice*. Bios Scientific Publishers, Oxford.

Harris, J. (1985) *The Value of Life. An Introduction to Medical Ethics*. Routledge Kegan Paul, London.

Institute of Medicine, Committee on Assessing Genetic Risks (1994) *Assessing Genetic Risks: Implications for Health and Social Policy* (eds L.B. Andrews, J.E. Fullarton, J.A. Holtzman, A.G. Motulsky). National Academic Press, Washington DC.

King, M. (1997) *A Better World for Children.* Routledge, London.

Lewontin, R.C. (1991) *The Doctrine of DNA. Biology as Ideology.* Penguin, London.

Lifshitz, F. and Moses, N. (1989) Growth failure. A complication of dietary treatment of hypercholesterolemia. *Am. J. Dis. Child.* **143**: 537–542.

Michie, S., McDonald, V., Bobrow, M., McKeown, C. and Marteau, T. (1996) Parents' responses to predictive genetic testing in their children: report of a single case study. *J. Med. Genet.* **33**: 313–318.

Muller-Hill, B. (1988) *Murderous Science.* Oxford University Press, Oxford.

Newman, R.T. Browner, W.D. and Hulley, S.B. (1990) The case against childhood cholesterol screening. *JAMA* **264**: 3003–3005.

Parsons, E.P. and Atkinson, P. (1992) Lay constructions of genetic risk. *Sociology of Health and Illness* **14**: 437–455.

Parsons, E.P. and Atkinson, P. (1993) Genetic risk and reproduction. Sociology Review **41**: 679–706.

Parsons, E.P. and Clarke, A. (1993) Genetic risk: women's understanding of carrier risk in Duchenne muscular dystrophy. *J. Med. Genet.* **30**: 562–566.

Ramsey, M. (1994) Genetic reductionism in medical genetic practice. In: *Genetic Counselling: Practice and Principles* (ed A. Clarke). Routledge, London.

Thelin, T., McNeil, T.F., Aspegren-Jansson, E. and Sveger, T. (1985) Psychological consequences of neonatal screening for alpha-1-antitrypsin deficiency. *Acta Paediatr. Scand.* **74**: 787–793.

Wertz, D.C. and Reilly, P.R. (1994) Genetic testing for children and adolescents: who decides? *JAMA* **272**: 875–881.

Wertz, D.C. and Reilly, P.R. (1997) Laboratory policies and practices for the genetic testing of children: a survey of the Helix network. *Am. J. Hum. Genet.* **61**: 1163–1168.

The genetic testing of children: some legal and ethical concerns

Sheila A.M. McLean

1. Introduction

The rapid advances in molecular biology promise a revolution in disease identification. Changes in the perception and incidence of illness and disease have helped to focus attention on genes as a major source of disease. As Pullen says,

'improvements in living standards and advances in medicine have led to a gradual decline in the incidence of nutritional deficiencies and infectious diseases in childhood ... others have been uncovered which are largely, or even entirely, genetically determined' (Pullen, 1990).

Genetic screening, on an individual or community-wide basis, can now detect a number of conditions, and will doubtless identify many more. The importance of the genetic component in health cannot be underestimated. For example, Wexler suggests that '...gene defects underlie 3000 to 4000 different diseases, and this is before one considers polygenic aetiologies or aetiologies in which there is interaction between genes and environment' (Wexler, 1991). The capacity to detect 'faulty' genes is, however, by no means uncontroversial, for a number of reasons.

First, with some exceptions, genetic information describes possibilities rather than certainties, yet this is poorly understood, at least outside the genetics community. The shortfall of information has been commented on in numerous fora. As the Nuffield Council on Bioethics put it:

'A broad public understanding of the scientific basis of medical genetics is essential if informed public policy decisions are to be taken about the introduction of genetic screening programmes. Such programmes ... have both an individual and a public dimension' (Nuffield Council on Bioethics, 1993).

However, despite the potential problems of so doing, the temptation may be to screen for everything that we can unless the consequences of so doing are adequately weighed, including the caveat that the progress of an identified condition

The Genetic Testing of Children, A.J. Clarke (ed.).
© 1998 BIOS Scientific Publishers Ltd, Oxford.

may well be affected by external factors and even if not, it remains relatively unpredictable in some crucial ways. As Hubbard and Wald point out,

> 'Genetic predictions, whether they involve testing or screening, are based on the assumption that there is a relatively straightforward relationship between genes and traits. However, genetic conditions involve a largely unpredictable interplay of many factors and processes' (Hubbard and Wald, 1993).

Second, there is a troubling impetus to identify social characteristics with genetic markers. The investigation into the human genome will doubtless provide a further incentive to proceed down this path. Thus,

> 'Genetics frequently provides an explanation or excuse for individual behaviours or traits, as well as for group differences. The genome project will enhance the tendency to give genetic explanations for individual and group differences in two ways. First, research will suggest genetic correlates for a wide range of human traits and behaviour. Some may prove to be spurious, but others will withstand scrutiny. Whenever such a genetic correlate is suggested for some ethically, legally, or economically consequential outcome, there will be a temptation to explain it as fundamentally – that is exclusively or exhaustively – genetic, and hence outside the individual's responsibility or capacity to control. ... Second, in the initial rush of findings of genetic connections to important human traits and behaviour, both scientists and the public may become too eager to embrace genetic explanations for a vast range of ethically significant phenomena' (Murray, 1991).

Science, then, may seem to be poised to decide the nature/nurture debate on the side of nature, encouraging the shedding of both personal and societal reponsibilities. After all, if people are 'made' the way they are, there is little if any incentive either for them to seek to change or for communities to offer them rehabilitation.

Third, and perhaps most critically, the fact that a genetic basis can be found for a particular condition is arguably lacking in utility if there is no method available to prevent its onset, to ameliorate its symptoms or to cure it. This last is of particular significance where screening programmes are instituted or when individual testing is contemplated. As Berg notes, 'One may well argue that there is no need to seek knowledge about one's prospects concerning future health unless the information can be of practical use. Thus, since there is no realistic prospect of therapy on the horizon, it could be argued that there is not much use in finding out if one possesses the gene for Huntington's chorea. No preventive or therapeutic intervention is available to the (at-)risk person, who could doubtless argue (for) a "right not to know"' (Berg, 1993). Of course, there may be some future benefit to be gained from refuting the argument from a 'right not to know'. For example, the gathering of data through community-based screening programmes may help in achieving more information about the condition itself. As Maddox says, 'Understanding does not always presage more effective treatment, but the promise is high' (Maddox, 1991). Equally, however, it could be argued that screening has not yet been shown to have these kinds of benefits, and that this is particularly true when the programme is essentially epidemiological rather than preventive or curative.

These problems affect all identifications of genetic constituents. No matter one's biological age, we are all vulnerable to the possibilities for discrimination, devaluing and distress which can follow from knowledge without therapy. However, it is likely that the consequences of such diagnostic capacities will be more profound for those who are relatively immature and whose understanding of the implications of the knowledge will be even less informed than that of adults. Explaining genetic information has been said to be difficult even between geneticists; it must be even more problematic in respect of non-qualified adults and intuitively it seems even less likely to be understood in all its complexities by the young.

Of course, all screening is open to challenge. Although there is what Stone and Stewart (1994) have called 'a second wave of evangelistic screening fervour...' for screening, its efficacy is in doubt. What genetic screening may achieve, in the absence of a concomitant cure, is the generation of more and more of the 'worried well' together with increasing numbers of frightened possible sufferers. Reports of young women having double mastectomies in the hope of averting a genetically indicated risk of breast cancer show the possible effects. It is not known, perhaps not even knowable yet, whether or not this radical procedure will in fact prevent the development of cancer, yet for some it is the logical outcome of genetic diagnosis.

In the Western world, much emphasis is placed on the status of childhood. Children are generally perceived as vulnerable, in need of adult protection. Yet we are at pains to give them such rights as we feel they can handle (for discussion, see McLean, 1995). Thus, laws make it unlawful to engage in sex with a young woman, but the law has also conceded that, where sex is occurring, children should have a right in certain circumstances, to the provision (in private) of contraception (*Gillick* v. *West Norfolk and Wisbech Area Health Authority*, 1985). This apparent ambivalence in fact represents a marriage between the twin aims of respecting and protecting children.

Undoubtedly, these same aims should permeate all aspects of medical care of children. But genetic information is different. As the Danish Council of Ethics (1993) points out:

'One of the peculiar aspects of genetic screening is that the things being screened for – gene and chromosome changes – can be passed down from one generation to the next. As a corollary to this, the detection of such a change means immediately having to regard the parents and any siblings or children of the person concerned....as belonging to a high-risk group....Therefore, genetic screening can very easily affect relatives of the people in question, genes being shared with relatives'.

The rationale for finding out the child's genetic make-up in this situation is not that described above. It may be a reflection of the attribution to the child of a right to know, but it can scarcely be described as protection. In fact, unless a corresponding right not to know is acknowledged, the result of screening in childhood may be to defeat our legitimate desire to protect children.

For the purposes of this chapter, it will be possible only to consider post-natal genetic testing and screening, although it is clear that problems also exist in respect of testing which occurs pre-implantation or pre-natally. Not the least of these problems is the likelihood that adverse test results will lead to non-implantation or pregnancy termination, choices which are not unlawful but which provoke considerable moral and ethical dispute. It must, however, be borne in mind that although these questions are not directly tackled, some of the conclusions reached in respect of post-natal testing will have a resonance for the debate surrounding its pre-natal use.

Childhood genetic testing, for the purposes of this chapter, will be defined as including presymptomatic testing for both childhood-onset conditions and for late-onset conditions. Equally, any genetic testing will be included, whether it is carried out as a screening test in the community as a whole or whether it is offered to individuals in specific families. This broad definition is employed since – without therapy – arguably any genetic diagnostic intervention has purposes and consequences which are unusual in medicine. Now, clearly, when tests are undertaken for certain other non-genetic conditions, it is known that a positive result may also be untreatable, but the reason for distinguishing genetic testing is not simply the absence of a cure. Genetic diagnosis would still be unique even if a cure were available, because of one simple but important fact. The diagnosis of a genetic condition in a child has ramifications for all of those related to that child. The rationale for testing, then, may well not relate specifically to that child, but may serve familial purposes, posing tensions between the interests of different people. Thus, 'when genetic testing of one person can benefit another family member, privacy and autonomy interests of the former may collide with the relative's interests in protecting her health or planning her future' (Suter, 1993).

This competition between privacy and responsibility, between autonomy and the rights of others, is one which has the potential to threaten our current understanding of the hierarchy of human rights and interests. This issue is important also for two further reasons. First, it calls into question any consent which – in the case of children – will often be given by someone else, usually a parent. In other words, and this will be discussed later, the rationale for the provision of parental consent may reasonably be thought to be in need of challenge, or at least of close scrutiny, since it may be designed, at least in part, for reasons which go beyond the individual child. Second, if consent is given by the child him/herself, the knowledge that others in the family may be affected places a heavier burden on the child than would knowledge of a non-heritable condition. Should the child share the information? Is there a moral obligation to warn others of the risks which they too run? Moreover, what are the implications of identifying gene carrier status?

Finally, it must be noted that where the authority to test comes from a third party, it flies in the face of the generally accepted rule that counselling should always be a pre-requisite of a valid consent. If the issue is the welfare of the child, then counselling the parents, for example, would not meet the standard required, especially if there is reason to believe that genetic knowledge may cause harm to the individual concerned.

The genetic testing of children, therefore, is even more complex than that of adults. Both the rationale for, and the sequelae of, testing are infinitely more problematic in these cases, even if some good comes out of the results. I have indicated that I will address two different kinds of testing, but in order to make the picture as clear as possible, I will also exclude testing for conditions for which therapy is available. The rationale for this is that the fundamental problems which I have already raised effectively disappear where treatment is available. Indeed, if treatment is available, then it would arguably be irresponsible not to provide the necessary tests and to commence treatment. Equally, the burden of knowledge which the child bears, and the decision whether or not that information should be shared, are no greater than they are in any other case of illness if family members can also be treated.

2. Presymptomatic testing for childhood-onset disorders

It is argued by some that this kind of testing presents different problems, and requires different solutions, from those involved in testing for late onset conditions. The reasons given are in themselves informative. In their response to the Report of the Clinical Genetics Society (1994), for example, the Genetics Interest Group (GIG) argues that '...there are valid reasons to test in the case of disorders for which there is no presymptomatic medical intervention' (Dalby, 1995). These reasons, they say, include

'...possible freedom from anxiety; facilitating open relationships; and the parents' need to secure the best environment they can for themselves, the child who will develop the disorder, and other children in the family' (Dalby, 1995).

These are laudable aims. Yet, a point of interest is that few of them actually concern the child *per se*. And, as Davis (1995) has pointed out, '...we cannot assume that making use of present patients for the good of future patients is ethically legitimate, particularly since the patients concerned are in no position to volunteer....' Nor can we assume that easing the job of parents is more important than the voluntariness of agreement to participate. Of course, where the child is able to give his/her own consent, then it might be said that we should permit this in the interests of respecting children rather than patronizing them. This, of course, is a valid point.

However, it is a general rule of law that, for consent to be valid and therefore worthy of respect, it should be free from external pressures. It would be difficult to be certain that a child, living within the ambit of a concerned family, would in fact be freely seeking such information, both because of the likely family pressure to know and because of the fact that for consent to be valid the child must understand the nature and consequences of what they are doing. If genetic information is as complex as has been suggested, one would have to be very rigourous in assessing the level of understanding, especially where there is room for any doubt about the true voluntariness of the request for screening in the first place. This would not rule out the capacity of children to provide consent to screening, but it would demand careful assessment of any purported agreement.

Moreover, to satisfy the protective role of the law, in cases where the child is insufficiently competent to give a consent, the rationale for testing becomes more dubious. As I have noted, the majority of the reasons given in favour of testing by the GIG relate to the family, or the parents, rather than the specific child. To be sure, the child may be the recipient of benefit from an informed family, but parental rights to consent on behalf of their children are legally constrained.

Parents have the authority to agree to medical intervention in respect of their child subject to the general rule of law that the intervention should be 'in the best interests' of that child. Although 'best interests' is a somewhat vague term, it nonetheless presumes a balance which falls in favour of benefit rather than burden. Even if, as they argue '...children can cope with information about themselves from an early age...' (Dalby, 1995), in legal terms this is not the issue. The 'best interests' test requires evidence that burdens do not outweigh benefits. The burdens of assimilating, even coming to terms with genetic information may not be wholly insurmountable, although they must be great.

It has been said that

'The advance of genetic knowledge has already increased the range of medical and procreative opportunities, and the choices raised by their advent can be discomfiting. Genetic screeners worry that the publicity given to screening programmes may cause needless apprehension among people whom the roll of the genetic dice has favoured, and that genetic information may lead to unrelievable anxiety among those whom it has not....The revelation of genetic hazard has been observed to result not only in repression, but in anxiety, depression, and a sense of stigmatization' (Kevles, 1985).

If this is so, then what are the benefits?

Of course, the benefits may be a better-adapted family, but this is by no means an unequivocal outcome of genetic knowledge. Although we would wish that children were treated, if anything perhaps, better when such knowledge is acquired, there is no guarantee that this will be the case. The psychological impact on both child and family may be negative rather than positive. In addition, even if these benefits do accrue, they relate to more than merely the child and the rationale for parents seeking the information in the first place will doubtless also relate to the welfare of the rest of the group. Ultimately, we may wish to concede that parental consent in such cases is valid, but from this brief description of the constituents of a valid consent, it can surely be argued that the validity of the consent needs considerably more scrutiny than is usual. We cannot presume that parents are best equipped to decide on what is in the best interests of a particular child, merely because they are his or her parents. And where the harm which may flow from diagnosis is so clear, the law should engage in a real, as opposed to theoretical, balancing of benefits and burdens.

3. Testing for late-onset disorders

Although many of the arguments in this situation are similar to those in the previous section, there is rather more consensus here, although its basis is unclear. For those who would argue that testing for childhood-onset conditions is permissible, the rationale seems to be the support which could potentially be provided to the child. Yet, these same commentators are generally opposed to testing for late-onset conditions. However, the benefits which are claimed to follow from other testing might equally be argued to be present here. In the view of the Genetics Interest Group, for example, '...it would be unethical to acquire knowledge that a child will be affected as an adult unless there is early treatment available' (Dalby, 1995). This is in direct agreement with guidelines drawn up many years ago by the World Health Organisation, but it still begs some questions.

If parental consent is valid in the first case, it is no less valid in the second, unless there are some significant differences. But what would these differences be? If the rationale underlying the permission of parental authorization of testing is that parents can perhaps help their children to adjust, then why should the time frame within which that adjustment takes place be relevant? Indeed, where the child faces a longer future, might early knowledge not enable the family better to support the child? If the 'best interests' of the child are served by knowing about a condition which will manifest itself soon, why is it the case that the same interests would not be served by having a longer time to come to terms with the likely onset of disability or disease?

The fact that these questions are difficult to resolve points to the reason for dubiety about childhood testing at all, and enhances concern about the legal standing of parental consent. Indeed, it is tempting to assert that the value of testing is not in fact primarily for the child, but rather for the parents and/or extended family. This is not in itself an ignoble aim, but it does fly in the face of the rationale for allowing parents to consent on behalf of their child when the benefits do not actually accrue to that child.

4. Conclusion

The legal position is, of course, only one perspective, but it is an unavoidable one. Inevitably there are matters of clinical and psychological import in these decisions, but ultimately any procedure carried out without a legally valid consent amounts to an assault. What has gone before highlights some real concerns about the extent to which parental consent in these cirumstances is sufficiently valid to justify the genetic testing of children who cannot consent on their own behalf. Yet, if there is a value in families knowing this information, it is plausible to argue that the earlier the information is found out the better. This means that, very often, the child would be undoubtedly characterized for legal purposes as incapax, and therefore unable to offer their own consent. Third party agreement would, therefore, be necessary, yet its standing is challengeable.

It is the difference between genetic and other health care information which poses this dilemma. We must avoid turning our responsibility for protecting children into another form of paternalism. However, we must also bear in mind that the central issue is the best interests of the child, not the interests of those who are related to him or her. And at this stage, it is also necessary to look to the wider picture. To date, what has been under consideration is the effect of the information gained on the family group. But genetic information also has a wider social and political impact. As has been said,

'Individuals do not live in a vacuum, although many of the philosophies underlying the so-called liberal Western tradition seek to emphasize the uniqueness of the individual by valuing concepts such as autonomy and integrity. These ethical principles safeguard the person against unwarranted intrusion into the private sphere of life, but they are nonetheless defeasible in the face of greater needs or more powerful principles. The fact that genetic information is of critical importance for the community as well as the individual places that individual's privacy and other rights in potential conflict with the greater good of the community, either immediate (i.e. family) or distant....' (McLean, 1994).

Whether identifying psychological traits or clinical conditions, genetic information carries within it the potential to be used in a manner which generates or reinforces discrimination. Although other groups may also be the victims of this based on medical and other conditions, the scope in genetics is much wider, both for the individual and for his/her family. The person who suffers, or who will suffer, from a known condition becomes vulnerable to discrimination in employment, insurance and health care, to name but three areas. Nor is it enough to say that the information once gained will be retained in confidence. No system for storing information yet devised is foolproof. Moreover, and more worryingly, once information is known, in the absence of strict privacy laws, there is nothing which prevents employers or insurers from seeking to have access to it and nothing which prevents them from building plausible arguments as to why they should have this access. Thus,

'Although knowledge about genes offers benefits, these must be balanced against possible harms, like stigmatization, anxiety due to ignorance or knowledge of genetic status, and discrimination. Because society's resources have limits the cost of this knowledge must also be taken into account, as must the aims of researchers and the value of the research they propose' (Robinson, 1994).

In conclusion, therefore, there may be some relatively strong arguments for childhood detection of genetic conditions, but there are also strong ones against. These arguments relate both to the immediate well-being of the child and to his or her future. As was famously said in a different context, parents may be free to make martyrs of themselves but they may not make martyrs of their children (*Prince* v. *Massachussets*, 1944). This, then, is an argument for caution. We must evaluate what benefits are to be gained from genetic testing and to whom these benefits will accrue.

As was said earlier, genetic testing is not in itself uncontroversial, nor is it universally regarded as good. When it is done without the consent of the person who will be most immediately affected, it is even more problematic. As has been noted,

'No matter what way a person comes to terms with such information – which may be either a blessing or be terrifying – in all societies where maximum priority is attached to the right to self-determination with regard to information there is a consensus that the decision for or against such predictive diagnostics and the result thereof must remain a private matter' (Schmidtke 1992).

Thus, particularly when there is no therapy available, it can be argued strongly that the impetus for testing must come freely and after counselling only from the individual concerned. This would hold true whether or not the condition will likely manifest itself sooner rather than later. And when current legal tests require best interests or welfare to dominate third party treatment of children, there is real reason to examine closely the validity of proxy, even parental, consent.

Of course, all of these arguments may be turned on their head when therapeutic potential can hold its own with diagnostic capacity, but for the moment this seems some long way off. As Friedmann (1990) points out, '...there remains a serious gap between disease characterization and treatment'. The task for the present is not to be seduced by bare knowledge, however sophisticated, into reneging on our commitment to respect children. But this respect may include a protective function – protection from their use as vehicles to cement families, to provide information for others or from future discrimination. Arguably, genetic knowledge is so significant that it should never be obtained without the real and free consent of the individual him/herself. This would not disbar the mature, understanding child from finding out their genetic inheritance, but it would require that the decision is not taken in advance of their developing legal capacity where there is no direct benefit to them of having the knowledge, and indeed, where there may be real reason to believe that harm could accrue. The fact that we can identify such information means that

'...advances in genetic knowledge...will gradually affect every person with or without the capacity to choose genetic screening and testing. Information about the molecular contents of virtually every gene harmful to human health can be – and gradually will be – used for testing. Human beings will be genetically laid bare and vulnerable as never before.' (Fletcher and Wertz, 1991)

In the absence of convincing arguments to the contrary, then, it is necessary to err on the side of caution. By permitting the mature young person the right to seek such knowledge, assuming that the search is conducted on the basis of a genuine consent, we respect the person. However, our protective role in respect of children demands that – unless we know that a good will follow, for example, therapy – we should not be overly willing to permit testing to take place where the child's consent cannot be obtained. In fact, when coupled with the serious doubts raised here about the validity of any purported parental consent in these circumstances, we may even conclude that the use of children in this way would be unlawful.

References

Berg, K. (1993) *Confidentiality Issues in Medical Genetics: The Need for Laws, Rules and Good Practices to Secure Optimal Disease Control.* Council of Europe, CDBI-SY-SP (93)3, Strasbourg, 5 October 1993, p. 5.

Clinical Genetics Society (1994) Report of the working party on the genetic testing of children. *J. Med. Genet.* **31**: 785.

Dalby, S. (1995) Response of the Genetics Interest Group to the Clinical Genetics Society report: 'The Genetic Testing of Children'. *J. Med. Genet.* **32**: 490–491.

Danish Council of Ethics (1993) *Ethics and Mapping of the Human Genome*, p. 57.

Davis, J. (1995) Genetic testing for familial hypertrophic cardiomyopathy: ethical issues. *BMJ* **310**: 858.

Fletcher, J.C. and Wertz, D.C. (1991) An International Code of Ethics in Medical Genetics Before the Human Genome is Mapped. In: *Genetics, Ethics and Human Values: Human Genome Mapping, Genetic Screening and Therapy*, (eds Z. Bankowski and A. Capron), xxiv CIOMS Round Table Conference, 97, p. 97.

Friedmann, T. (1990) Opinion: The Human Genome Project – Some Implications of Extensive 'Reverse Genetics' Medicine' *Am.J. Hum. Genet.* **46**: 408.

Gillick v. West Norfolk and Wisbech Area Health Authority [1985] 3 All ER 402 (HL).

Hubbard, R. and Wald, E. (1993) *Exploding the Gene Myth*. Beacon Press, Boston, MA, p. 36.

Kevles, D. (1985) In: *The Name of Eugenics: Genetics and the Uses of Human Heredity*. Penguin, Harmondsworth, pp. 297–298.

Maddox, J. (1991) The case for the human genome, *Nature*, 352: 12.

McLean, S.A.M. (1995) Genetic screening of children: The UK Position, *J. Contemp. Health Law Policy* **12**:113.

McLean, S.A.M. (1994) Mapping the human genome – friend or foe? *Soc.Sci.Med.* **39**: (9) 1221.

Murray, T.H. (1991) Ethical issues in human genome research *FASEB J.* **5**: 55.

Nuffield Council on Bioethics (1993) *Genetic Screening: Ethical Issues*. London, para 8.3, p. 75.

Prince v. Massachussets (1944) 321 US 158.

Pullen, I. (1990) Patients, families and genetic information. In: *Family Rights: Family Law and Medical Advance*, (eds E. Sutherland and R.A. McCall Smith). Edinburgh University Press, Edinburgh, p. 42.

Robinson, A. (1994) The ethics of gene research. *Can.Med.Assoc.J.* **150**(5): 721.

Schmidtke, J. (1992) Who owns the human genome? ethical and legal aspects *J. Pharm. Pharmacol.* **44** (Suppl. 1): 205.

Stone, D. and Stewart, S. (1994) *Towards a Screening Strategy for Scotland*. Scottish Forum for Public Health Medicine, Glasgow, p. 45.

Suter, S.M. (1993) Whose genes are these anyway? Familial conflicts over access to genetic information, *Mich. Law Rev.* **9**: 1854.

Wexler, N.S. (1991) Disease gene identification: ethical considerations. *Hospital Practice* **145**: 145.

2

Talking to children – and talking with them

Priscilla Alderson

When the princess kissed the frog, he turned into a prince and they lived happily ever after. Well, quite happily – as told in the sequel. The frog had an identity crisis. Was he a frog or a prince? He set out to find his true identity (Scieszka and Johnson, 1992). Like all great stories, the frog prince deals with the eternal questions.

How much are we trapped by biology, limited and defined by our bodies, our genes, and how much can we live beyond these apparent limits?

How much can we achieve by our own efforts and with other people's help, and how much do we have to rely on other powers – science, genetics or magic?

How much does our state at birth determine who we are and will become?

How much does our behaviour, even our appearance, reflect the attitudes of other people, such as love or rejection, and their beliefs about us?

There are many stories of ugly lonely 'beasts' who, when they are loved turn into handsome heroes. How much are we really our 'worst' or our 'best' sides?

1. Introduction

This chapter is not an instruction manual, as if talking with children, like baking a cake, will turn out right if you follow the recipe. Instead, children are very varied and unpredictable agents in their own right, partners in discussions even when they refuse to talk. This chapter is mainly about stepping back and looking at some of the underlying complications which have to be dealt with at some level, especially when talking with people who are very intimately affected by genetic conditions. The main areas covered in this chapter consider communication in relation to risk, contradictions, timing and content, children's competence, and methods and barriers when talking with children.

The Genetic Testing of Children, A.J. Clarke (ed.).
© 1998 BIOS Scientific Publishers Ltd, Oxford.

2. Words and meanings

Everyone tells stories – their versions or accounts of raw experience. Science is often seen as a way of escaping from stories and telling neutral 'truths' or 'facts'. But like all other accounts, scientific ones have to use language, which is loaded with complex meanings and values as much as fairy stories are. Genetic 'information' or 'facts' about a genetic condition are complicated by:

- uncertainties – about how pronounced the condition might be
- speculation – about the way a child's life might be affected by the genetic condition or by other influences
- value judgements – about which lives are worth living and who counts as a human being, such as in the efforts put into developing and marketing the 'triple test' (maternal serum screening for Down's Syndrome).
- anxiety – about risk and responsibility, control and blame, and sometimes irrational guilt about genetic events which are beyond our control
- the underestimation of human resourcefulness and the strength of community when we jointly confront and live with difficulty and disability – as we almost all have to do in old age
- 'geneticization' (Lippman, 1994) – a great (over)emphasis on biology and technology – genes, mapping the genome, antenatal screening and selection, hopes of genetic engineering, methods of controlling and preventing 'problems' – with inadequate attention being paid to relevant social concerns
- related to geneticization is the tendency to inflate the importance of genetic influences on health – many genetic disorders hardly affect daily life, many are extremely rare, and with many conditions only a few of the people that have them are severely affected. Yet it is often implied that genetic disorders are extremely numerous, and generally very severe, and each affects large numbers of people. Among the general public who do not belong to genetic interest groups or work in genetics but have quite serious genetic disorders, how many of them think of themselves as having faulty genes like defective machines? How many carriers think of themselves as carrying a time bomb waiting to explode in later generations? Do people generally perceive themselves in these consciously technical, mechanistic terms?

All these complications can affect the way people talk about genetics with children – and with anyone else of any age. Most of this chapter applies to all age groups. The complications raised by genetic issues may be highlighted when talking with children but they are not unique to them.

3. Risk in talking with children

As discussed in a later chapter (Childhood, Genetics, Ethics and the Social Context) certain strong themes in the way we think about genetics resonate with many other aspects of life, including the task of talking about genetics with children. Some of these themes can heighten concern about risk. There is the feeling that 'we ought to get it right', that complex information should be both clear and correct, and that information will raise the child's anxiety. Yet at the same time there is also the feeling that anxiety should somehow be managed and relieved. Besides coping with these contradictory aims, the adults' hopes to 'get it right' are inevitably accompanied by worry about 'getting it wrong', about being incorrect, or about giving too much or too little information for the child's good.

Subconsciously, other trends in genetics might increase an informant's anxiety: quests for high standards of research and knowledge, and for the 'perfect child'; the assumption that some pregnancies should be terminated because coping with the consequences would be too hard for the parents and the potential child; the desire to protect people from suffering; a lack of confidence in people's ability to cope with difficulty. Those who are strongly guided by such assumptions are likely to see talking with children as risky.

Adults with greater confidence about sharing knowledge with children are less worried that there is a 'correct' method. They are less anxious about taking risks and making mistakes, trusting that they will learn from children's responses and with them work out appropriate ways of sharing knowledge, recovering from blunders, and together coping with confusing, uncertain and distressing information. They have confidence in children's resilience and resourcefulness.

Decades ago, when four (fictional) children wanted to go camping and sailing on their own, their father agreed saying, 'Better drowned than duffers, if not duffers, won't drown' (Ransom, 1930). Today he would probably be imprisoned for negligence. Western society is becoming increasingly anxious about risk, blame and litigation; it is almost impossible to have an accident when every event can be seen as someone's 'fault' (Green, 1992). If we consider a spectrum of risks associated with different activities (*Figure 1*), there has been a progressive shift of many activities towards the careful, cautious or fearful categories since at least the 1970s.

The shift along this spectrum towards fearfulness has been accompanied by increasing ascriptions of responsibility and potential blame for any 'accident' that may occur.

When talking with a child, adults need to reflect on how they feel about and portray, for example, the possibility of the child as a carrier passing on a genetic condition to the next generation: as fearful and to be avoided at all costs, or as requiring caution, or as a risk that is worth taking and acceptable whatever happens. When parents say that one benefit in warning their teenage son about genetic risk is that it will inhibit his sexual relationships, is this really in their son's best interests, now and in the long term? Should he be controlled by fear, rather than by his own sense of responsibility? What might the effects of this be on his present and future sense of identity? Do parental warnings that particular activities are to be feared serve the interests of the parents or those of their child?

4. Contradictions

Modern genetics is caught within social ideals which contradict and yet also sustain and reinforce one another (*Figure 2*). People grappling with these conflicts can feel that their aims and values are being pulled apart by opposing forces, yet they are also curiously reaffirmed by these pairs of opposites. This can make

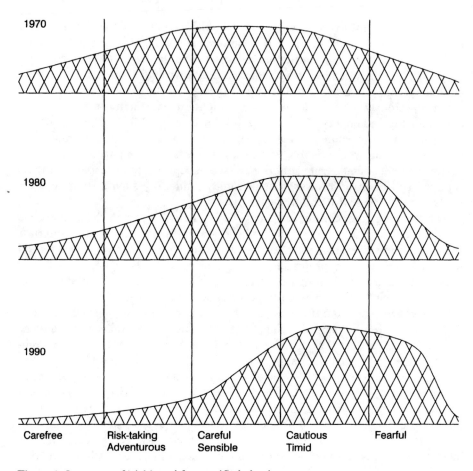

Figure 1. Spectrum of 'riskiness' for specific behaviours

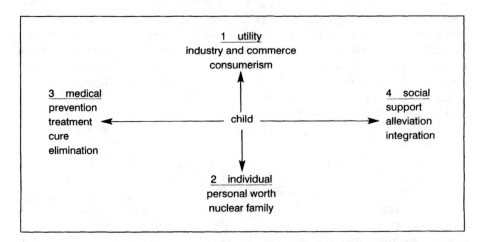

Figure 2. Contradictory ideals

explanations about genetic issues still more complicated, such as in the choice of words and emphases and the presentation of options.

The first opposition is between social utility and personal worth: some people increasingly measure things and people for their utility, their commercial worth, and the way they contribute to or detract from efficient industry. Impaired people in this light are seen as a net loss. If consumerism is applied to the sphere of reproduction, it encourages people to expect the 'perfect' child and to reject the 'imperfect' fetus. As one expectant couple in the US is reported as saying, 'If he can't have a shot at being president, we don't want him' (Rapp, 1988).

The second dimension of opposing views relates to how a child with a genetic condition can be seen as either a case of disease to be rooted out, or as a person requiring social support. It is said that one reason for the rising divorce rate is that people are encouraged to expect higher rewards than ever before from close relationships as a haven from a hard world; mothers are exhorted to keep their children clean, safe, busy, stimulated and achieving (Rose, 1990). Mutual satisfaction within families is highly valued, sometimes in terms of parents expecting to be rewarded by their children with expressions of affection and with high achievements, and children expecting love to take the form of expensive gifts.

The contrast along this social versus medical dimension involves both contradiction and reinforcement. Medical technology enables us to terminate pregnancies affected by genetic conditions. Indeed, some doctors now talk of 'eliminating' a genetic defect. This can lead to less tolerance of disability and disfigurement, and higher expectations of treatments and cures. It is advanced technology, however, that also produces the aids and practical means of support, which alleviate disabilities and enable severely disabled people to lead fulfilled lives, integrated into society in ways that were previously not imagined. The contradictions arise in the ways technology is used – to exclude or to aid inclusion, to 'eliminate' people with cystic fibrosis or to alleviate their condition and greatly extend their lives. 'Integration' includes the global interest in disability rights, equal opportunities and the tolerance and celebration of diversity, all of which can be aided by new technologies.

In the past few decades, possibilities have expanded and values have become more diverse. So, for example, a child with Down's syndrome who might scarcely have been noticed as different in many earlier societies, can now be seen as either a misfit, being too great a burden on the state, or else as a highly valued, loving and loved person (Goodey, 1991). Such a child can be portrayed as either a problem for health professionals to prevent and to treat (with plastic, cardiac or ENT surgery, speech therapy, and so on), or as someone with much to offer when fully accepted into society (Alderson, 1990). This acceptance can be actively promoted by, for example, effective treatment of the heart conditions which affect many children with Down's syndrome, and which would be treated without question in other children. When the medical and the social perspectives complement one another in this way, the same principles are applied equally to everyone and there is positive allowance for the differences between individuals.

These four positions, or perspectives, are pulling apart from each other, and yet they can reinforce one another. Explaining the more severe genetic conditions to a child involves making sense of these contradictory forces, as far as is possible.

5. Timing and content

Talking with children affected by a genetic disorder raises questions about what and when and how to tell them. It also involves listening to the children and learning from them. Each child, and each child–parent relationship, is unique. There is a limit to how helpful general ideas can be when talking with the individual child – so much depends on the specific relationship and on how freely and comfortably the child and adult talk about many other things.

It may be assumed that parents should have an orderly plan, gradually unfolding information as the child 'develops', but life is not like that. Some children understand something long before the textbooks say that they can, others want to talk years after that 'stage'. Children ask unexpected questions at awkward times; parents have to think of quick responses. One mother described how she was preparing her daughter, who had a mucopolysaccharide disorder, for major surgery. Her daughter asked, 'Might I die?'. The mother said that she hesitated, and she knew that the hesitation expressed far more than any follow-up words, so she decided to explain more than she had expected to. She worried that her young daughter would not cope well with the alarming news. The girl wrote a story about heaven; although distressed at the time, she did seem to come to terms with the information (Alderson, 1993). Apart from deciding to initiate discussions, adults are perhaps best advised on the timing and extent of their talking by the child's own questions and other cues.

Genetics involves words loaded with good and bad news: 'bad' genes or 'incorrect, impaired, wicked, naughty' genes. 'Make it right' or . . . 'better', and so on. How can the 'right' words be found? One way is to find out the words which the child or teenager uses, their questions and concerns, what they think their friends will or do think, their 'ascertainable wishes and feelings' in the words of the Children Act, 1989. This includes trying to sense how much they want to talk, when and whether they want to talk. Sharing complex and potentially distressing knowledge with people of any age is not easy. The main difficulties arise when adults try to 'talk down' to children.

There are barriers to communication, especially in busy clinics, such as limitations of time and space, and the lack of confidence and skill of some adults. There are problems of language with young children, people with learning difficulties and those who speak little or no English. Communication goes beyond words in tone of voice (optimistic, anxious, nervous or authoritarian), facial expression and body language which can be more expressive than words. Careful listeners pick up many cues from pauses and unspoken responses when giving and receiving information. More important than practical barriers are those of attitudes – prejudice that it is unwise or unkind to inform children or to trust them to make decisions, the 'adult alliance', which assumes that adults must be in control and that to defer to children is to betray both this alliance and also children's reliance on adults. There is anxiety about intruding on children's privacy on the one hand, or burdening them with unwanted anxieties on the other (Clarke, 1994; Clinical Genetics Society, 1994; Dalby, 1995; Michie, 1996).

How can reticent children be respected without being excluded from discussions? Geneticists and counsellors may have different agenda from their clients, while assuming that they have the same concerns and order of priorities. All these barriers can affect discussions with adult clients as well as children. Attempts to create special skills for talking with children risk belittling and demeaning them. It is more worthwhile to practise sensitive ways of relating to each individual to find appropriate ways of sharing information with them as clearly as possible whatever their age or circumstances.

6. Competence

Talking with children is frequently discussed in terms of assessing their competence to understand and to make decisions, such as agreeing to a predictive genetic test. Some people advocate waiting until young people are 18 years old and therefore 'competent'. Others believe that some children want to have answers, to sort out misunderstandings, and sometimes to make decisions at a much younger age. There might seem to be a contradiction between the two views, raising the age for involving children in one case and lowering it in the other. They are reconciled when attention is paid to competence as long as this is not identified with a specific age.

Competence to make decisions means understanding the relevant information, having the wisdom or discretion to evaluate it in the light of one's best interests, and having the confidence to act with some independence and, if necessary, to take responsibility and accept blame. Few if any adults fully understand, or have complete discretion and independence, or indeed want to have such extreme capacities. Competence is a relative not an absolute quality, and higher standards should not be expected from children than from adults, or from the parents who may decide for them. As mentioned earlier, genetics is complicated with many uncertainties.

Assessments of competence usually attend wholly to the child's capacities. Yet the context also needs to be assessed. The competence of most people depends in part on how clearly they are informed by health professionals, and how much their discretion and independence are respected by all concerned. Besides relationships between the individuals, the more general social context affects competence, such as how far experts understand the condition and the relevant testing or treatment, how new or risky the treatment is, how much time and space is allowed in the hospital for quiet discussion, the beliefs held by the relevant adults about children's abilities, and many other factors.

Assessments of the child must therefore be moderated in the light of the specific setting, for example whether the parents are used to sharing knowledge and decisions with the child or not. This moderating can be done positively or negatively. If the parents tend to keep secrets and to reserve power to themselves, a negative approach would assume that the child is immature and incompetent. The context would be reflected back into the assessor's view of the child. In a positive approach, the assessor would consider whether, despite the constraints and perhaps initial signals that the child does not and cannot want to be involved, this is not inevitably the case. Further discussion might clarify the child's wishes about being involved and the parents' willingness to reconsider. The context can be used either as a further constraint, or to help move towards a fairer and more careful assessment. The two approaches partly depend on whether the assessor works in psychological traditions of taking the individual as a somewhat static unit of analysis, or else sees people within a network of dynamic relationships and experiences.

When the competence of younger children is respected, they are treated with the same respect accorded to adults. (1) Adults cannot be tested against their will in order to inform or benefit others. (2) Their results are confidential. (3) An adult who very much wants to have a test which is often done for other people would not be refused.

Those who want to set different standards for non-urgent interventions for children and for adults should be expected to justify this; the onus is on them. Respect for children can expose them to the dangers of unwanted tests and interventions. It is important to check that they are not being unduly pressured to consent, and that they understand any risks and disadvantages – as in counselling adult patients. Many adults feel ambivalent about tests for serious conditions, and undertake them partly to inform their relatives. So to look for complete conviction and freedom from family pressures in the child would be as unrealistic as to expect this in all adults. It is a question of a reasonable balance between the child deciding as an individual and as a member of a family, and this will not always be easy to discern.

One very important factor is whether children do or do not have knowledge of the relevant genetic condition based on personal experience, perhaps of a close relative suffering severe problems from the condition. If they do not have such direct personal experience then however mature and intelligent they are, their understanding and wisdom are likely to be much more limited. Children with experience of serious illness or disability and treatment, even if they are well below average at school, can have high levels of competence concerning decisions about their health care. During interviews with children in schools for physical, sensory, emotional and learning difficulties, we have found children who are far more competent and aware than might be anticipated from generalizations made about their disabilities (Alderson and Goodey, 1996). This finding has two important implications for talking with children about genetics. A much broader range of children may be able to be informed than has previously been supposed, and overly pessimistic descriptions and decisions about some genetic disorders must be reconsidered.

One unpublished description of a working class family, in which two younger children have severe genetic conditions, assumed that the older sister, aged 16, knew nothing about the disorders and should not be tested for her carrier status, because the family did not know the numbers of the affected chromosomes. But, of course, the sister had profound experience of the effects of the disorders which were far more relevant to any decisions she might make than formal genetic knowledge would be. There are other valuable sources of knowledge, such as the mass media, and their impact is discussed in Chapter 19 on the social context.

Competence has four levels (Alderson and Montgomery, 1996), and the first two have no threshold of age or ability. One-year-old children can exchange information and views, resist or accept interventions, and react positively to explanation and persuasion and negatively to coercion. The levels of competence are:

- to be informed;
- to express views;
- to influence a decision;
- to be the main decider about proposed treatment or care.

The United Nations Convention on the Rights of the Child, 1989, and the Children Act 1989 advocate the first three levels, the third one in accordance with the child's ability. Law Lords in the *Gillick* case, 1985, advocated the fourth level for children who can understand the relevant information, make wise decisions and are competent to do so in the judgement of the doctor treating them (for a discussion of this case and subsequent rulings and the current state of English law see Alderson and Montgomery, 1996; also see the Age of Legal Capacity Scotland Act 1991, s.2 (4)).

Assessments of competence are difficult and contentious. There are no clearly agreed methods and criteria. Psychological testing tends to be biased by outdated developmental theories which under-estimate children's abilities. A solution to these problems is to assume that school-aged children are competent to influence personal decisions affecting them, and that anyone who thinks that their informed views should not be taken seriously would have to demonstrate that the child is not competent.

7. Conclusion

Some geneticists are concerned about the technical imperative, the belief that it is better to do something than nothing, better to act now than to wait and see. They have reservations about the rush to have genetic tests for later-onset disorders for which there is no effective treatment or prevention. Some are concerned about pressures either to give alarming information to children, or to withhold it from them leaving them in frustrated uncertainty. Because of these pressures, it is not helpful to make rules about specific ages when children should be informed or involved in decisions. So much depends on individual cases.

Instead of being a 'how to' manual, this chapter is intended to help people to reflect on underlying questions and contradictions, and to think about what they know and do not know before deciding what to do. This is in the hope that by unravelling some of the questions, they may find it easier to respond to children's enquiries and to respect their interests. Through talking with children, we can learn from them how to do this, and we can reduce problems of coercion, fear, ignorance and resentment. Talking can help to clarify mutual understanding while, whenever possible, adults and children work together towards the best, or the least harmful, decision for that child.

References

Alderson, P. (1990) *Choosing for Children: Parents' Consent to Surgery*. Chapter 5. Oxford University Press, Oxford.

Alderson, P. (1993) *Children's Consent to Surgery*. Open University Press, Milton Keynes.

Alderson, P. and Goodey, C. (1996) Research with disabled children: how useful is child-centred ethics? *Children and Society* **10** (2): 106–116.

Alderson, P. and Montgomery, J. (1996) *Health Care Choices: Making Decisions with Children.* Institute for Public Policy Research, London.

Clarke, A. (1994) *Genetic Counselling: Practice and Principles.* Routledge, London and New York.

Clinical Genetics Society (1994) Report of the Working Party of the Clinical Genetics Society on the Genetic Testing of Children. *J. Med. Genet.* **31**: 785–797.

Dalby, S. (1995) Genetics Interest Group response to the UK Clinical Genetics Society report 'The genetic testing of children'. *J. Med. Genet.* **32**: 490–491.

Goodey, C. (1991) *Living in the real world: families talk about Down's syndrome.* Twenty-One Press, Newham.

Green, J. (1992) Some problems in the development of a sociology of accidents. In: *Private Risks and Public Dangers* (ed S. Scott). Avebury, Aldershot, pp. 19–33.

Lippmann, A. (1994) Prenatal genetic screening: constructing needs and reinforcing inequalities. In: *Genetic Counselling: Practice and Principles* (ed A. Clarke). Routledge, London and New York, Chap. 7, pp. 142–186.

Michie, S. (1996) Predictive testing in children: paternalism or empiricism ? In: *The Troubled Helix: social and psychological implications of the new human genetics* (ed T.M. Marteau and M. Richards). Cambridge University Press, Cambridge.

Ransom, A. (1930) *Swallows and Amazons.* Jonathan Cape, London.

Rapp, R. (1988) Chromosomes and communication: the discourse of genetic counselling. *Med. Anthropol. Quarterly* **2**: 143–157.

Rose, N. (1990) *Governing the Soul: the Shaping of the Private Self.* Routledge, London.

Scieszka, J. and Johnson, S. (1992) *The Frog Prince Continued.* Penguin, Harmondsworth.

Reflections on genetic testing in childhood

Dietmar Mieth

1. Pragmatic ethical issues concerning diagnostic tests on children

Ethical aspects of genetic testing of children can be divided into three different sets of questions that are closely interrelated:

(1) General issues raised by diagnostic testing in childhood
(2) Issues raised specifically by genetic testing in childhood
(3) The broader, social context of genetic testing.

In addition, this chapter discusses the issues raised by a concrete example of genetic testing in childhood – screening infants for cystic fibrosis (CF).

1.1 Diagnostic tests that do not employ genetic techniques

Our first topic deals with general questions concerning diagnostic tests in childhood. Such tests are subject to points of consideration, criteria and norms that will also be relevant in the context of genetic tests. Given the widening range of application of genetic tests, the significance of guidelines for medical practice in this area will grow and they may even come to be legally enforced. In many cases it may be possible to transfer the arguments and conclusions from the general field of diagnostic testing to the more specific field of genetic testing. It is also possible, however, that medical practice in the context of genetics may reveal ethical difficulties and problems that also arise in the wider, more general context but which had not previously been recognized or sufficiently emphasized. This is particularly true in the context of predictive diagnostic tests carried out on currently healthy, so far unaffected individuals; such tests will often employ genetic techniques but some non-genetic tests may generate similar information.

Diagnostic tests can have the following purposes

- they may be employed for therapeutic reasons after symptoms of a disease have been found

The Genetic Testing of Children, A.J. Clarke (ed.).
© 1998 BIOS Scientific Publishers Ltd, Oxford.

- they may be employed to prevent a disease, for example through certain pro-phylactic measures such as diets
- they may be used to generate information that the individual may wish either for his/her own sake or for reproductive planning, lifestyle decisions, etc.
- they may be employed for the purpose of research.

Diagnostic tests on children are generally employed for the same reasons as diagnostic tests on adults. If we assume that children are persons who do not have the capacity to give informed consent on their own, then they are seen to need the caring voice of their parents or the community to stand in on their behalf. Balancing the decision between someone else's possible interests and his or her right to autonomy is a theoretical as well as a practical problem in itself. Cognitive and psychological development in children varies greatly: the capacity of the individual child to assess the meaning and consequences of a test will have to be considered. The child's autonomy must be respected as far as possible, but beyond this it may still be difficult to determine what is in the best interests of a person on whose behalf one has to assume responsibility and make a decision.

Another general ethical question arises in relation to informing the child of the test results (if the test is carried out). Should the child be informed of the result of specific tests? When should he or she be informed? Who else will be informed? The conflict between responsibility and respect for the child's autonomy will be easier to resolve in cases where the test itself is not harmful and the proposed therapeutic or preventive benefit is substantial. However, parents may find themselves in situations of tragic conflict when the risk involved in diagnosis is high yet the benefit of a possible therapy is uncertain. Similar to these are cases where diagnostic tests are employed to prevent a disease, or are preliminary to introducing certain preventive measures such as a diet. The balance between responsibility and respect for autonomy will generally be attainable if the tests are likely to cause little or no harm in comparison with the outcome for an affected child who is not treated promptly. The decision becomes more difficult, however, when the risk of harm from the test is significant, if the statistical risk of disease is minimal or if the benefit of early diagnosis is small.

If the state introduces vaccinations or diagnostic tests for an entire population or group at risk, such as children, the likely benefit has to be great as, for example, in the case of epidemics of dangerous infectious disease. The rule of thumb for all these cases is to integrate the child's own view as far as possible, although this may be overruled by other considerations such as the overall health of the child, the protection of others, the curability rate of a disease, and so on. In contrast to adults' decisions, the child's opinion is only one factor among many. The overall responsibility lies not in his or her hands but in the hands of the parents or the state.

The use of a diagnostic test solely for purposes of information seems highly problematic in the case of testing in childhood. While an adult person may have a reasonable interest in informing himself or herself about his or her likely future health without strictly medical indications, the same cannot be presupposed for a child. On the contrary, his or her right not to know has to be respected, even if this becomes relevant only in the distant future. Cases have to be very specific and exceptional to overrule this ethical right of a person.

Diagnostic tests carried out for the sake of research are also difficult cases. They should certainly not be regarded as ethically acceptable if the testing procedure causes harm. Furthermore, the test has to be of some potential benefit to the child and the consent of the parents has to be obtained. These are the most important guidelines. They have to be adapted for use in each specific field of application.

1.2 How are these issues to be considered when the diagnostic tests employ genetic techniques?

The fact that a test provides evidence about the individual's genetic constitution is ethically relevant not because of the new technology that it employs but because of its effects on society and on human behaviour. Human genetic technology changes our understanding of reality through its precision and its rapid development, but even more so through the new perspectives it offers and the new expectations that it raises. As a result the structure of individual and societal life becomes more complex, and has to be reinterpreted. The changes caused by new technologies are not only scientifically relevant but have to be considered against the background of the society's hopes, expectations and even strong beliefs that certain health problems will be solved by these technological developments; these beliefs are often reinforced by the active and public involvement of committed individuals motivated by the failure of conventional medical approaches.

The most dramatic difference between conventional and genetic diagnostic tests is the number of persons potentially affected by the result. For if a genetic disease or disorder is diagnosed, other family members or relatives may be affected as well – in the present or the future. Therefore, the nature of decision-making also changes: how can we take the interests of other family members into consideration? How is their right to privacy to be weighed against their possible interest in knowing their genetic status, and especially their possible interest in preventive or prophylactic measures (where applicable) ? Up to now ethicists or medical professionals have been quite vague in their formulations:

'Confidentiality of the results of the test is an ethical imperative. Genetic data should not be released, except with the free and informed consent of the woman or couple. **If the genetic data are relevant to the interests of other family members, the woman or couple should be strongly recommended by the genetic counsellor to allow the release of such data to these family members**'. (Group of Advisors, 1996).

Apart from this there are other considerations calling for ethical evaluation and criteria:

- the question whether a biochemical test should be replaced by a genetic test;
- the question of whether a test is relevant for families and should be employed for families, that is where one member of a family is affected and there is a risk that blood relatives may also be affected by, or carriers of, the same disorder, or whether population-based screening should be implemented;
- the question of using diagnostic tests when rational, curative therapeutic measures are not available: how can this be weighed against the 'right not to know'?;

- the related question whether the availability of symptomatic treatments or special diets that can alleviate the symptoms of a disease should weigh in favour of permitting the use of genetic tests;
- the question whether forgoing reproduction is an option for high risk groups;
- the question of the ethical significance of nonaction in relation to the ethical evaluation of the consequences of a certain action;
- the question whether a predictive test should be permitted when a successful therapy is possible at the time of the appearance of symptoms;
- the question of the social discrimination and stigma resulting from genetic tests; how can these be weighed against the possibly helpful manipulation of a genetic disorder by lifestyle measures after an individual has been informed about their personal risk of the disorder?

There are still more questions that need to be answered by the relevant specialists – for example the safety of genetic testing, the personal and financial burdens of testing and the accuracy of the tests – which will be relevant facts for making decisions about the permissibility of a new test.

1.3 Genetic tests on children are to be placed in the context of more general questions that arise with human genetic technology: how are technical and practical medical developments related to general societal trends?

These questions can be stated more concretely:

(a) Genetic tests are – at least partly – commercial goods. To what extent do economic and political pressures influence their actual application? Developments in genetic testing take place within the context of national political goals relating to research, development, technology, and productivity (Koch and Stemerding, 1994).
(b) The ethical priority of the individual's rights and the protection of the family against larger institutions is expressed in the following norm: for new tests to be acceptable they must be of value to affected individuals or families or high risk groups rather than to society as a whole. To what extent, however, does society rule through directing the interests of individuals? (Mieth, 1993).
(c) Protecting the confidentiality of private data is a more general societal problem, not restricted to the area of genetics; it is one part of the protection of the private sphere. However, in the face of the genetic revolution, it is conceivable that the state may have to set more and more severe limits on some individuals, in order to protect the privacy of others (Gierer, 1990). At what point is the intention of the one principle – the protection of the private sphere in order to insure individual freedom – perverted by limiting the freedom of others?
(d) The increase of genetic knowledge without an increase in therapeutic measures is met by a euphoric change of paradigms on the part of society: the hope that genetics will succeed in the therapy of hereditary and multifactorial diseases, or at least offer dramatic advances in their treatment, where conventional pharmacological or biochemical measures have failed to bring

about revolutionary changes. However, to what extent are these hopes merely wishful thinking, ignoring the fundamental failure of human intellectual growth and social progress to take human contingency into consideration? Who still expects limits to human achievement in the long term? For example, physicians and medical experts argue that progress in diagnostics will sooner or later lead to progress in therapies. This statement is problematic in two ways: first, it expresses an optimism that may be totally unfounded; second, it ignores the situation of those individuals who are given diagnoses with no prospect of therapy. These individuals cannot be allowed to become a 'quantité négligeable' on the basis of speculative utilitarian calculations.

(e) There is a societal means for solving problems, namely, to have 'licensing bodies' decide, case by case, on difficult ethical problems. These determine the indications and specific conditions that permit the medical practice in question. Such a problem-solving procedure, however, will to some extent remove ethical problems of decisive importance for the future from public discourse.

It is impossible to answer all these questions competently and immediately. Therefore, one question has been singled out for further consideration. One presupposition is necessary here: that genetic tests that do not lead to likely therapeutic benefits may only be permitted for adults who are able to articulate and assert their right not to know. For children, tests can be employed only with those diseases that can be treated effectively at the time of onset of symptoms. But even in these cases one more condition must be fulfilled: the treatment of symptoms must have better chances of being successful when a disorder is known before the symptoms of a disease are detectable – because otherwise there is no reason to have the information before the onset of the disease. With this presupposition we can focus our discussion on genetic testing for CF.

1.4 The case of cystic fibrosis

Testing newborn infants for CF has in the past been possible within affected families by means of sweat testing (a biochemical test carried out on the sweat of the siblings of affected infants) and, at the population level, by biochemical screening of blood spots taken at 5–7 days of age. Population screening was evaluated in several areas but was then mostly discontinued both because of difficulties with implementing the test and because of the lack of a decisive advantage to affected infants who were diagnosed early. By combining the biochemical test with molecular genetic testing, an improved method of screening has become available. However, there is still no causal therapy, although developments in treatment are improving the prognoses for affected individuals who are diagnosed early.

This is not the appropriate context in which to make a decision about the appropriateness of introducing a population screening test for CF, but we can usefully raise some of the questions that will need to be considered in coming to such a decision.

● The accuracy of tests is reduced if only a proportion of the relevant, disease-causing mutations in a population can be detected: what percentage of mutations must be identifiable for testing to be worthwhile? (cf. Larsen, 1995) How does the cost-benefit analysis of genetic testing compare to that of biochemical testing, considering psychosocial as well as financial outcomes?

- What criteria do we have to make a decision about introducing a new genetic test, either to substitute for a conventional test or where no other test is feasible? If the reason for testing is research or scientific progress only, for example to gain information on mutations or variants and so on, then the patient obviously is only the subject – not the beneficiary – of the procedure.
- To allow the introduction of genetic tests on children, we need much information from tests carried out on healthy (unaffected) and on affected individuals. Of course safety and accuracy have to be guaranteed *lege artis*, but predictive genetic tests on children will only make sense if harm at the time of the onset or clinical diagnosis of the disease can be avoided by earlier testing. For that to apply, there have to be reasons to begin therapies or quasi-therapies before the onset or clinical diagnosis of the disease. If no harm is to be expected when therapy starts at the usual time of onset or diagnosis, it is unlikely that predictive testing in childhood could be justified.
- Unless these concerns have been adequately addressed, why should screening be introduced at all? It would be possible to limit tests to families with previously affected children that are actively seeking this information. It is not unlikely that personal stigmatization and social discrimination can result from knowledge of specific items of information, for example in reproductive matters. This may be especially true when screening identifies some healthy carrier infants as well as truly affected infants, as in newborn screening for CF.
- The introduction of genetic screening tests for CF is clearly accompanied by certain commercial, professional and instutional pressures. There is a long list of corporations with financial interests in the provision of testing services and of test kits for use in public and private laboratories. Lene Koch has – in my opinion successfully – made efforts to prove the existence of pressures to introduce and establish these tests in the context of a screening programme to identify healthy carrier adults (Koch and Stemerding, 1994).

It seems imperative to agree with Redemann's statement (1993)

'The introduction of genetic screening for other diseases as a new diagnostic and preventive tool of modern biotechnology is still in its infancy and needs especially more public information and the improvement of counselling, education and methods applied, before it can be introduced on a broad basis. The same is true for screening in CF. As long as there is still no unanimity reached about how, when, where and if at all to screen for CF, there is no need to rush to a decision, as the experience of the past shows that, for the prognosis, treatment is rather more crucial than genotype. We need to gain more information first, to be able to make a responsible decision on the issue of population screening for this disease, otherwise we are no better than any fictitious Vogon from outer space carelessly destroying Earth.'

With the genetic testing of children we face the need to resolve difficult practical issues while respecting several sets of interests and duties: the interests of affected persons which we have to respect, the general interest in scientific progress and development in which different motives interact, and ethical rights and duties that apply to everyone. If these ethical rights require more information before a decision can be reached about implementation and the development of more counselling services before tests are widely introduced, then scientific progress will have to be slowed down or perhaps even stopped in certain situations. For it will take time, for example, to meet such conditions as those laid down in the WHO Guidelines for Pre-Test Counselling 1995 (Fletcher *et al.*, 1995) as 'minimum conditions'. These guidelines should be transferred to the issue of genetic tests on children. They are not meant as ethical conditions, and they do not address the question of education and competence, but in the future such professional ethical counselling will be a prerequisite for the establishment of genetic tests.

2. Fundamental ethical issues concerning genetic tests on children

New possibilities in therapy not only raise new questions but old questions as well, although often in a novel way. These questions concern specific test procedures and specific personal circumstances, and they also surface in other contexts such as genetic counselling or counselling in the form of assistance with decision-making. Without making claims of comprehensiveness, the following fundamental ethical questions can be applied to the genetic testing of children:

(1) The question of whether personal decisions have been reached in an ethically valid manner; this also implies the question of who decides what and when? and what form does the accompanying counselling take?
(2) The question of the limits to our interest in knowledge, which is usually justified although it may frequently be in conflict with other goods.
(3) The question whether an anticipated risk – specifically an increased risk of an adult-onset disease in a potential, future child – restricts the right of potential parents to have their own child (Kielstein 1996).

The first question thematizes the discussion of counselling concepts based on 'free and informed consent' not only with regard to comprehensive criteria (for example, not to be directive or coercive) but also with regard to the procedures followed in the course of this counselling – before testing, in parallel with testing or following on afterwards. The individual steps in this counselling and decision-making process have been outlined by Haker (1998). Decision-making in familial or couple relationships as well as the balancing of parental decisions with the possibility of the child's 'assent' need to be considered more thoroughly in psychological studies.

The second question addresses the limits to our thirst for knowledge, limits that can (for reasons of responsibility or moral identity) or must (due to higher goods) be established. This encompasses a number of important related questions. Is it morally right for parents to acquire knowledge about their children, who cannot yet give their own free and informed consent? that is:

● about diseases which are untreatable and causing symptoms which are untreatable and unrelievable?

- about diseases which announce themselves early in symptoms and which are equally or at least adequately treatable from the moment of the appearance of these symptoms?
- about incurable diseases, whose clinical course cannot be modified and which usually first manifest at a point in time (as in the case of Huntington's chorea) when the parents' advisory responsibilities and potential parental burdens no longer play a role or have been taken over by the (now adult) child? (Clinical Genetics Society, 1994; Harper and Clarke, 1990).
- acquired with the help of those types of diagnostic tests which are invasive and a burden to the patient? Must we not respect the limits specified as 'minimal risk and minimal burden' in the Human Rights Convention on Biomedicine in relation to persons incapable of consent, even though these limits were too narrowly linked with clinical trials of profit to third parties?
- with regard to carrier status for single gene (Mendelian) recessive diseases, in which it is the responsibility of each individual potential carrier to make their own decision about carrier testing? Knowledge of this type of carrier status may be important for making informed reproductive decisions, but is not otherwise relevant to the carrier's health. Or should we allow the individual's competence to be subsidiarily supported by parental knowledge without their consent? If so, should not this knowledge be limited to knowledge acquired through counselling in contrast to knowledge acquired through testing?

All of these questions are essentially concerned with the limits of parents' rights in relation to the prospective rights of children. Even more radical is the question of limits in relation to the prospective rights of future persons, before conception is intended. This is the question whether an anticipated risk to a potential child violates or revokes the right of potential parents to produce their own offspring. This is to be determined on the basis of prognoses and probabilities as well as on the basis of demonstrated capabilities for dealing with extremely difficult situations. Genetic tests on children, included in the long-range planning of family life, are not, like pre-natal diagnostics, potentially selective. Admittedly, the question of the alternative (adoption), does arise here. In this context an ethically questionable detail also comes to light: the interests of adoption agencies in test results or the interests of adoptive parents. Here the line must be drawn between health purposes and commercial interests.

The notification of test results is a question that falls in the realm of psychology, since the assessment must inevitably be empirically determined. Only the 'whether' of the right to information and the corresponding responsibility to inform can be asserted on ethical grounds, not the 'how' and the 'when.' The fundamental ethical issues must continue to be debated (Clarke and Flinter, 1996; Kitcher, 1996).

References

Clarke, A. and Flinter, F. (1996) The genetic testing of children: a clinical perspective. In: *The Troubled Helix: social and psychological implications of the new human genetics* (eds T.M. Marteau and M.P. Richards). Cambridge University Press, Cambridge, Chap. 7, pp. 164–176.

Clinical Genetics Society (1994) Report of the Working Party on the Genetic Testing of Children. *J. Med. Genet.* **31**: 785–797.

Fletcher, J. *et al.* (1995) *Guidelines on ethical issues in medical genetics and the provisions of genetic services.* WHO Hereditary Diseases Program. WHO, Geneva, Table 7, p.62.

Gierer, A. (1990) Einstieg in fremdes Bewusstsein – neurowissenschaftliche, erkenntnistheoretische und ethische Grenzen. In: *Ethik in den Wissenschaften* (eds K. Steigleder and D, Mieth), Attempto, Tubingen, pp. 65–68.

Group of Advisors on the Ethical Implications of Biotechnology (1996) Opinion No. 6, Ethical Aspects of Prenatal Diagnosis, 20 February 1996, Brussels, Belgium. n.2.8.

Haker, H. (1998) Entscheidungsfindung im Kontext pranataler Diagnostik. In: *Beratung als Zwang, Schwangerschaftsabbruch, genetische Diagnostik und Aufklarung und die Grenzen kommunikativer Vernunft* (ed M. Keltner). Leske & Burlich, Opladen, in press.

Harper, P.S. and Clarke, A. (1990) Should we test children for 'adult' genetic diseases? *Lancet* **335**: 1205–1206.

Kielstein, R. (1996) Clinical and ethical challenges of genetic markers for severe human hereditary disorders. In: *Changing Nature's Course. The Ethical Challenge of Biotechnology* (eds G.K. Becker and J.P. Buchanan), Hong Kong, pp 61-70. (Rita Kielstein lists the alternatives of responsible parenthood: not having children, not getting married and adoption.)

Kitcher, P. (1996) *The Lives To Come: the Genetic Revolution and Human Possibilities. (To Test or Not To Test?).* Penguin Books, Harmondsworth pp. 65–86.

Koch, L. and Stemerding, D. (1994) The sociology of entrenchment: a cystic fibrosis test for everyone? *Soc. Sci. Med.* **39**: 1211–1220.

Larsen, W. (1995) Psychological impact of screening for cystic fibrosis. In: Human Genome in Europe: Scientific Ethical and Social Aspects. Académic Royale de Medecine de Belgique, Brussels, Belgium, pp. 50–55.

Mieth, D. (1993) Justified interests. In: *Ethics of Human Genome Analysis* (eds H. Haker, R. Hearn and K. Steigleder). Attempto, Tubingen, pp.272–289.

Redemann, B. (1995) Psychological impact of screening in cystic fibrosis. A patient's point of view. In: *Human Genome in Europe*, l.c. (see Larsen, 1995) 55–58, p.56.

4

On the receiving end of genetic medicine

Christine Lavery

1. Introduction

Knowledge and understanding of the genetics and clinical manifestations of even some of the rarest inherited conditions continues apace. Whilst recognizing that there are more than 5000 identified genetic diseases, it is principally through the eyes and experiences of families whose children are affected by the mucopolysaccharide (MPS) and related diseases (MPS) that we examine attitudes to the science of genetics.

2. MPS diseases

MPS diseases are rare inherited lysosomal storage disorders, in which there are deficiencies of specific enzymes. They embrace a wide range of conditions which cause progressive physical and, in many cases, severe mental disability. At present there is no cure and death usually occurs in childhood. At least one baby born every ten days in the UK will be diagnosed with one of these conditions.

It is therefore of little surprise that, on learning that their child or children have an MPS disease, parents have to learn quickly and to embark on teaching and informing a wide range of professionals involved in the care of their child. Some professionals, particularly doctors, are surprised to find that a vast majority of parents, even those who may be considered generally less educated, search out information and become the expert not only in their own child and the condition that affects him or her but also in the wider issues around genetics.

In recent times it has often become possible to carry out genetic tests for carrier status on the healthy siblings of an affected child, where it is known they may be at risk of carrying an MPS disease – especially when the affected child's genotype is known. As a result, genetic testing of children for carrier status is an issue that has been much debated within the Society for Mucopolysaccharide Diseases; this debate culminated in the Society undertaking a pilot survey to investigate attitudes within MPS families and to record some of their experiences. (*Tables 1 and 2*)

The Genetic Testing of Children, A.J. Clarke (ed.).

Table 1. Diagnosis of the affected child of 53 parents in survey

Type of Disability	Number (%) in sample group
Hurler/Hurler Scheie/Scheie	11 (25)
Hunter[a]	19 (43)
Sanfilippo	13 (30)
Morquio	1 (2)

[a] Pattern of inheritance for all diseases is autosomal recessive except Hunter which is X-linked.

Table 2. Parents' reporting of carrier testing in their unaffected children

Number (%)	Decision
17 (39)	Have had carrier testing performed
20 (45)	Have not had carrier testing performed
6 (14)	Did not know carrier testing was available
1 (2)	Did not answer question

We believe this small pilot study reinforces the argument in favour of the right to test children for carrier status in serious childhood – onset diseases. As our survey clearly demonstrates (*Tables 3 and 4*), the ages at which children have been tested has been wide-ranging.

Brothers and sisters growing up in a family of one or more MPS children cannot ignore the differences. These youngsters become carers the moment they can walk and talk. As these children develop they take on the burden of emotional worry alongside their parents; they cannot ignore the decline and eventual

Table 3. Diagnosis of affected child in the families of 18 unaffected siblings in survey

Type of disability	Number (%) in sample group
Hurler Scheie/Scheie	4 (22)
Hunter	3 (17)
Sanfilippo	9 (50)
Morquio	1 (5)
Mannosidosis	1 (5)

Table 4. Number and ages of siblings when carrier tested

Age When Tested	Number (% of those tested)		(% of total)
Not tested	7	–	(39%)
Tested under 5 years	7	(64)	
5–8 years	2	(22)	
9–12 years	0	–	
13–15 years	2	(22)	
All Tested	11	(100)	(61%)
Total	18	–	

demise of their much loved brother or sister. With these anxieties come the questions, 'Will I develop MPS?', 'Will my children have MPS?'. In the words of one nine year old:

'The teacher asked me to spell miscellaneous. I told him I had not learned that yet but I can spell Mucopolysaccharide'.

The survey showed not only that there was a wide range of experiences between families, with carrier testing being performed at different ages, but there was also a wide range of attitudes towards testing. There was a diversity of views between parents, and no consensus view between parents and children. The majority of siblings favoured testing before puberty; this allows time to come to terms with a positive carrier test result and to work out how to deal with the information. We, at MPS, agree with the Genetic Interest Group's response to the Report of the Working Party of the Clinical Genetics Society, 'Genetic Testing of Children', that the child must be put to the fore, with parents helping their children to make informed decisions. We know that children are astute, and that quite young children can be involved in the decision-making process in relation to carrier-testing. It is vital to maintain a balance in the debate for and against carrier testing. Whilst a decision to proceed with carrier testing is serious, the implications should not be exaggerated.

In the case of autosomal recessive diseases, being a carrier does not increase the risk of an unborn baby being affected unless the partner has the same defective gene. However, such is the impact of living in an MPS family that many children in a wide age group believe they are entitled to carrier testing. One sibling expressed her view at a recent International Symposium on MPS:

'I remember every moment of what it was like to be sister to Adrienne, who passed away 5 years ago at the age of 16 years. MPS still and always will affect my life. Then there is the question of carrier status and the professional attitude to carriers of autosomal recessive diseases. Statistics have no meaning when you are the person facing the risk. How can you guarantee I won't be the one?'

Over recent months, more families have become aware of the wider debate about genetic testing in childhood and feel that professionals are closing ranks 'against' them by making genetic tests less accessible at a time when science is beginning to make answers available in this area. The strength of feeling within the MPS Society in favour of the right for genetic testing for carrier status to be made available to children over the age of 8 years has increased significantly.

It is a cruel irony that many families, who could benefit from the recent breakthroughs in genetic carrier testing for MPS using mutation analysis, are seeing these opportunities closed to them as professionals debate restricting genetic testing in those under 16 years. These are the same families that, despite caring for their MPS son(s) or daughter(s), have for years been using every last ounce of energy to raise funds for the scientific research that has led to these breakthroughs.

It was recently suggested that the siblings under 16 years of age at risk of carrying a defective MPS gene were likely to be stigmatized if found on genetic testing to be carriers. Stigmatization, however, does not come with the knowledge that an MPS sibling is a carrier. If stigmatization is going to show itself at all it will do so when the sibling wants to bring a friend home for tea, because they then have to decide whether to explain that their brother(s) or sister(s) cannot talk or walk, make funny noises, look very different, throw food around and are doubly incontinent.

3. Conclusion

In conclusion, the MPS Society urges professionals not to view genetic testing in isolation but to see it as part of a whole package of care for a family where one or more children are affected by the disease but the lives of the other brothers or sisters are dominated by it as well. 'The Answer' may not exist as a specific single policy applying to all, but it may rather be for brothers and sisters to be able to make their own informed choices as and when they are ready.

5

Testing children for balanced chromosomal translocations: parental views and experiences

Chris Barnes

1. Introduction

The increase in the number of genetic disorders for which testing is now possible has seen a corresponding increase in the amount of discussion about the genetic testing of children (Ball *et al.*, 1994; Bloch and Hayden, 1990; Clinical Genetics Society, 1994; Dalby, 1995; Harper and Clarke, 1990; Marteau, 1994). Diagnostic tests with proven medical benefits to the child are seldom controversial. Additionally, a general consensus against testing healthy children for late-onset untreatable disorders such as Huntington's disease has emerged (International Huntington Association and The World Federation of Neurology, 1990).

However, there is far less agreement about the genetic testing of children in a number of other situations (Bradley *et al.*, 1993; Ryan *et al.*, 1995; Thelin *et al.*, 1985). Genetic testing to determine whether a child is a healthy carrier of a single-gene disorder or a balanced chromosome rearrangement has been particularly controversial. A Working Party of the UK Clinical Genetics Society (CGS, 1994) concluded that testing children to determine their carrier status, where this would be of purely reproductive significance to the child, was probably best avoided. This view was in keeping with that of the Nuffield Council on Bioethics (1993): namely that the genetic testing of children which would not be immediately beneficial should normally be deferred until the subjects can give valid consent. One central anxiety about the genetic testing of a healthy child is the damage that could ensue from possible 'post-test' changes in parental attitudes. Genetic counsellors also worry that knowledge itself has the potential for causing harm, with an appreciation that, 'confronting a child with distressing and unsolicited information about their "defective" genetic status can be defined as harmful' (Montgomery, 1994).

The Genetic Testing of Children, A.J. Clarke (ed.).
© 1998 BIOS Scientific Publishers Ltd, Oxford.

The principal arguments for and against testing apparently healthy children have been comprehensively described and discussed elsewhere (Clarke and Flinter, 1996; Michie, 1996). In practice, parents often seek guidance from genetic health professionals about the wisdom of having their children tested, but there is little evidence as to whether 'carrier' testing in childhood is harmful or beneficial to the child or the family, and few attempts have been made to look at the experiences of families in which children are known, or potential, carriers. In particular, there is a paucity of formal studies of parental views and experiences of carrier testing in childhood. Karyotyping healthy children who are potential carriers of chromosomal rearrangements has been performed for many years but, without published studies of the psychosocial effects of such testing, 'no information is available as to whether the testing process causes or prevents harm in practice' (Clarke and Payne, 1995). Here, some findings of a study (Barnes, 1996), designed to explore some of the views and experiences of 'carriers' of balanced chromosomal translocations about the testing of their healthy offspring for these inherited rearrangements, are described and discussed.

2. Materials and methods

The study population consisted of carriers of balanced chromosome translocations who were also parents of carrier children and/or untested children, ascertained from the relevant databases of the South Thames Regional Genetics Centre [East] (STRGCE) based at Guy's Hospital in London, and the North West Regional Genetics Centre (NWRGC) based at St. Mary's Hospital in Manchester. A questionnaire, together with an introductory letter was sent to parents from the STRGCE dedicated database and the NWRGC family register who met the study criteria. The overall response rate to the questionnaire was 60.7% (105 of 173 parents) and data from 103 parents was finally used, excluding two parents who proved to be ineligible. The questionnaire responses were coded and stored in system files, and data analysed using the Statistical Package for the Social Sciences (SPSS/PC and SPSS for Windows).

3. Results

3.1 Personal information about study group parents

Around a third ($n = 54$) of the study group, and a quarter ($n = 26$) of respondents were men. Seventy-seven of the 103 respondents (74.8%) were translocation carrier mothers. The average age of respondent parents was 36 years (range 23–57 years) compared with 37 years (range 21–58 years) for non-respondents. The average age when parents had been identified as carriers was 30 years (range 14–52 years).

3.2 Pregnancy loss

Predictably within a group of translocation carrier parents, the average rate of pregnancy loss was high, with over 70% of respondents (or their partners) having lost at least one pregnancy and over 30% losing three or more. Most losses were first trimester miscarriages, although 15 pregnancies had ended in therapeutic termination because of the presence of a known or probable unbalanced translocation.

3.3 Parental knowledge of carrier status before having children

Most parents (87.4%) were unaware of their own translocation before having their first child. Only one parent had been tested as a child. Nearly a third of parents with carrier children had no knowledge of their own carrier status until shortly after the result of a pre-natal test.

3.4 Knowledge and views of parents about their translocation

Even though all respondents had received genetic counselling, over a quarter (26.2%, $n = 27$) acknowledged a lack of awareness of the consequences of an unbalanced pregnancy. A higher proportion of carrier fathers (34.6%) than carrier mothers (23.4%) lacked this knowledge. Of the parents who claimed awareness of the effects of an unbalanced translocation, many were able to give only very general details of the effects. This raises the possibility that many of those parents who said they knew about these effects may in fact have had little in-depth knowledge.

Nearly half of parents (43.6%, $n = 45$) could not remember the likelihood that their particular translocation could give rise to the birth, rather than the conception, of a liveborn, affected (chromosomally "unbalanced") baby. Over one-fifth (21.6%) of parents believed that the risk of viability associated with an unbalanced form of their particular translocation was 50% or greater. Another 31.4% of parents estimated this risk to be between 25% and 49%. This suggests that many parents either seriously overestimate the risk of viability, or make no distinction between the risk of an unbalanced pregnancy at conception and at term.

In answer to the question "How much do you feel that your own life has been affected by 'being a carrier'?", the mean score (Likert-style scale) of a translocation carrier mother was three times the equivalent score of a carrier father (4.0 as opposed to 1.3). There was a noteworthy correlation between the estimation of the effect of a parent's own carrier status on his or her life, and the number of pregnancy losses suffered, suggesting that pregnancy loss has a significant effect on parental views of and attitudes towards the translocation. As some children will directly experience parental distress caused by pregnancy loss, the effects of such intrafamilial experiences on a child's own attitudes to their actual or potential carrier status warrants further exploration.

3.5 General information on children of study group parents

The 103 parents came from 93 kinships and had had a total of 239 children. Two hundred and thirteen (89.1%) of the children were still alive, of whom 211 were the genetic offspring of the carrier parent. Nearly two-thirds of children ($n = 144$) had been karyotyped, and 85 balanced translocation carriers (all still living) were identified. Seventy-six living children remained untested. The parents of three children did not know if they had been tested, and five tested children had parents who were unaware of the test result. Nearly two-thirds (65.3%, $n = 94$) of tested children (nearly 80% of the 85 carrier children) had been karyotyped prenatally. Only 18.8% of carrier children ($n = 16$) were tested post-natally before the age of 16, 11 of whom were found to be carriers as the result of a test primarily performed to establish the carrier status of the child. These 11 children represent 12.9% of the total number of carrier children in this study. The study design did not permit a detailed exploration of the decision-making processes for the parents of these 11 children in the 'post-natal carrier group'. However, parents gave some insights into the difficulties and dilemmas that may be involved.

'H. was tested within a few months of having lost our second child and therefore I was under a great emotional strain. I probably felt unable to cope with the additional worry of having H. tested, but we felt we needed all the facts of what faced us'.

Because of the small number of children in this study (11) whose carrier status was established as the result of a 'true' carrier test, information about parental feelings on learning that a child is a translocation carrier as a result of such a test was therefore limited. However, data from this study suggest that parents can suffer marked levels of distress.

'Although we knew that there was a possibility that she was a carrier, it came as a big surprise and shock when the results were positive. The only comfort I felt was in knowing that I could warn her in advance, and she wouldn't have to suffer six miscarriages like I did in order to know about this problem'.

Forty-four carrier children were tested as a result of a pre-natal test where one parent was known to be a carrier. In such circumstances, news that the fetus was 'only a carrier' may have been expected to have brought extreme relief. However, parental reactions in this situation were unpredictable. Half of these parents experienced a 'low' level of relief, with over a quarter having a 'high' level of guilt. Around one in five parents described a 'high' level of worry and/or sadness on hearing that the pre-natal test had excluded an affected fetus but gave a 'carrier' result. Three parents experienced a high level of 'devastation'.

'The relief that I didn't have to terminate was eaten up by the guilt that she was a carrier. I still feel I've got a lot ahead of me'.

Unfortunately, the study did not explore to what extent such feelings persisted during the course of the pregnancy, or were altered by the birth of the 'healthy' baby. However, data suggest that professionals may well underestimate the levels of anxiety, guilt and sadness associated with a carrier result in their efforts to give reassurance with the 'good news' result of an unaffected fetus.

3.6 Informing children of their carrier status

Over three-quarters of carrier children (77.6%, $n = 66$) had not been informed of their translocation. However, the vast majority of this uninformed group (80.3%, $n = 53$) were under 6 years old. All carrier children aged over 12 ($n = 9$) had been told about their chromosome rearrangement. Parents of seven of the 19 children who had been told of their translocation said that their child had been informed before the age of 6 years: one as young as 2 years old. The gender of the carrier child did not appear to be a factor in determining whether or not they were told about the translocation. As two of the children had been tested and informed of their carrier status after the age of 15 years, only 17 children (20.0% of carriers) learned that they were translocation carriers as 'true' minors, a number too small to make any significant conclusions about the effects of this information from a parental viewpoint.

Various explanations had been used to impart translocation information. Some parents attempted to describe the actual rearrangement of the chromosomes.

> 'I explained about sets of chromosomes, and how they split and are joined up again with their equal number, and that one pair had a chunk missing but it had joined up to one of the others, so all the pieces of the jigsaw were there, even if differently arranged'.

Some of the carrier children learned of the translocation after they had asked a direct question, such as one about the death of an affected sibling, or their status as an only child. Others used the opportunity of a conversation about a related issue, such as difficulties surrounding childbirth, tests in pregnancy or the existence of a genetic disorder as a way of introducing the subject. Sometimes, the subject was introduced by reference to the implications of the translocation for future pregnancies.

> '[I said] that her body was slightly different to others but it would only affect her when she came to have babies, and that mummy was the same and it was nothing to worry about'.

> 'It was explained that we were carriers and that in future his partner would have to have tests to see if the baby would be like his [affected, 'unbalanced'] brother'.

3.7 Overall effect on child of knowledge about carrier status

Parents were asked to comment on the child's reaction(s) to news of their carrier status. None of the children tested were said to have had any negative reaction to the news. Overall, there is no evidence from this study that childhood testing in translocation families causes harm, and no parent regretted having their carrier child tested.

3.8 Parental plans to tell 'uninformed' carrier child about translocation

Of the 66 children who had not been told they were carriers, parents planned to impart this information in almost all cases (93.9%, $n = 62$). However, many parents had only a broad idea at what age to tell the child. This suggests that parents may not have thought about the optimum age for divulgence in any detailed manner, or that chronological age is only one of many factors that are regarded as important when parents consider this issue. It may be that a child's emotional and intellectual maturity are of prime importance to parents in their plans to talk to their child about the translocation and its implications. Some parental comments in this study support this hypothesis. For some parents, a delay in informing their child may be related to their own uncertainty and incomplete knowledge about the nature and implications of the translocation.

3.9 Parental plans to karyotype untested child

Seventy-six children remained untested, 72 of whom were still minors under 16 years of age. Parents of more than half (54.2%, $n = 39$) of the untested minors (two-thirds of parents of the over-fives) had given the matter of testing considerable thought.

'I have thought of nothing else. I pray my child is not a carrier'.

However, parents of untested children were not specifically questioned about the degree of anxiety resulting from their child's potential carrier status, so it is not possible to know whether the frequency of parental thought about testing was in any way related to anxiety about this issue. Thirty parents (41.7%) had been offered the chance to discuss the possibility of having their child karyotyped with a professional; 23 parents accepted the offer. However, two parents reported considerable dissatisfaction with the discussion, and parents of seven children told of professional advice they had received.

'The genetic counsellor suggested that our child should be tested on reaching early teens'.

'We didn't have a choice – we were told that she would not be tested'.

'I was advised against testing at his age as it could cause him anxiety when he was too young to really understand'.

'I think I agreed with the genetics department that there was no need to test her until she was older'.

Only 11.1% ($n = 8$) of untested minors had parents who had requested a carrier test for their son or daughter. Of these parents, five had been told that the child was too young to be tested and one test was 'refused' for an unspecified reason. Two considered their parental rights to have been infringed.

'I feel I have the right to know'.

'We feel we are the ones that should make this decision'.

Another two parents felt 'unhappy' or 'annoyed' with the response to their request.

3.10 Discussions with untested children about potential carrier status

Two-thirds of untested children under 16 years (68.1%, $n = 49$) had a carrier parent who had never talked to them about their potential carrier status, although half of these children were under 6 years old. Only seven children (9.7%) had held frequent conversations about the possibility of being a translocation carrier with their carrier parent.

3.11 Anticipated parental feelings if untested child proves to be a carrier

For 40.3% of parents ($n = 29$), a high degree of sadness was associated with thoughts about their child's potential carrier status and nine parents predicted a 'high' level of 'devastation' in the event of a positive carrier test.

3.12 Parental plans about testing child

This study suggests a high level of parental interest in childhood testing, with the parents of 63.9% ($n = 46$) of untested minors wanting their child tested before the age of 16. Given this figure, it is probably surprising that more parents had not requested that their child be tested. However, nearly half of the 46 parents wanted their child tested after the age of 10 years, so many may have been waiting for the child to reach this age before raising the matter with an appropriate professional.

'Testing should be done as early as possible from the child's point of view, because children are not afraid of things as much when young, and it also enables the result to be known about before puberty'.

3.13 Reasons behind decision about testing

The carrier parents of the untested children were also asked to select reasons why they were in favour of, against, or unsure of their child being karyotyped in childhood, from a given list of commonly cited reasons. The majority of the 46 parents who were in favour of their child being karyotyped (84.8%) wanted the child to avoid the 'shock' of learning about the translocation in later life. Parental anxiety was a factor for three-quarters, and two-thirds considered that 'being a parent' put them in the position of knowing what was best for their child. Over half (51.3%) of parents in favour of testing acknowledged the possibility of childhood pregnancy as a reason for testing.

The parents of eight children were opposed to testing. They were unanimous in their choice of two reasons: respect for the autonomy of the future adult and the 'pointlessness' of testing for a status with only reproductive implications. Sixteen children had parents who were undecided about testing, over half of whom were worried about possible harmful effects on the child. The need for further discussion before making a decision about testing was felt by 62.5%.

4. Discussion and conclusions

This study explored some of the views and experiences of 103 carriers of balanced chromosomal translocations about the testing of their healthy children for these inherited rearrangements. A number of areas were explored, and the main conclusions which can be drawn from this work are outlined below.

A high proportion of parents in this study were either totally unaware of the effects of a chromosomally unbalanced pregnancy, or were able to provide only very general details of the nature of such effects. Also, more than half of parents appeared to believe that the risk of viability associated with an unbalanced form of their particular translocation was 25% or greater. Clinicians involved with translocation families need to consider the possibility that deficiencies in parental knowledge and understanding may seriously hamper the transmission of accurate information to carrier (or potential carrier) children.

In this study, many 'childhood' carrier tests were performed pre-natally. The status of almost 80% of carrier 'children' had in fact been established during a pregnancy. In a substantial proportion of these tests, the parental translocation was unknown and the fetus had been karyotyped for an unrelated reason (e.g. a maternal-age associated risk of Down' syndrome). Thus, although more than half of all carrier children were tested pre-natally because the pregnancy was at a known risk of an unbalanced rearrangement, a substantial proportion of fetal carrier findings were totally inadvertent and had to be disclosed for the proper management of the pregnancy. This fact is often overlooked in the wider discussions about the ethics of childhood carrier testing. Once a familial translocation has been identified in this manner, it can be difficult to explain professional reservations about testing minors to a parent requesting a carrier test for other children.

As the parents studied here were themselves born before the advent of pre-natal karyotyping, their own carrier status would not have been discovered inadvertently as a fetus. This will be in contrast to the next generation of carrier parents where the presence of a familial translocation is more likely to have been established some time before pregnancy. The extent to which this will alter parental views and experiences of their own and their child's carrier status remains to be seen, but would seem to be an important subject for future studies in this area.

Data on feelings experienced by parents on learning news of a carrier child's status, after a test primarily performed to establish that status, were limited by the small numbers involved. However, this study suggests that parents of children tested in this way can suffer marked levels of distress. After pre-natal tests on known carriers (or partners of known carrier men), parental reactions to the 'good' news of a balanced (i.e. unaffected carrier) fetal karyotype were not predictable, with half of these parents experiencing little or no relief, and several suffering substantial levels of guilt, worry and sadness. This study suggests that levels of parental distress associated with the pre-natal identification of the fetus as a (healthy) carrier may well be significantly underestimated.

Whether or not such distress in known carrier parents could be prevented or ameliorated by giving pre-natal results as 'affected' or 'unaffected' is not known, but may be worth consideration if this is discussed beforehand and the parents are agreeable.

The majority of untested children in this study were unaware of their potential carrier status. As half of these 'uninformed' untested children were under 6 years of age, it seemed reasonable to conclude that many parents had decided that their 'at risk' but healthy offspring were too young for this to be of any consequence to the child. However, some parents of known carrier children had told their child about the translocation at a very young age, despite the reproductive implications of the translocation being many years ahead. Some of the factors that go towards helping a parent decide whether, when and how to transmit this information have been highlighted by this study. However, comparatively little is known about the probable complex interplay of such factors, and how they relate to the transmission of other intrafamilial information. Health professionals, parents and their offspring – both as children and as potential future parents – may well benefit from a better understanding of these processes.

This study also demonstrates the need for more open discussion between parents and professionals about the issue of childhood carrier testing. A high level of interest in childhood testing was expressed by the parents of the untested minors, with two-thirds wanting their child to be tested before the age of 16. Despite this, around half of the untested children had parents who had had no 'formal' opportunity to discuss the matter of testing their child with a professional. A few parents had directly asked for their child to be karyotyped but had experienced the professional 'advice' they received about this to be detrimentally directive. It seems clear that many parents could benefit from a forum whereby their views and concerns relating to childhood carrier testing could be discussed. However, if parents perceive a genetic counselling service to have rigid and judgmental 'rules' about such testing, this would be likely to discourage parents from seeking help. Parents may then become increasingly isolated in any anxieties, fears or misapprehensions they may have. They may also attribute professional caution around childhood carrier testing to an exercise in medical power and an attack on parental rights. This would be likely to harm, rather than help, families who are trying to cope with an often difficult genetic situation. Genetic counselling centres therefore need to consider ways in which parents can be offered suitable opportunities for open discussion about their carrier and untested children.

It must also be remembered that the study was designed to assess parental views and experiences. It was not designed to examine the impact of carrier testing on children from their viewpoint, and the limitations of assessing parental perceptions, as opposed to those of the children themselves, must be acknowledged. It is also difficult to judge what a parent may mean by 'informing' a child of his or her carrier status. In this study, seven carrier children were said to have been informed of their translocation before the age of six, one child being as young as 2 years old. Although chronological age is only a broad guide to intellectual and emotional capacity, there would seem to be a clear difference in the abilities of a toddler and an 11-year-old child to make sense of information they may be given. Some parents recognised that the age of their child was, or could be, an important factor in a general 'matter-of-fact' attitude to their carrier status.

'He just accepted it without much question but he is too young to realise the far-reaching effects'.

This study has provided insights into some of the views and experiences of carrier parents when thinking about their children who are, or may be, carriers of inherited translocations. There was no evidence that being tested (or not being tested) caused harm to a child, and no parent regretted having their carrier child tested. However, the number of minors who underwent carrier testing as a child (as opposed to a fetus) is small. Larger studies are needed to look in more detail at the effects of testing and non-testing, both from the point of view of the 'at risk' child and the family.

When policy guidelines about the genetic testing of children are devised, professionals need to consider the opinions and learn from the experiences of parents such as the subjects of this study, who may well be the best judges of their own child's potential to deal with information about an inherited condition. However, the protection of a child from harm remains the predominant consideration, with the preservation of the child's autonomy as a future adult continuing as a powerful concern. As a result, there is clearly a great need for studies on the child's perceptions of carrier testing and on both the short-term and long-term effects of such testing.

References

Ball, D., Tyler A. and Harper, P. (1994) Predictive testing of adults and children. In: *Genetic Counselling: Practice and Principles* (ed A. Clarke). Routledge, London, pp. 63–86.

Barnes, C.A. (1996) Testing children for balanced chromosome translocations: a study of parental views and experiences. *Unpublished M.Sc. thesis,* University of Manchester, Faculty of Medicine, Department of Medical Genetics.

Bloch, M. and Hayden, M.R. (1990) Predictive testing for Huntington's disease in childhood: challenges and implications. *Am. J. Hum. Genet.* **46**: 1–4.

Bradley, D.M., Parsons, E.P. and Clarke, A.J. (1993) Experience with screening newborns for Duchenne muscular dystrophy in Wales. *BMJ* **306**: 357–360.

Clarke, A. and Payne, H. (1995) Genetic testing in children: the relevance for adoption. In: *Secrets in the Genes* (ed P. Turnpenny). British Agencies for Adoption and Fostering, UK, pp. 176–187.

Clarke, A. and Flinter, F. (1996) The genetic testing of children: a clinical perspective. In: *The Troubled Helix* (eds T. Marteau and M. Richards). Cambridge University Press, Cambridge, pp. 164–176.

Clinical Genetics Society (1994) *The Genetic Testing of Children.* Report of a Working Party of the Clinical Genetics Society. Clinical Genetics Society, Birmingham and J Med Genet **31**: 785–797.

Dalby, S. (on behalf of the Genetic Interest Group) (1995) GIG response to the UK Clinical Genetics Society report 'The Genetic Testing of Children'. *J. Med. Genet.* **32**: 490–491.

Harper, P.S. and Clarke, A. (1990) Should we test children for 'adult' genetic diseases? *Lancet* **335**: 1205–1206.

International Huntington Association and the World Federation of Neurology (1990) Ethical issues policy statement on Huntington's disease molecular genetics predictive test. *J. Med. Genet.* **27**: 34–38.

Marteau, T.M. (1994) The genetic testing of children. *J. Med. Genet.* **31**: 743.

Michie, S. (1996) Predictive genetic testing in children: paternalism or empiricism? In: *The Troubled Helix* (eds T. Marteau and M. Richards). Cambridge University Press, Cambridge, pp. 177–183.

Montgomery, J. (1994) The rights and interests of children and those with mental handicap. In: *Genetic Counselling: Practice and Principles* (ed A. Clarke). Routledge, London, pp. 208–222.

Nuffield Council on Bioethics (1993) *Genetic Screening. Ethical Issues.* Nuffield Council on Bioethics: London.

Ryan, M.P., French, J., Al-Mahdawi, S., Nihoyannopoulos, P., Cleland, J.G.F. and Oakley, C.M. (1995) Genetic testing for familial hypertrophic cardiomyopathy in newborn infants. Clinicians' perspective. *BMJ* **310**: 856–857.

Thelin, T., McNeil, T.F., Aspegren-Jansson, E. and Sveger, T. (1985) Psychological consequences of neonatal screening for alpha-1-antitrypsin deficiency. *Acta Paediatr. Scand.* **74**: 787–793.

6

Identifying carriers of balanced chromosomal translocations: interviews with family members

Anita Jolly, Evelyn Parsons and Angus Clarke

1. Introduction

1.1 The background to this study

Genetic tests are widely used in medical practice today as a clinical aid which can be invaluable in establishing the diagnosis of the clinical disease affecting a sick individual. Additionally, they can be used to identify healthy individuals who carry faulty genes or chromosomes associated with genetic disease, but who are not themselves clinically affected. The children of these affected or carrier individuals may inherit either the disease or the carrier state.

For some inherited conditions, such as balanced chromosomal translocations, it has been common practice in the past to test children born to known carriers to discover if they are also carriers. There is some unease surrounding this practice, however, as it is feared that it may cause long-term psychological harm in the child. It also removes the child's right to make his or her own decision about carrier testing when older and it removes his or her right to confidentiality. There is no set of consensus guidelines as to when children should or should not be tested for genetic abnormalities, but a report from the UK Clinical Genetics Society on testing in childhood (1994) attempted to provide a series of recommendations for clinical practice. These were based upon reports of professional practice and principles of law and ethics – but not upon empirical evidence from families, as there was next to none (Marteau, 1994a). This study sets out to gather some evidence.

The consequences of testing can vary dramatically and are influenced by social and emotional factors as well as economic resources (Marteau, 1994b). Identifying an individual as a carrier of a genetic disease can have profound psychosocial risks, namely stigmatization, discrimination and an adverse effect on self-image (Haan, 1993). In addition there is some evidence that this knowledge has been frequently associated with anxiety (Zeesman *et al.*, 1984) and a less optimistic perception of future health (Marteau *et al.*, 1992a).

Stigmatization is the labelling of a person or group as having undesirable characteristics (Haan, 1993). It has been shown in the past that mass screening programmes can produce such effects. One such program set up in 1966 in Orchomenos, Greece, was aimed at detecting persons with sickle cell trait; this resulted in 7% of the population asserting that they would actively avoid relationships with persons carrying the trait (Stamatoyannopolous, 1974). Such stigmatization caused people to conceal their carrier status. Genetic testing can also have a negative effect on self image. In a study done on the stigmatizing effect of the carrier status for cystic fibrosis (Evers-Kiebooms *et al.*, 1994), it was found that carriers had significantly less positive feelings about themselves than non-carriers. Both carriers and non-carriers attributed significantly more negative feelings to other carriers of the CF gene, than they did to non-carriers. Such negative feelings do not seem to lead to sustained increases in anxiety in identified carriers, however, and may reflect a different psychological process (Bekker *et al.*, 1994).

Genetic discrimination refers to discrimination directed against an individual or family based solely on an apparent or perceived genetic variation from 'normality'. Discrimination has been recorded in many social institutions in North America, especially in the health and life insurance industries (Billings *et al.*, 1992); it has also been practised by employers against healthy workers who happen to carry a recessive disorder such as sickle cell trait (Bowman, 1991).

There are several reasons why children are tested for genetic disorders. There may be direct implications for the health of the child, which could be immediate or which may arise in later childhood or adult life. There are sometimes therapeutic or preventive measures that could influence the course of the disorder if started early. There may be important decisions regarding education and later career, which will be altered by the presence or absence of a disabling disease. There may also be genetic implications for the child transmitting the disorder in later life. Finally, many parents feel that knowledge of a future serious disorder in their child is preferable to prolonged uncertainty (Harper and Clarke, 1990). Genetic tests may be carried out on children in the hope of reducing family anxiety. By only considering this immediate benefit, however, parents and professionals may overlook longer-term psychosocial consequences.

Although there is no information available about specific psychological damage caused to children by genetic carrier testing, the impact of such testing could be great. Misunderstandings by the parents or child could lead to serious misconceptions about the future (Wertz *et al.*, 1994). Positive test results may reinforce already latent feelings of unworthiness (Page *et al.*, 1992). Siblings, formerly united by a bond of risk, may suddenly find themselves separated by the outcome of genetic testing, grappling with issues of 'survivor guilt' (Wertz *et al.*, 1994). In siblings of people with cystic fibrosis, worries about carrier status are related to fear of intimacy and of interpersonal relationships (Fanos and Johnson 1995a,b). Other potential hazards include harm to the parent–child relationship and harm to the child's self-esteem (Wertz *et al.*, 1994).

This study set out to explore the long-term psychological consequences of testing children for balanced chromosomal translocations. We chose to use qualitative research methods to examine: the impact of carrier status test results on the family, the children's understanding of their carrier status when they were first given the information and the broader significance to families of their experiences of genetic counselling.

2. Methodology

Leninger (1985) claimed that qualitative methods were of value in discovering new and unknown knowledge both to generate theories and hypotheses and to provide fresh perspectives on known areas. One of the main strengths of qualitative research methods is the gathering of rich narratives that have a quality of 'undeniabilty' (Smith, 1975). Another reason for choosing a qualitative approach is that it enables the investigator to study the topics of interest from the perspective of the subject, not the researcher (Duffy, 1987). As Spradley (1979) says, 'the ethnographer seeks to learn from people, to be taught by them.' A further potential advantage is that qualitative research is interactive and a personal relationship is forged between the researcher and respondent, which can lead to large amounts of data being disclosed if a good rapport is established.

A qualitative approach was chosen as most appropriate for this study because of its exploratory nature, because it would have been inappropriate to pre-determine the issues and topics that would be important to the research subjects, because the design had to be observational and retrospective rather than experimental and prospective, and because the expected number of participants was too small for statistical analysis of questionnaire responses.

The principal methods employed in this study were in-depth interviewing and the review of documents. Interviews were conducted by a single researcher in the home of each research subject (*Table 1*). The medical records of most participants were also reviewed to gather background information and to verify the genetic test results. Comprehensive field notes were also compiled, detailing who was present at the interview, in which room in the house it took place, what body language the participants used and where each participant sat.

The aim of this study was to record family experiences of genetic testing and counselling and to explore the social or emotional impact of these experiences on the individuals involved. Although retrospective in nature, the study was thought to be valid because these experiences would be likely to be well remembered in so far as they had been emotive. That they also encompassed women's memories of pregnancy and childbirth would make them likely to remain vivid for many years (Simkin, 1991). Reliability was further increased by letting an independent judge scrutinise the coding categories so that only confirmed coding assignments were retained for the analysis.

Table 1. Family members in this study

Name	Genetic Status	Type of Test	Age when tested
Mrs Waters	Carrier	Blood	Adult
Claire	Carrier	Blood	2 years
Ian	Carrier	Amniocentesis	Pre-natal
Mrs Connor	Carrier	Blood	Adult
Lucy	Carrier	Blood	6 years
James	Carrier	Blood	4 years
Elizabeth	Carrier	Amniocentesis	Pre-natal
Mrs Parry	Carrier	Blood	Adult
Sian	Carrier	Blood	3 years
Rhian	Affected daughter		
John	Carrier	CVS	Pre-natal
Mrs Callaghan	Carrier	Blood	Adult
Richard	Carrier	Blood	4 yeers
Amy	Non-Carrier	Amniocentesis	Pre-natal
Mrs Williams	Carrier	Blood	Adult
Elen	Carrier	Blood	5 years
Thomas	Carrier	Amniocentesis	Pre-natal
Mrs Grant	Carrier	Blood	Adult
Louise	Carrier	Blood	6 years
Matthew	Carrier	Blood	1 month
Mrs Ford	Carrier	Blood	Adult
Gareth	Carrier	Blood	2 years
Mrs Brown	Carrier	Blood	Adult
Sharon	Affected daughter		
Phillip	Carrier son	Blood	12 years
Sandra	Daughter-in-law		
Robin	Carrier son of Phillip and Sandra	Amniocentesis	Pre-natal
Mrs Hobbs	Carrier	Blood	Adult
Mark	Affected son		
Julia	Carrier daughter	Amniocentesis	Pre-natal
Mrs Moore	Carrier	Blood	Adult
Luke	Carrier	Blood	6 years
Sarah	Carrier	Blood	3 years
Kevin	Carrier	Amniocentesis	Pre-natal

2.1 The sample population

Subjects were ascertained by sorting through the cytogenetics records held in the Institute of Medical Genetics in Cardiff relating to the years 1971–1985. Individuals aged at least 10 years were eligible for the research if they had been identified as carriers of balanced chromosomal translocations by testing as young children or in fetal life.

The reasons why such testing was performed included maternal concern about the risk of Down's syndrome in the fetus, a history of recurrent miscarriage or infertility in the family and a history of a cytogenetic anomaly affecting another family member. If a parent was found to be a carrier of a balanced chromosomal rearrangement after a carrier fetus had been identified, this would have given reassurance that the fetus would probably be clinically unaffected but it would raise the possibility that other children in the family might also be carriers. Usual practice in these circumstances in the past has been to test the other children regardless of their age or ability to consent.

There were 54 potential subjects identified from our laboratory records, and we identified 17 families for whom the relevant departmental clinical records were available. One family was excluded from the study on the grounds of a long and complicated medical history, which could have been a confounding factor. Out of the remaining 16 families, seven were excluded because they lived outside South East Wales.

Families were contacted by writing to the mother of each known carrier child after obtaining permission to do so from their general practitioner. When consent had been given, the mother was contacted by telephone and a convenient date for a home visit was arranged; whether or not the woman's partner was present was left to the discretion of the family. Eight mothers replied and consented to a visit by the researcher. To increase the sample, an additional two families were recruited in the same way but through tracing Cardiff hospital record numbers where no clinical genetics records had been found.

The group of subjects defined as the 'children' in this study consisted of the individuals in these 10 families whose carrier status was tested either as children under the age of sixteen or pre-natally. At the time of the interviews, they ranged in age from 11 to 27 years. Parental permission was sought to interview the children after their own interview had been completed. Some agreed immediately and a date for a second home visit was then arranged. Some parents wanted to consult their children first, however, and they were telephoned a few days later to see if the 'children' could be interviewed.

Only one mother said that her children did not want to be interviewed, saying that they 'just were not really interested'. In one family, a date to visit her son was arranged through the mother. When the interviewer arrived for the interview, however, she was told that the son had gone out and 'was not really fussed in taking part'. In one further family the son walked in on the tail-end of the interview that was being conducted with his mother. Although he seemed interested in finding out about the translocation he was not keen to be questioned about anything else and switched on the television set. The interview with the mother was completed, and he was then asked if he would like to speak to the researcher on a subsequent occasion; he declined.

In total, 12 parents (aged between their early 30s and late 50s) and 11 children (aged between 11 and 27) were interviewed. The parents consisted of two couples (the natural mothers and fathers of the tested children) and eight mothers alone. The children came from seven of the ten families included in the study. In four families just one child was interviewed, either because it was only this child who had inherited the balanced translocation, or because their siblings were too young to be interviewed or had grown up and moved out of the area. In one family the carrier child had grown up and was married and had a son himself; in this case both he and his wife were interviewed. In the remaining two families, groups of two and three children were interviewed collectively. Although the families were not asked to give written details of occupation, most mentioned the jobs they or their husbands held.

2.2 The interviews

'An interview is a conversation with a purpose.' (Kahn and Cannel, 1957, p. 3149).

According to Cannel and Kahn, an interview is a conversation 'initiated by the interviewer for the specific purpose of obtaining research-relevant information and focussed by him on content specified by research objectives of systematic description, prediction or explanation' (Cannel and Kahn, 1968). Interview styles differ greatly. For this project we chose to employ a semi-structured interview, in which the interviewer has worked out a set of questions in advance but is free to modify their order based upon his or her perception of what is most appropriate at the time – he or she can change the way questions are worded, give explanations, leave out particular questions which seem inappropriate with a particular interviewee or include additional ones. While specific, well-defined areas were to be probed, it was imperative that the respondents did not feel constrained to only one topic, thus enabling them to speak of anything that seemed pertinent to their experiences.

All interviews took place in the subject's home and – with the subjects' consent – were audio-taped. The interview techniques were modified for the younger participants who found it hard to concentrate on what was being asked; they were encouraged to read from the interviewer's aide-memoire list of questions and topics and were helped to pronounce and understand the words. The children were then asked to think about what issues were being raised and whether they thought the question was 'good or bad' and how they felt about it. They were then allowed to read the next question out. This technique was not pre-planned, but rather it was developed in the field as this problem was encountered. Mothers sat in on three of the interviews with the children and spoke occasionally. In the remaining four, the children were interviewed on their own. Field notes were written within half-an-hour of leaving the interview.

It was appreciated that this study might raise issues that could be addressed by referral for genetic counselling and hence the subjects were informed of the help that would be available if it was requested.

2.3 Analysing the data

All tapes were transcribed by the interviewer. Transcription was completed before the analysis was undertaken. The transcripts were studied and recurrent themes were identified; the corresponding sections of text were coded. Of the many themes that emerged (>40), only four were selected for analysis in this chapter – coping, emotions through pregnancy, stigma and the children's perception of the genetic disorder and the testing process (the theme of 'age and understanding'). These themes were chosen because they illustrated issues that were raised by many of the subjects and they gave information about the consequences of genetic testing.

Confidentiality has been protected by coding the transcripts and using pseudonyms in this report. Only the interviewer ever listened to the audio-tapes and anonymity has also been ensured for health professionals mentioned by the interviewees.

3. Coming to terms with being a carrier – how family members coped with learning about their carrier status

When a person learns of an abnormality in his genetic constitution, he is liable to experience stress associated with a disruption of his psychological equilibrium (Falek, 1977). This stress can be profound because genes are seen as part of the self and thus genetic disorders cannot be externalized, as is common in other diseases where the illness may be depersonalized and objectified. If genes malfunction, all of the cells in the body are affected and the entire person may be experienced as defective (Kessler, 1979). Falek identifies five stages in the coping sequence which are based around the four preliminary phases of response commonly observed in studies of individuals under stress and these are followed by a return towards equilibrium: (i) shock and denial; (ii) anxiety; (iii) anger and/or guilt; (iv) depression; (v) psychological homeostasis.

A more flexible approach to understanding such adjustment processes is that of Lazarus and Folkman (1984). They argue:

> 'We define coping as constantly changing cognitive and behavioural efforts to manage specific external and/or internal demands that are appraised as taxing or exceeding the resources of the person.' (Lazarus and Folkman, 1984).

They claim that critical life events are not inherently deterministic, but are situations subject to individual definition. This is an interactive approach, which views coping as a continually changing process of appraisal, negotiation and adjustment.

Many participants in our study appeared to have adjusted to their carrier status over the passage of time. However, there was evidence that this psychological homeostasis had been achieved using both the approaches described above.

Julia Hobbs took the news of her carrier status calmly, saying:

> 'But it didn't shock me really. I don't know why, I can't explain why it didn't, but it didn't shock me.'

She was aware, however, that this information could be a bombshell:

> '...well you go to the doctors or counselling and they will sort of tell you something and then they'll ask you questions, you know, little questions like, 'how is this going to affect you, are you sure that you can cope with this?' and blah, blah, blah and if they keep asking you these questions and you've only just heard, it can be really awful.'

There is a potential for many different responses to genetic information.

3.1 Initial shock

Richard Callaghan had only recently been told of his carrier status and he seemed slightly bewildered by the news, describing himself as being 'shocked' and 'astonished' by it. He seemed to have some difficulty in adjusting to the information and had not really taken it all on board, saying that he wasn't ready to answer the questions because it had all happened so suddenly.

Richard appeared to be going through the early stages of the coping sequence as described by Falek, and he had not fully accepted what he had been told. When asked whether he would have bothered to get tested if it had not already been carried out, he said:

'Yer, I probably would to see if it's true or not, just in case it's a load of bollocks or not'

suggesting that he might be trying to deny his carrier status.

Many of the parents who were told that they were carriers experienced anxiety. This is illustrated by Mrs Moore:

'Well I was a little bit worried when I found out. I didn't know what it was, but other than that once we was explained the basic things like, I wasn't taking in what he [the counsellor] was saying to me.'

Some parents also perceived these concerns in their children:

'I think it was a worry to him [her son], but er I think once he knew that things could be done for him, you know, I think then it wasn't quite so bad.'

Finding out that her husband was a carrier was very distressing for Mrs Sandra Brown, who was pregnant when she was told. Falek states that when a person is told of their genetic abnormality they are initially only able to take in a little of this new information, if any, because they are in shock. This was so for many of the interviewees and is typified by Mrs Brown:

'And er she [the genetic counsellor] talked to us and explained this and that, but er at the time I don't think we really understood what she was because there are so many outcomes rather than cut them down to say well there is this option, she explained why the chromosomes were like they were and why your sister was like she was, but of course you're meeting all these people and you're thinking "What the hell does this mean?", like. You know, so we was none the wiser even after we was told really.'

Mrs Williams had her amniocentesis for maternal age and only knew that the test could detect Down's Syndrome and spina bifida. She was unprepared to be told of any other abnormality:

'So I knew myself that there was something wrong. So er I saw the counsellor again and he told me there was no spina bifida, no Down's Syndrome in my little boy and then I didn't want to know what it was.'

Mr Grant said that his daughter 'sobbed and sobbed' when she was told she was a carrier, although his wife then went on to say that when she mentioned the translocation to her daughter recently she had been very dismissive of it:

'Well I did mention it to her when we had the letter and she just said, "Oh there's quite a few people I know with chromosome defects". She was quite matter of fact about it.'

The Grant's daughter, like Richard Callaghan, appeared to experience Falek's first stage of coping, manifesting shock. Her subsequent dismissiveness could indicate a move to calm acceptance or could be an attempt at denial.

3.2 Calm acceptance

Gareth Ford was also only recently told by his mother that he was a carrier, but she described his reaction as:

'He didn't bat an eyelid. He just said "Oh right" and left it at that.'

Mrs Ford felt that her son had not been affected by the news and this was reinforced by Gareth Ford himself, who said that he 'wasn't bothered' by it. Mrs Ford herself was also calm about the family translocation: 'I just never thought about it [the translocation], to be honest. You know, you just forget you've got it, don't you? You know I haven't talked about it for years.'

This sentiment was echoed by Mrs Brown (snr) who said that the translocation was the 'least of my worries now'. Her son's anxiety did not last for long; he said that the knowledge 'just went quickly out of my mind'.

Although Mrs Water's children had been told when they were young, when she mentioned it to them again she said that it was like telling them for the first time as it 'hadn't really sunk in before'. However, she said:

'My daughter seems to have taken it in her stride and isn't too alarmed ...'

The reactions of both Gareth Ford and Mrs Water's daughter indicated that they did not have much difficulty in coping with knowledge of their carrier status, although this may be partly due to the fact that any repercussions from it would only become apparent when they started to have children. Thus it did not seem relevant to them at the time, although it might become relevant in the future. Gareth Ford alluded to this by saying:

'Well it doesn't affect me until I have children, anyway.'

These individuals did not seem to be following Falek's model of coping. They assessed the translocation and attached little importance to it at that time in their lives. Hence, they were not upset by the news. This accords with the more interactive approach to understanding individuals' reactions to such information, where the individual is an active agent who has the capacity to negotiate their own response.

Mrs Conner's children also were unconcerned about being carriers and Lucy Conner perhaps spoke for them all when she said, 'I mean I feel completely indifferent really', when talking about carrying the translocation. Her mother confirmed this, saying that 'it all seems very remote from them at this minute', explaining that they all had 'much more important things on their minds' such as GCSEs and University.

3.3 'At a bad time'

Information is harder to cope with if the individuals are trying to come to terms with other devastating news. This is shown by the Parry family who were told of

Mrs Parry's carrier status directly after their newborn daughter, Rhian, had died as a result of the family's chromosome problem:

> 'You had so much to take on board, to gather in as much information about her and they were telling us about the chromosome as well and with Rhian, so it was quite a while after that we got everything straight in our own minds. It took longer I think than if we'd just been tested.'

3.4 Family experiences and contexts

Anxieties about being a translocation carrier arose in part from the personal experiences and understandings of family members. Where no adverse affects had been encountered, as in the Connor family, very little importance was attached to the carrier state. Lucy Connor said:

> 'I think it has so little effect on our lives anyway, well it just has no impact on our lives, does it?'

Mrs Waters also felt that it was of minor significance, especially in relation to recent serious illness:

> 'Now I know I keep saying about the breast cancer, but that's something different. It's something that she [her daughter] knows that I've had and I've been very ill with it and I've had to have a lot of bad treatment. That's something tangible, that's something she can see. But with this other thing [the translocation], it's not you know, there are no repercussions from it as far as we know.'

When a family had had a distressing experience related to the translocation, however, the anxiety was much more pronounced. Mrs Hobbs, who had a son with Down's Syndrome, spoke of being very concerned for the future when her carrier daughter may decide to have children – she felt 'very anxious, very, very anxious'. These different perceptions on the same state of health are not unusual. In a study done by McQueen (1975) it was found that just under half the people attending a screening programme for Tay Sachs disease felt that to be a carrier would matter a great deal, whereas the remainder attached varying degrees of importance to this.

Mrs Connor compared her genetic anomaly with other genetic disease:

> 'it's not as if it's of critical importance ... It would be different if you were carrying a chromosome that was linked to, I don't know, cystic fibrosis in the family or something ... It would be essential, well at least more important to have a test (amniocentesis) done if the disease were serious.'

By comparing it to a genetic disorder that she perceived to be much worse, Mrs Connor was allaying her own fears about the translocation. Such a comparison was also made by Julia Hobbs, who said:

> 'I mean I'd prefer not to have it. Anybody would prefer not to have it, but er I still think it's the case that I have got it and it's not the end of the world. I mean I could have something really drastic like leukaemia or something, I could be dying, so it's not going to kill me.'

Julia had come to terms with her carrier status by relating it to a much more debilitating condition, which helped her put it into perspective. For these two individuals the impact of the translocation was subject to their own personal definition of the disorder in relation to other diseases. This again is more easily appreciated in the context of the interactive approach to understanding family reactions. If the translocation then went on to cause a serious problem in these families, the translocation would be redefined and more importance would be attached to it. Mrs Waters actually said, 'Well I mean it's everything in perspective, isn't it?' when referring to the minimal effect that the translocation had had on her family.

Most of the subjects spoke in fatalistic terms of acceptance when describing their present day feelings of the translocation:

'Well it doesn't bother me. Like you say, that's Nature's way. I got it and there's nothing I can do about it' (Luke Moore – carrier son). -

4. The tentative pregnancy

4.1 The emotional strain of pre-natal testing

There is no cure for an unbalanced translocation. The only intervention medicine can offer those couples who wish to have a child is that of pre-natal diagnosis and the selective termination of affected pregnancies.

The advent of pre-natal diagnosis was heralded by the medical profession as a technological breakthrough that gave women more reproductive choice. This makes the assumption that pregnancy is somehow made easier because the chances of having a 'negative outcome' – a mentally or physically handicapped child – can be minimized. In a study looking at the opinions and attitudes of women towards pre-natal diagnosis (Tymstra *et al.*, 1991) the respondents had a strong positive attitude towards the diagnostic procedures, especially if treatable abnormalities could be detected. Rothman (1988), however, argues that 'pre-natal diagnosis **changes** women's experiences of pregnancy' [our emphasis] and does not make it easier, regardless of the eventual outcome. As Mrs Parry points out:

'It made us very nervous and, although it was helpful that we knew that everything was all right, it was still in the back of your mind then, 'is it, isn't it?' until everything's over. We found it very worrying, all the way through the pregnancy, very apprehensive.'

A tentative pregnancy is one in which the normal feelings of elation and pleasurable expectation of motherhood cannot be enjoyed to the full. One reason for this is that the mother may fear that there might be something wrong with her baby, a concern that is both generated and allayed by the offer of pre-natal testing. All the mothers in the study group, who had undergone the procedure of amniocentesis, described it with strong, emotive language using words such as 'horrendous', 'horrible' and 'terrible'. Many identified the period between having the test done and getting the results as the most worrying time during the pregnancy. This is typified by Mrs Waters:

'because the amniocentesis was awful really. I had to have it done twice and each time I had to wait three weeks and it was so stressful.'

Mrs Williams described her feelings as being 'all churned up inside' as she waited for her results and Mrs Connor and her husband were 'counting the hours', saying:

'... er it's very upsetting to ... to be carrying a child and to feel fine and to think that everything's going all right and then to have these doubts thrown at you. So, yes we had an agonizing few days, it was awful really. But you've got to carry on – I had two other little children to look after so you've got to carry on, but I was just waiting for the days to go so I could get the results of the test. So I mean it was awful.'

In a review of the literature, White-van Mourik (1994) found that women, who had undergone a termination for fetal abnormality detected by amniocentesis, often experienced strong negative emotions. The majority of such terminations involved planned and/or welcome pregnancies and thus the intervention entailed the loss of a wanted child. This distress is added to because of the usual timing of the diagnosis and subsequent termination, the second trimester. Maternal–fetal attachment begins as early as ten weeks in the pregnancy and increases significantly as the pregnancy progresses (Caccia et al., 1991), thus it is well established by mid-gestation. Attachment grows significantly once the results of a pre-natal test are known to be normal. Mrs Brown's husband added that it was only after they received their result, which confirmed a healthy baby, that they could 'start enjoying the pregnancy' thus backing Rothman's (1988) idea that participating in the test, and the fear that accompanies it of an unfavourable test result, casts a shadow back over the early months of the pregnancy.

It would be logical to think that obtaining a favourable result (for both amniocentesis and chorionic villus sampling CVS) would be a source of comfort and relief to the parents. Previous research has shown that positive emotions increase significantly as the amniocentesis procedure unfolds (Harward, 1992). Indeed this was the case for many mothers whose minds were 'put at ease' by it. As Mrs Connor said:

'Um and having that actually gave me a lot of peace of mind after that because we knew then that she (her daughter) was going to be all right.'

4.2 Seeing is believing

Mrs Waters did not experience such relief from a favourable test result, however, either immediately after receiving the result or when her son was born:

'You know, when I saw him, well I should have felt tremendously relieved that it was all over and I had a normal son, but it wasn't like that. So I think that um I know there's the expression better to be safe than sorry, but I think that the whole thing caused a lot of distress.'

Neither Mrs Parry nor Mrs Hobbs could dispel their fears of not having a healthy baby throughout their respective pregnancies. As Mrs Hobbs put it:

'So there was this issue that even though I'd been told that she (her daughter) was a carrier like me and that there was nothing else wrong with her, I was not convinced until I had actually set eyes upon her.'

Her anxieties were relieved only when she could see for herself that her daughter was all right even though technology had given her this information earlier. Pre-natal testing at best only gave partial relief – it was not enough to calm her fears. The same was true for Mrs Parry:

'It's not until you actually see the baby – the worry's still in there.'

Even after the baby has been born, the memories of amniocentesis linger on and shape future events. For Mrs Waters the trauma of the whole process and the con-comitant stress that it caused gave her a very real problem, in that:

'... the outcome of that [the amniocentesis] was that it took me a long time to actually bond with my son – that is how much distress it caused me.'

Such adverse affects of pre-natal testing have been reported by Marteau and col-leagues, who found that women who had undergone amniocentesis held less pos-itive attitudes to the baby after the birth and were significantly more worried about the baby's health (Marteau *et al.*, 1992b).

Although only one mother in our study experienced such bonding difficulties, others have pinpointed the fear of amniocentesis as a tangible reason for limiting their family size. This is true for Sandra Brown:

'You know I think I'd be pregnant tomorrow it's just this that is holding us back you know, well holding me back more than anything. It's like I say to you, it's OK for him to say 'let's have another one' but it's me that's got to go through it all you know. He's just got to put it there. You know it's me that's got to go through all the tests, it's me that's got to have all the amniocentesis tests, it's me that has got to think well at the end of the day you could take my baby away, you know.'

Her husband, the carrier, sympathized with this dilemma and also worried about having more children saying:

'If I was to have more children in that case that's what I was thinking about most, because if I was to have another child she'd have to go through all that tests again.'

Pre-natal diagnosis, then, can generate anxieties that are not always completely relieved by a negative diagnosis. Concerns include fears about the baby's health and maternal stress. Such anxieties can exert their effect throughout the preg-nancy and may even extend beyond, to disrupt mother–baby bonding and affect future reproductive decisions.

5. Feeling 'different'

5.1 Evidence of stigmatization

When two strangers meet each will categorize the other and subconsciously attribute characteristics felt to belong to this assigned category to the other; in other words first impressions are important and do enable one to attach a 'social identity' to a stranger dependent upon their appearance. The term 'social identity' not only encompasses a person's structural attributes, such as occupation, but also personal ones, such as honesty, which are associated with the other more tangible traits (Goffman, 1968).

Goffman argues that stigmatization occurs when there is a discrepancy between what society expects of an individual, in terms of physical appearance, behaviour and beliefs (i.e. an imagined or virtual social identity), and what attributes that individual can actually be proved to possess – those real attributes falling short of the expected ones. Thus the term stigma refers to an attribute that is (deeply) discrediting.

Two classes of stigmatization exist. The first deals with the discredited stigma – here the 'differentness' is known about already or is evident on the spot. The second concerns the discreditable stigma, where the difference is neither known about by those present nor is immediately perceivable by them. The discredited individual, possessing a visible discrepancy between the expected and the actual social identity, is likely to experience prejudice and awkwardness during social interactions with those members of society who do not depart negatively from the particular expectations in question, who can be termed 'normal'.

None of the subjects interviewed spoke of experiencing a discredited stigma, which was to be expected as the genetic variable that causes them to 'deviate from the norm' is invisible. As Julia Hobbs says:

'Because I mean it just doesn't crop up and when I do tell people they're quite surprised because it's not something that does show. Um so er I don't think it makes us any different at all.'

Here Julia is showing how she does not attach any stigma to her carrier status and is quite willing to talk about it to people outside her family.

A different response was obtained when Richard Callaghan was interviewed, soon after learning about his carrier status:

'I just seem the odd one out like, the stranger, like.'

This stigma was also felt by Mrs Williams and Mrs Moore, although it was only present transiently at the time the translocation became known:

'Well I felt different in the beginning, but I'm no different to anybody else.'

'Well I felt a bit weird at first when I knew about it.'

This is consistent with the findings of Wolff et al. (1989) who found that most of the translocation carriers in their study reported a phase after the diagnosis was made when they had predominant feelings of stigmatization. Whereas a discredited stigma can be attached to an individual with no prior knowledge of that person being required, a discreditable stigma can only be known to those who have direct knowledge of the person being stigmatized. Goffman states that the discovery of such discrediting information will spoil personal identity. As two people get to know each other they become familiar with each other's personalities and biographies – this is personal identity, the 'identity peg' upon which is hung all the facets of an individual's appearance and lifestyle that make him unique

(Goffman, 1968). Therefore, if information perceived to be detrimental is then discovered about one party it may spoil their personal identity.

Mrs Moore talks of this potential:

> 'Perhaps people would think 'Oh there's something wrong with her (meaning herself) like', you know. I don't know because I haven't actually mentioned it to anybody. So I don't know what their reaction would be.'

Mrs Waters spoke of the reactions of her parents-in-law, 'And then my husband's parents found it too embarrassing to talk about really'. Similarly, when Mrs Williams was asked if she had received any adverse reactions, she spoke of some people 'not understanding'. Both women had experienced felt stigma when they revealed their carrier status.

Although many people did not experience more than a mild stigma, if they experienced any at all, they alluded to their carrier status in terms of being less than 'normal'.

This is illustrated by Mrs Waters, who was not embarrassed by her translocation within the family and had no problems passing this information onto her children. She 'didn't find it hard to say that there was something in the family' but later went on to talk about how she felt her children would not wish to mention the translocation to anybody, unless it was really necessary, because:

> 'You have got that potential that there is something wrong with you.'

This sentiment was echoed by Mrs Ford, who spoke of herself in these terms:

> 'Like they (the doctors) knew that there was something wrong with me. They knew I had that ... that chromosome.'

Here these two people seem to be admitting 'flaws' in their personal identities, which made them feel less than perfect and thus susceptible to discrimination.

Mrs Williams tells of how her son's carrier status was incorporated into his personal identity:

> '...he's always grown up knowing he was a little bit different, not different, but that there was something different, you know.'

5.2 Information Control

A discreditable stigma is not evident and is therefore unobtrusive in face-to-face interactions, and so a stigmatized individual will try to conceal the source of his stigma from as many people as possible. Goffman calls this 'passing' (Goffman, 1968) and when aiming to achieve this, an individual exerts information control, whereby he is very selective as to whom he reveals his stigma. Such information control was exhibited by several of the translocation carriers, including Mrs Connor when she spoke about her children:

> 'I've told them (her children) and they know that you are coming here today, so it hasn't been a secret or anything. But I've kept it private in the family. I know this is confidential and I'm speaking to you in confidence and er it's confidential in the family really. I mean you don't go around talking about it, do you?'

Richard Callaghan gives his reasoning for not telling people:

'I'm not quite sure if it's based on only me and my mother who've got this. It's only us two who should be having a discussion about it and not involving the rest of the family because it might worry them ... I like to keep it personal see, just in case they worry a bit or they stay away, they might be wary of it. So I don't like people interfering anyway. If I tell people about my personal life then they start to worry about it then and stay away from me.'

Here Richard is aware that he might be stigmatized by being known as a translocation carrier, and he minimizes this risk by keeping the information as secret as possible.

Luke and Sarah Moore also share this feeling:

Sarah (reading from my list of questions): 'Have you told anyone outside the family?'
Luke: 'No, I haven't told anyone.'
Interviewer: 'Why might that be?'
Luke: 'Well I don't know.'
Sarah: 'Because they will laugh at you. People will laugh when you tell them.'
Int. 'Why would they laugh?'
Luke: 'They don't keep it a secret, do they?'
Sarah: 'No, it'll spread all round the school.'

Here these children are expressing a fear of being different from their peers which may make them objects of ridicule; this would be very painful to young adolescents who try their best to 'fit in'. The only way they can control this is by not telling anyone at all. All the interviewees in this study had discussed the translocation to some degree with other family members. This was in part due to necessity, so that others of childbearing age could be informed of their risks, and thus was mentioned on medical advice.

For others this topic was discussed amongst family members as a means of adjusting to the information. This was true for Mrs Moore, Mrs Ford and Mrs Williams;

'Well I talk to my sister who hasn't got it and she understands. Because we're a close family really.'

'Me and my sister used to sit down sometimes and talk about it.'

'But er as for other family members, well they were away, but they rallied round, well we've always been a close family anyway.'

It was easy for this to be discussed within the family because often many relatives were also carriers who could share emotions and experiences. Goffman refers to this as the concept of 'the wise'. Family members, who were not carriers, were trusted by those who were because they were emotionally close to their carrier relatives and could therefore understand and empathize with them. Goffman terms these people 'sympathetic others' who were wise to the effects of being a carrier (Goffman, 1968). In addition, many of the interviewed carriers had also divulged their carrier status to their 'close friends':

'I've told my friends about it, but the teachers don't know anything, it's just my close friends who know about it' (Sian Parry).

'My friends know, I've told my friends that I've got this chromosome thing' (Julia Hobbs).

'I would tell close friends' (Mrs Moore).

'Oh well I did talk to the odd friend about it' (Mrs Connor).

These people did not feel that they had to hide this knowledge as it would not spoil their personal identities to such an extent as to disrupt any close friendships. As Matthew Grant says:

'Um a few people know, best friends and so on, but it's not something that I tell everyone ... If er they [new acquaintances] sort of wanted to know you a bit, then it would be a part of you that you couldn't get away from and so it would be something that they would have to be told.'

As close friendships develop, people are almost obliged to reveal more about themselves and so this information becomes progressively more difficult to hide. However, friends who are privy to this information can also become part of the information control process, as Sian Parry illustrated:

'I didn't mind them [her friends] telling their parents, but I asked them not to tell anyone outside like any of their other friends. I want to keep it with my close friends.'

When Sian was asked how her friends reacted, she said:

'It was OK. They understood it, they just took it all in.'

It is normal not to want everybody to know your intimate details and thus such information control may not be due solely to these individuals feeling stigmatized by their carrier status. There was also evidence that some subjects did not feel any stigma, saying that they felt 'no different' from anybody else (Mrs Williams, Mrs Ford, Sian Parry and Phillip Brown). As Julia Hobbs states:

'And er I still wouldn't say that it's anything that's made me feel any different to anybody else, because it's not something that should ... Like I said, my friends know and they haven't treated me any differently or treated me as if I've got something odd about me ... they don't treat me any differently, they've just taken it in and thought 'Right we can't see anything wrong with her, there's nothing different about her and she's been like this for the last five years that I've known her', so it doesn't really affect them and they are close to me.'

Mrs Parry, Mrs Williams and Mrs Hobbs all spoke of the benefit that sharing this information would bring to others:

'If anyone was in the same boat, I would tell them.'

Mrs Parry told how she discussed her experience of CVS with those of her work colleagues who were worried about pre-natal diagnostic procedures and Mrs

Hobbs recounted anecdotal stories to her students when she lectured about women and medicine.

6. The age of consent

6.1 When can a child decide?

The families interviewed in this study had generally not asked their children if they wished to undergo genetic testing, with the exception of the Brown family. For some, consent could not possibly have been obtained as their carrier status was determined pre-natally; in the case of those tested as young children, it had been left to the parents or physicians to make the decision. The European charter for children in hospital states that 'children and parents have the right to informed participation in all decisions involving their health care. Every child shall be protected from unnecessary medical treatment and investigation.' (Alderson, 1993a,b). This can be extrapolated to children receiving medical intervention in the primary care environment. However, the children in this study were not tested because they requested it but rather because the genetic counsellor involved thought that it would be prudent to do so or because the child's parents wanted to know for their own peace of mind:

'I just wanted to know and the only way I was going to know was by getting the test done' (Mrs Ford, speaking about her son).

The United Nations Convention on the Rights of the Child advocates the right of every child to self determination, dignity, respect, non-interference and the right to make informed decisions. However, these rights did not appear to be the principal consideration of the adults involved in the decisions in these families. Rather, genetic testing was carried out because it was thought to be in the best interests of the child in relation to their future reproductive decisions. Consideration did not seem to have been given to the potential harm that could be done by tests of genetic carrier status.

6.2 When is a minor sufficiently competent to decide about testing?

Piaget, working in the 1920s, stated that children met cognitive milestones in a set sequence in the same way that they reached motor milestones, and until the appropriate cognitive milestones had been passed children would be incapable of reasoned decision making. He inferred that children from 18 months up to seven years interpret reality exclusively on the basis of their own perceptions and experiences, without being able to consider anyone else's point of view. Their understanding is based on their own intuitions in an egocentric fashion that prevents objectivity. Subsequent research, however, has shown that children as young as three years can demonstrate objectivity when asked to solve puzzles that are put to them in an age-appropriate way (Donaldson, 1978).

In Piaget's third stage of cognitive development, that of concrete operations, children can reason in terms of concrete realities and direct personal experiences. They do not yet have the ability to think about abstract ideas. Such abstract thinking is achieved in the stage of formal operations, near the age of twelve. At this age reasoning also becomes more logical and less egocentric than in childhood. Thus most teenagers who have reached this stage are capable of discussing hypothetical situations and problems. Adolescents then proceed to develop new cognitive stages using processes of assimilation and accommodation.

Piaget's theories have been criticized. Some studies have failed to demonstrate any consistent pattern of cognitive development during adolescence (Keating, 1980). Cognitive development does not unfold simply at its own internal rate, but is also affected by personal experience and other external factors (Alderson, 1992a). There is variation among children of the same chronological age as to the level of their understanding and their ability to contribute to decisions about their present and future health. This was demonstrated by Julia Hobbs who said:

'I had to grow up a lot quicker than some children because of Mark' (Julia's brother, who was affected by Down's syndrome).

Anne Solberg, studying hundreds of Norwegian 12-year-olds, found that external factors relating to home and family affect a child's maturity. Children, whose parents expected them to be responsible, respond well and are trusted with more and more adult responsibilities (Solberg, 1990). Social class, family income, ethnicity, the number of children in the family, the parents' age and education and other factors in the home and family can all affect children's decision-making (Alderson, 1992a).

The young people in this study demonstrated a transition from one level of cognitive thinking to the next. Most of the children were told about their carrier status when they were still fairly young:

'I mean when my youngest daughter was born [who was a carrier], my eldest child was six in any case and she's always been bright. I don't suppose I went into chromosomes at the age of six' (Mrs Connor describing at what age she told her daughter, Lucy).

At this stage, although the children seemed to have understood what was said, their comprehension was far from complete:

'So er I don't really think I understood it [the translocation] fully, but as I've gotten older I've understood it more. But I don't think I was confused about it really' (Julia Hobbs).

Sian Parry's parents told her about her carrier status when she was 10 years old and she is now 13:

'Well I understood it [the translocation], but I didn't think about it again, so I needed to be told about it again'.

As the children grew older, the information they had been given gradually took on more meaning:

'I think that over the years I began to understand what was said and as I got older more things were explained and I started understanding more' (Matthew Grant).

'It didn't bother me too much when I was first told, but as I'm getting older, it's starting to sink in a bit' (Sian Parry).

Julia Hobbs had been told 4 years before being interviewed for this study, when she was 12:

'I did understand what it [the translocation] was when I was first told, but I don't think it sort of er as I've gotten older, because I've grown up with Mark and so I've learnt what he's like, but as I've gotten older that's when I started realizing, you know, the pressures that people with handicapped children are under. I don't think I really understood how much of a problem it could be until, you know, maybe a couple of years ago, when I started realizing how much of a problem it can be looking after a handicapped child and then I thought it would be my turn.'

This passage illustrates that, as she was progressing through the stage of formal thinking, Julia began to understand for the first time that she too could pass on this genetic disorder to her own children. The healthy siblings of a sick child will at some stage begin to appreciate that they too may be vulnerable to a genetic disorder or may subsequently transmit it to their children.

Only two mothers did not tell their children when they were young. Mrs Callaghan had simply forgotten about the translocation, but Mrs Ford explained:

'I just thought I would tell him when he was older and could understand more.'

Young people can make personal decisions, even before they reach the stage of formal thinking, because they can utilize their concrete operational skills. Research has shown that even intelligent adults do better when they use concrete operational thought in decision making (Dulik, 1972; Kuhn & Brannock, 1977). This is because it is easier to think about real problems than abstractions. It has also been shown that young children employ certain aspects of formal thinking well in advance of the age predicted by Piaget (Keating, 1980). Decisions are also based on feelings and needs, both of which a child can express (Alderson, 1993b). However, it would not be easy for a young child to make an informed decision about the abstract matter of consenting to carrier testing, which has no immediate, tangible benefits or harm. This was recognized by many parents who felt that even though their children had understood some aspects of the translocation, they could not have any idea about its implications. The age at which such maturity is attained is specific to each child, but for many it will start around early adolescence.

Mann et al. (1989) argue that competent decision making in adolescence is reliant upon nine factors: choice, comprehension, creativity, compromise, consequentiality, correctness, credibility, consistency and commitment. They state that by the age of 15 many adolescents have achieved a reasonable level of competence in most of these components. They claim that younger adolescents (12–14 years) are less able to create options, identify a wide range of risks and benefits, foresee the consequences of alternatives, and gauge the credibility of information from sources with vested interests. This latter concern is of importance in relation to genetic testing. If a young person is given the autonomy in this context, they must be mature enough to have at least some emotional independence from their parents – establishing such independence is, of course, one of the tasks of adolescence (Havighurst, 1953).

Certain barriers to achieving competence in decision making during adolescence do exist (Mann *et al.*, 1989). These include attitudinal constraints (e.g. beliefs about the proper age for making decisions), peer group pressures to conformity, breakdowns in family structure and functioning and restricted legal rights to make important personal decisions. Alderson (1992b) found that in a survey of 8–15 year-old students in the London area, the average age at which the respondents deemed themselves able to independently consent to surgery was 16.6 years. Alderson believes that the freedom to make medical decisions was sometimes equated with being somehow 'grown-up', such as being old enough to go out to work or leave home. Thus she cautioned that opinions about appropriate ages for consent could reveal more about public beliefs than about any particular child's needs or abilities (Alderson, 1992a).

From the late 1960s to the early 1990s changes in the law on consent have progressively increased the rights of those under 18 to self determination – always dependent on their capacity to understand the implications of their decisions. The case of *Gillick* v. *West Norfolk and Wisbech Area Health Authority* (1985) established that children under 16 years could give legally effective consent to medical treatment, independent of their parents' wishes, provided they had sufficient understanding and intelligence.

6.3 Why would children want to make a decision to undergo testing when young?

There seemed to be no resentment of the way the 'children' had been treated when younger. Only Julia Hobbs gave any explanations why she thought the genetic testing of children could be a problem.

> 'I'm lucky because I've got a very good relationship with my parents, but I can see how it would be difficult if you didn't and I can also see that if you didn't get on with your parents you may be angry with them for making that decision [to get tested] for you.'

None of the group of 'children' felt that their genetic make-up had led other family members to treat them differently in any way. Talking about her carrier son, however, Mrs Brown (snr) did say:

> 'I thought more about Phillip, you can understand, worried about him a little bit more then because he's got to go through it all.'

Mrs Brown also had a daughter with an unbalanced form of the translocation and thus her concern could be attributed to natural anxiety arising from the desire for her son not to go through the same experiences as she had.

Many reasons were given by the 'children' as to why they thought that childhood testing for this particular chromosomal anomaly was beneficial. Phillip Brown said:

> 'I'm glad I had the test done when I was younger, I think. Because – knowing me – I would have waited and waited until it was too late and something might have happened.'

Later in the conversation Phillip and his wife Sandra explained further:

> Int.: 'When will you tell your son that he is a carrier?'
> Sandra: 'Oh I'll tell him as soon as he's after the girls' (laughs).
> Phillip: 'As soon as he's a teenager I should imagine.'
> Sandra: 'Yes as soon as he starts er you know ...'
> Phillip: 'Having girlfriends.'
> Sandra: 'You know because these days I think you just have to talk to them. You know you see 12- or 15-year-old girls pregnant, 12-year-old boys as the father, they're kids of 12, it's like being brother and sister. But er as soon as I can see, you know ...'
> Phillip: 'As soon as he's mature enough, we will tell him as soon as that.'

Richard Callaghan, who was not told of his carrier status until he was 17, felt that he would have liked to have been told earlier:

> 'Well I think I would have rather got it over and done with when I was a young age ... er I could understand what it was all about then, than have it done now, because it's too much pressure on me now. I'd rather have had it over and done with at an earlier stage.'

Matthew Grant was happy to have grown up knowing he was a carrier:

> 'I think that it was probably best knowing then, before, because it has helped me become aware of certain things. Because if for instance I found out about it at the age of sixteen, it would have come as a huge shock. It probably would have affected me more than it already had.'

When asked why it would have been more of a shock, he replied:

> 'I think by that age you probably have certain ideas and certain plans and then you realize this and it stops those plans.'

A major reason for testing minors is to relieve parental anxieties. Julia Hobbs gives this as a reason why she would want her own children to be tested:

> 'But I think I would get them (her future children) tested for my own peace of mind, because I know that I'd worry otherwise and if I'm worrying about things then it may put a strain on the family. So I would get it done because it would make life easier for everyone.'

Julia may not mind that her parents knew her carrier status because she has a very supportive family. Matthew Grant also gives another reason why he thinks that it could be advantageous for parents to know:

> 'I would probably get my children tested when they were young, so I would know as well as them, so when the time comes for them to find out I've known for a while and so when they ask you questions about it, you are more prepared to answer them.'

Although these statements support the notion that it is important to keep the knowledge of the translocation alive in the family, it is not essential for children to be tested for them to be aware of the issues.

7. Responses to genetic counselling

7.1 Transfer of information

One of the goals of genetic counselling is to provide individuals and families with the information they will find helpful in making reproductive decisions. Following genetic counselling, therefore, a couple should be fully cognizant of the facts surrounding their genetic diagnosis. Even though individuals will often have an increased awareness of the specific genetic disorder in question after such counselling, however, this knowledge is often only marginally accurate (Kessler, 1989). Many of the families in our study seemed to have been confused by the information they received.

'You know, I don't even know what it's called' (Mrs Callaghan).

'We were certainly very confused about everything, about the risks that we were under of, you know, of having an abnormal child and the risks of having one that would miscarry. … So really at the time we were very, very confused and very distressed because, you know, the counsellor put the chances so high at one time um and I started to feel I didn't know what was going on' (Mrs Waters).

'Now that's what our Mam was getting confused with because she went along to the doctor whatever, and I think that our Mam took it the wrong way and thought that Dad was unbalanced' (Mrs Moore).

There was also substantial dissatisfaction with the amount of information received:

'I wouldn't say it was upsetting, but I would say if we could have had more information. It was er it was very basic really. [Pause] It seemed to be all percentages – you've got a one in four chance of this and it seemed to be geared all round percentages. I think we would have liked more information about it' (Mrs Parry).

'Well I wanted to know all this and that about it you know. But like I say I still don't know that much about it now' (Mrs Moore).

'… nobody's really sat down with us and explained you know. I know that I am a carrier of this chromosome, same as my mother and it can affect the children if you are going to bear children. It's basically all we know really, we haven't really nobody's really sat down and explained any details' (Phillip Brown).

Luke Moore also highlighted that lack of relevant knowledge could cause difficulties:

'Well if you don't know much about it, it might be difficult to discuss it with a partner' (Luke Moore).

The task of accurately relaying genetic information is hampered by the fact that little is known about genetics by the lay person. In a study on a sample population that was slightly more educated than the general public, it was found that only 38% had any knowledge of the relatively common genetic disease, cystic fibrosis (Decruyenaere et al., 1992). Therefore it was not surprising that none of the subjects in this study were familiar with the consequences of an unbalanced translocation, unless they had had some previous experience of it in their family. The authors concluded that people are insufficiently acquainted with genetics and recommended education of the general public.

7.2 Recall of information

Incorrect recall can also be a problem. Mrs Williams, for example, had incorrectly remembered her daughter as not being a carrier, when in fact she was. It is common practice now for the information imparted during genetic counselling to be written down in a letter that is sent to the families. Only the Waters' family received any written communication, but Mrs Waters found this to be very helpful:

> 'Not only did she [the genetic counsellor] explain things clearly to us that day, but I asked for her to write a letter so that I've got ... I mean if you're not really familiar with these terms they can go in one ear and out the other, can't they?'

7.3 Communication in the wider family

Another goal of genetic counselling is to enable interested family members to make informed reproductive decisions. This requires a willingness on the part of specific individuals to pass on information to other members of their families. A variety of processes can obstruct this flow of information within a family. Ayme and co-authors, for example, found that over half of the potential carriers of a balanced chromosomal translocation within a group of families had not been tested (Ayme *et al.*, 1993). Either the index members had not passed on the knowledge of their carrier status to all their relatives, or they transmitted it in such a way that their relatives had not perceived a need to explore it. These families had all been counselled 'non-directively'. A more active promotion of the diffusion of information within families may be appropriate, especially where the carrier parent was ascertained through a history of recurrent miscarriage or the birth of an affected, chromosomally unbalanced child (Wolff *et al.*, 1989).

As all of the families in this study were initially counselled some years ago, they were actively told by the counsellor or their GP to inform other family members:

> 'I was told to mention it to my brother and his family' (Mrs Connor).

The result was that all the families discussed their carrier status with some, if not all, of their relatives.

> '... so with the result then that all the family had to go for tests. Um my mother's sister, she got one sister in Lincoln, all her family, my sister that's away, everybody had to go you know.' (Mrs Williams).

The assumption was made by the counsellors involved that all extended family members needed to know their genetic make-up and that such information should be viewed with paramount importance. This supports a claim made by Abby Lippman that in Western society today, 'genetics is increasingly identified as the way to reveal and explain health and disease, normality and abnormality' (Lippman, 1991). She states that:

'Today's stories about health and disease, both in professional journals and in mass circulation magazines, are increasingly told in the language of genetics. Using the metaphor of blueprints, with genes and DNA fragments presented as a set of instructions, the dominant discourse describing the human condition is reductionist, emphasizing genetic determination. It promotes scientific control of the body, individualizes health problems and situates individuals increasingly according to their genes.'

Lippman labels this tendency to explain physiological, physical and behavioural differences between individuals in genetic terms as 'geneticization'. She argues that geneticists, within the context of Western cultural beliefs, have created a need for people to know their genetic status, so that couples (and women in particular), are expected to do everything possible to produce a healthy child. Our study, however, produces evidence that for some people that need does not exist. Three subjects reported that their relatives did not want to discover their carrier status, because they could not see its relevance. For example, Mrs Waters said:

'I know there were some members of my father's family refused to be tested. They just thought it was a lot of nonsense.'

She went on to say:

'Well no, and um if you've been perfectly healthy all your life, your parents were healthy and your brothers and sisters were healthy and then someone comes along and says there's something wrong with your chromosomes, a lot of people just don't want to hear about it, they just don't want to know.'

Mrs Connor and Mrs Moore also found that not all the members of their families felt that it was of any importance:

'I don't know whether my brother has this chromosome actually. Um, by that stage they had had the two children that they wanted and they weren't going to have any more, so I don't know if my brother has the chromosome or not, um and we did talk to my husband's parents at the time, but it didn't really affect them in any way so I wouldn't be surprised if they've quite forgotten about it really. And of course it was mentioned to my parents, but we don't know which one of them that it came from, but it didn't matter then because they weren't going to have any more children' (Mrs Connor).

'We know that it originated from my dad, but they didn't want to know, my dad's side of the family. They didn't want to know because like they said, that they had had their children before. So I don't think it ever really came up with them' (Mrs Moore).

These family members did not feel that it was necessary for either themselves or their children to be tested, even though they were aware that it was a disorder that could be passed through the generations.

The majority of the subjects did not actively avoid speaking of the translocation to other members of their families, which is contrary to the findings of Suslak et al. (1983) and Spitzer et al. (1985) who found that translocation carriers often did not want to share the information about genetic risks with family members and that 'emotional genetic load leads to the denial of genetic risks in translocation families.'

Counsellees go to genetic counselling to obtain information particularly concerning genetic risk, aetiology and diagnostic issues (Sorenson *et al.*, 1981); and such information is frequently used to make decisions about reproduction and other important life choices. However, to be useful for decision-making, information first needs to be coded and stored in ways in which it can be accessed for later problem solving. To be coded and stored, information has to be transformed into personally meaningful items, which interface and, to a greater or lesser extent, integrate with previously stored information (Kessler, 1989). Thus, while many of the parents in this study had limited factual knowledge, they did know accurately about the various practical details involved in pre-natal testing. Mrs Moore, who confessed that her knowledge of the translocation was minimal, said:

> 'But like I say, you know, I know certain things about it, you know like the amniocentesis and that everything's all right, but er the counsellor told me that when I got pregnant I'd have to have this test done to make sure that everything's all right and that, but it don't affect me, like. I remember that.'

These practicalities were also relayed to her children. Thus the genetic information was remembered in a way that had personal meaning and not, for many, in an intellectual context. Many subjects acted upon their health professionals' beliefs that karyotyping the young children of recently identified carriers was the correct course of action to take. When Mrs Connor was asked if it was her idea to have her children tested, she replied:

> 'No not at all. My GP I think it was, or it might have been the Genetics Department, er said that it would be sensible to test the other two children um so I took them along to the surgery down here to have blood taken from them. Of course they were only quite little and it was a bit difficult to explain really.'

When the agenda is set so that it satisfies the doctor's criteria, the consultand will feel dissatisfied with the consultation, (as has been previously illustrated) and will not remember much from it. Such problems are compounded if technical jargon is used by the health professional. Mrs Waters, a qualified teacher, said:

> 'I mean if you're really not familiar with these terms they can go in one ear and out the other, can't they?'

Her experience also demonstrated that if a counsellee's preconceived thoughts are not recognized by the physician, an ensuing confusion will result because both parties have different internal agendas:

> 'I had expected something to be a bit wrong because I'd worried so much – having had a handicapped sister, but, when it did ... you know when they told us what it was ... and it was nothing to do with her, then it just didn't seem to make sense. That was very hard to accept that it wasn't connected ... It wasn't until we'd seen him (the genetic counsellor) a couple of times that we realized that the abnormality that he was talking about wasn't in fact Down's Syndrome as we thought it was. It was Patau's syndrome ...' (Mrs Waters).

8. Summary

We have gathered evidence about the remembered impact of identifying children as carriers of balanced chromosomal translocations. Individuals do experience a disruption of their psychological equilibrium when told that they carry the balanced translocation, but this is transient. Mothers who have undergone pre-natal testing report stress and anxiety. Although these adverse consequences were relieved for most by a favourable test result, some mothers experienced longer-term problems, including damage to the process of mother–baby attachment and fear of future pregnancies.

Pendleton and coauthors state that when illness is defined as an abnormal state by society, it becomes regarded as deviant. Medicine, indeed, has an important role in social control; those individuals seeking medical help may be labelled as deviant and suffer stigmatization. Therefore, individuals who are actively sought out by health professionals and are shown to be different, such as those identified as having a genetic abnormality through a family investigation, may also feel stigmatized. This was true for many of the interviewees, although the stigma was discreditable and not apparent to the outside world. These feelings also did not last for long.

Burn (1993) argues that as the human genome project gathers pace, we may be able to detect all major recessive genes in the near future and when everyone realizes that they are a carrier, there can be no stigma. This 'Trotskyist' solution to genetic discrimination and stigmatization might work in the long run – but great distress may be caused while these processes of social re-equilibration proceed.

The findings of this study must be interpreted in the light of its limitations. These include the fact that it was retrospective and that participants were describing events that occurred at least ten years ago – they were inevitably liable to inaccurate recall. Two of the interviewees alluded to this problem:

'It's easy to forget because it was so long ago.' (Mrs Williams)

'Oh my memory's so bad I can't think that far back.' (Mrs Brown, snr)

There could also be discrepancies between the subject's retrospective interpretation of events and the reactions they experienced at the time, as people will interpret their actions and emotions in ways that correspond with their present situations.

Most children exhibited no adverse emotions that could be attributed to knowledge of their carrier status. Only one child, Richard Callaghan, showed signs of discomfort during the interview. He had only recently learnt of the translocation and admitted that the news was a shock. However, this did not lead him to wish that he had never found out about the chromosome, rather that he had been told about it when he was younger. It is not imperative to test young children for a genetic disorder that will only have implications in later life, but it is important for knowledge of the disorder to be kept alive in the family.

Adolescents are capable of decision-making and are happy to do so in relation to genetic testing. Mitchell et al. (1993) studied attitudes of high school children in Montreal towards CF carrier screening. They found that, given the chance to learn and participate in a screening program, many regarded it as a positive experience.

Most participants understood and accepted the imperfect sensitivity of the DNA test. Cobb *et al.* (1991) also found that 86% of Scottish third-year University students favoured routine CF carrier screening, although education about the disease and test was essential before such an opinion could be given.

Chromosome testing of the children in translocation families could therefore be postponed until they are young adolescents. At this age most individuals will be able to give informed consent, following an in-depth counselling session with a clinical geneticist in which they would be educated about the translocation and its implications. Thus, carrier status would usually be known about before child-bearing became likely (if an individual did not want to learn of their genetic make-up, that would, of course, be their decision and they should not be pressured). In these circumstances, therefore, testing would generally not take place until an individual was fully competent to consent. The identification of carriers would, no doubt, continue to occur as a consequence of pre-natal diagnosis and carrier testing would also continue to be performed on some young children. In many families, however, the optimum approach may be for ongoing, open discussion of the family's genetic situation to be combined with a deferral of testing until the child is competent to make an informed decision on their own behalf.

References

Alderson, P. (1992a) Everyday and medical life choices: decision-making among 8-to-15 year-old school students. *Child: Care, Health and Development* 18: 81–95.

Alderson, P. (1992b) In the genes or in the stars? Children's competence to consent. *J. Med. Ethics* 18: 119–124.

Alderson, P. (1993a) European charter of children's rights. *Bull. Med. Ethics* 13–15.

Alderson, P. (1993b) Children's consent to surgery. Open University Press, Buckingham.

Ayme, S., Macquart-Moulin, G., Julian-Reynier, C., Chabal, F. and Giraud, F. (1993) Diffusion of information about genetic risk within families. *Neuromusc. Disord.* 3: (5/6) 571–574.

Bekker, H., Denniss, G., Modell, M., Brobrow, M. and Marteau, T. (1994) The impact of population based screening for carriers of cystic fibrosis. *J. Med. Genet.* 31: 364–368.

Billings, P.R., Kohn, M.A., de Cuevas, M., Beckwith, J., Alper, S. and Natowicz, M.R. (1992) Discrimination as a consequence of genetic testing. *Am. J. Hum. Genet.* 50: 476–482.

Bowman, J.E. (1991): Invited editorial: Pre-natal screening for haemoglobinopathies. *Am. J. Hum. Genet.* 48: 433–438.

Burn, J. (1993) Screening for cystic fibrosis in primary care. Family practice at last? *B.M.J.* 306: 1558–1559.

Caccia, N., Johnson, J.M., Robinson, G.E. and Barna, T. (1991) Impact of pre-natal testing on maternal-fetal bonding: chorionic villus sampling versus amniocentesis. *Am. J. Obstet. Gynecol.* 165: 1122–1125.

Cannel, C.F and Khan, R.L. (1968) 'Interviewing'. In: *The Handbook of Social Psychology* (eds G. Lindzey and E. Aronson) Vol. 2, Research Methods. Addison-Wesley, New York.

Clinical Genetics Society (1994) Report of the Working Party on The Genetic Testing of Children. *J. Med. Genet.* 31: 788–797.

Cobb, E., Holloway, S., Elton, R. and Raeburn, J.A. (1991) What do young people think about screening for cystic fibrosis? *J. Med. Genet.* 28: 322–324.

Decruyenaere, M., Evers-Kiebooms, G. and Van den Berghe, H. (1992) Community knowledge about human genetics. In: *Psychosocial Aspects of Genetic Counseling* (eds G. Evers–Kiebooms *et al.*) Wiley-Liss, New York.

Donaldson, M. (1978) Children's minds. Fontana, Edinburgh.

Duffy, M.E. (1987) Methodological triangulation: a vehicle for merging quantitative and qualitative methods. *Image* **19**: 130–133.

Dulik, E. (1972) Adolescent thinking á la Piaget: The formal stage. *J. Youth Adolescence*, **1**: 281–291.

Evers-Kiebooms, G., Denayer, L., Welkenhuysen, M., Cassiman J-J. and Van den Berghe, H. (1994) A stigmatizing effect of the carrier status for cystic fibrosis? *Clin. Genet.* **36**: 336–343.

Falek, A. (1977) Use of the coping process to achieve psychological homeostasis in genetic counseling. In: *Genetic Counseling* (eds H. A. Lubs and F. de la Cruz), Raven Press, New York.

Fanos, J.H. and Johnson, J.P. (1995a) Barriers to carrier testing for adult cystic fibrosis siblings: the importance of not knowing. *Am. J. Med. Genet.* **59**: 85–91.

Fanos, J.H. and Johnson, J.P. (1995b) Perception of carrier status by cystic fibrosis siblings. *Am. J. Hum. Genet.* **57**: 431–438

Gillick v West Norfolk and Wisbech Area Health Authority. [1985] 3 All ER 402.

Goffman, E. (1968) *Notes on the Management of Spoiled Identity.* Penguin Books, Harmondsworth.

Haan, E.A. (1993) Screening for carriers of genetic disease: points to consider. *Med. J. Aust.* **158**: 419–421.

Harper, P.S. and Clarke, A. (1990) Should we test children for 'adult' genetic diseases? *The Lancet* **335**: 1205–1206.

Harward, J.R. (1992) Emotion in women undergoing amniocentesis in the second trimester of pregnancy. *J. Perinatoly.* **12**(3): 257–261.

Havighurst, R.J. (1953) Human development and education.Longmans, Green and Co, New York.

Keating, D. (1980) *Thinking processess in adolescence.* Wiley, New York.

Kessler, S. (1979) The Psychological Aims of Genetic Counseling. In: *Genetic Counseling. Psychological Dimensions* (ed S. Kessler). Academic Press, New York.

Kessler, S. (1989) Psychologic Aspects of Genetic Counseling: VI. A critical review of the literature dealing with education and reproduction. *Am. J. Med. Genet.* **34**: 340–353.

Kahn, A. and Cannel, C. (1957) The Dynamics of Interviewing. Wiley, New York.

Kuhn, D. and Brannoch, J. (1977) Development of the isolation of variables scheme in an experimental and a natural experiment context. *Dev. Psychol.* **13**: 9–14.

Lazarus, R.S. and Folkman, S. (1984) *Stress, Appraisal and Coping.* Springer Verlag, New York.

Leninger, M. (1985) Nature, rationale and importance of qualitative research methods in nursing. In: *Qualitative research methods in nursing* (ed M. Leninger). Grune and Stratton, Orlando, FL.

Lippman, A. (1991) Pre-natal genetic testing and screening: ,constructing needs and reinforcing inequities. *Am. J. Law Med.* **17**: 15–50. Reprinted in: *Genetic Counselling. Practice and Principles.* (ed A. Clarke), 1994, Routledge, London.

Mann, L., Harmoni, R. and Power, C. (1989) Adolescent decision-making : the development of competence. *J. Adolescence.* **12**: 265–278.

Marteau, T.M., van Duijn, M. and Ellis, I. (1992a) Effects of genetic screening on perceptions of health: a pilot study. *J. Med. Genet.* **29**: 24–26.

Marteau, T.M., Kidd, J., Cook, R., Michie, S., Johnston, M., Slack, J. and Shaw, R.W. (1992b) Psychological effects of having amniocentesis: are these due to the procedure, the risk or the behaviour? *J. Psychomat. Res.* **36**: 395–402.

Marteau, T.M. (1994a) Invited editorial: *J. Med. Genet.* **31**: 743.

Marteau, T.M. (1994b) Towards an understanding of the psychological consequences of screening. In: *Psychosocial effects of screening for disease prevention and detection* (ed R.T. Croyle) University Press, New York.

McQueen, D.V. (1975) Social aspects of genetic screening for Tay-Sachs disease: The pilot community screening program in Baltimore and Washington. *Soc. Biol.* **22**: 125–133.

Mitchell, J., Scriver, C.R., Clow, C.L. and Kaplan, F. (1993) What young people think and do when the option for cystic fibrosis carrier testing is available. *J. Med. Genet.* **30**: 538–542.

Page, A., Eng, B., Chang, P.L. and Waye, J.S. (1992) Attitudes of high school students towards carrier screening for cystic fibrosis. *Am. J. Hum. Genet.*51(suppl): abstract 17.

Pendleton, D., Schofield, T., Tate, P., Havelock, P. (1984) *The consultation. An approach to learning and teaching.* Oxford University Press, Oxford.

Rothman, B.K. (1988) *The Tentative Pregnancy. Amniocentesis and the Sexual Politics of Motherhood.* Pandora, Hammersmith.

Simkin, P. (1991) Just another day in a woman's life? Women's long-term perceptions of their first birth experience. Part 1. *Birth* **18**(4) 203–210.

Smith, H.W. (1975) *Strategies of Social Research: the methodological imagination.* Prentice-Hall, London.

Solberg, A. (1990) Negotiating childhood. In: *Constructing and reconstructing childhood* (eds A. James and A. Proust). Falmer Press, Basingstoke.

Sorenson, J.R., Swazey, J.P. and Scotch, N.A. (1981) *Reproductive Pasts, Reproductive Futures: Genetic counseling and its effectiveness.* Alan R. Liss, New York. For the *National Foundation-March of Dimes.* BD: OAS XVII(4): 1–92.

Spitzer, K.H., Sanders-Fay, K., Mitter, N.S. and Gardner, L.I. (1985) Psychodynamics of emotional genetic 'load' in balanced carrier state: denial of genetic consequences by members of two kindreds with trisomy 10q2 and 11q2. *Am. J. Hum. Genet.* 37: 137A.

Spradley, J.P. (1979) The Ethnographic Interview. Holt, Rinehart and Winston, New York.

Stamatayannopolous, G. (1974) Problems of screening and counselling in the haemoglobinopathies. In: *Birth Defects: Proceedings of the Fourth International* (eds A.G. Motulsky and F.J.B. Ebling). Excerpta Medica, Amsterdam.

Suslak, L., Price, D.M. and Desposito, F. (1983) Transmitting Balanced translocation carrier information within families: a follow-up study. American *J. Med. Genet.* 20: 227–232.

Tymstra, T.J., Bajemal, C., Beekhuis, J.R. and Mantingh, A. (1991) Women's Opinions on the offer and use of pre-natal diagnosis. *Prenatal Diagnosis.* 11(12): 893–898.

Wertz, D.C., Fanos, J.H. and Reilly, P.R. (1994) Genetic Testing For Children and Adolescents. *JAMA.* 272: 875–881.

White-van Mourik (1994) Termination of a second-trimester pregnancy for fetal abnormality: psychosocial aspects. In: *Genetic Counselling. Practice and Principles* (ed A. Clarke). Routledge, London.

Wolff, G., Back, E., Arleth, S. and Rapp-Korner, U. (1989) Genetic counselling in families with inherited balanced translocations: experience with 36 families. *Clin. Genet.* 35: 404–416.

Zeesman, S., Clow, C.L., Cartier, L. and Scriver, C.R. (1984) A private view of heterozygosity: Eight-year follow up on carriers of the Tay-Sachs gene detected by high school screening in Montreal. *Am. J. Med. Genet.* 18: 769–778.

7

A retrospective study of genetic carrier testing in childhood

Outi Järvinen and Helena Kääriäinen

1. Introduction

1.1 General background

The recent advances of modern biotechnology have greatly enhanced our ability to identify persons carrying faulty genes. Genetic tests are now available for a growing number of inherited diseases for which there was previously no diagnostic test. Gene tests make it possible to detect genetic diseases in a fetus and to predict the future onset of a variety of diseases prior to any clinical manifestations. Genetic testing thereby offers many potential benefits to an individual, a family, or a whole population. Also, it helps to identify treatable genetic disorders at an early stage. Even more often, it shows that the individual at risk of a genetic diseases does not have the suspected gene and so relieves them of unnecessary worry.

New disease genes are mapped and characterized at an increasing rate, and simpler, cheaper methods are being developed for genetic testing and screening. This will inevitably lead more patients, relatives, and pregnancies to be tested.

As gene tests are being used more commonly and as they can sometimes be offered to the whole population, it is simultaneously becoming very important to study all possible aspects of genetic testing including their psychosocial consequences. In particular, gene tests performed in childhood have specific problems and features that have not been investigated so far.

2. Problems in testing children

As young children are not able to participate in deciding whether to take a genetic test or not, the decision must be made by their parents and doctors. In these difficult and often conflicting situations, many questions and worries arise. Should a

---

The Genetic Testing of Children, A.J. Clarke (ed.).
© 1998 BIOS Scientific Publishers Ltd, Oxford.

genetic test be performed whenever the parents request it? Or, on the contrary, should it always be postponed until the child is older?

The genetic testing of children is clearly appropriate when the onset of a condition occurs regularly in childhood, especially if there are useful medical interventions that can be offered (for example, diet, medication, surveillance for complications) (Clinical Genetics Society, 1994). In such situations the use of a molecular genetic diagnostic test can be regarded as being no different in principle from the use of any other diagnostic test. However, testing apparently healthy children for adult-onset disorders or to determine their carrier status for inherited disorders that might affect the health not of themselves but of their future children is generally regarded as problematic (Clinical Genetics Society, 1994).

The potential harms caused by childhood genetic testing have been claimed to include damage to the child's self-esteem, distortion of the family's perceptions of the child, loss of future adult autonomy and confidentiality, discrimination against the child's education, employment, or insurance, and adverse effects on the child's capacity to form future relationships (Clinical Genetics Society, 1994). However, advantages of such testing have been considered to be an opportunity for the child to adapt to their circumstances, for fostering openness within the family, and for reducing parental feelings of uncertainty.

3. What would be the optimal age for testing?

When an X-chromosomal disorder occurs in the family, the doctors as well as the family members know that the female relatives (healthy sisters or healthy aunts) might be carriers of this disease. In this situation, it is hard to know whether carrier testing should be performed in childhood or later in adulthood. Also, there are no clear rules about who has the right to decide whether the child should be tested. In two Finnish studies, as many as 37–74% of the patients or family members with X-linked genetic diseases considered childhood or the teenage years as the optimal age for carrier testing (Furu et al., 1993; Ranta et al., 1994). However, it can be argued that, when suggesting this, the parents were not aware of all the possible effects of the testing on their own and the child's future.

Can the child take part in the decision making? According to Dorothy C. Wertz, in many religious traditions, 11 or 12 years is the age when children are considered capable of making major decisions about life, death and eternity. Widely recognized theories of cognitive development postulate a trajectory in children's ability to comprehend external events. According to these theories, most children are unable to understand even simple health-related procedures before approximately the age of 7 years, although the age and the quality of comprehension vary widely among individuals. Ordinarily, children will not be able to understand the full implications of genetic testing until they reach the stage of formal operational thought, which may begin as early as the age of 11 years or may first appear several years later (Wertz, 1994). This means that testing in childhood takes away the individual's future right to make his/her own decision about testing as an autonomous adult and takes away his/her privacy.

4. The practice of carrier testing – past and present

Even before the advent of molecular genetic methodologies, the identification of children carrying faulty genes had been widely practised. For instance, when X-linked muscular dystrophy or haemophilia (A or B) is diagnosed in a boy, it has been customary in Finland for the paediatricians, haematologists or geneticists to perform carrier testing in his sisters. The method was to examinate serum creatine kinase in the sisters of the boys with Duchenne muscular dystrophy, and to perform coagulation tests to identify carriers of haemophilia A or B. The main aim was to ensure that testing had been offered to the whole family.

Recently, the practice of carrier testing in childhood has been strongly questioned and carrier testing is no longer performed routinely on children. Clinical geneticists are very reluctant to test children, although these tests are sometimes requested by paediatricians and parents. Nevertheless, research on the long-term consequences of genetic testing in childhood is lacking. For example we do not know what the families concerned think about the testing of children. Are they satisfied with the testing performed years ago? Have they ever had regrets about the testing? What, if anything, have the parents told their children about the previous testing? Have the parents told anyone else about the results? What do these tested children think about the testing? Has the carrier test caused any unforeseen psychosocial consequences?

As carrier testing in childhood has been carried out in the past, it is now possible to institute retrospective psychosocial evaluations of the impact of the testing on the children and their families so that future policies can be guided by evidence rather than conjecture and anecdote. In Finland, we have started to study the families in which carrier testing was performed in the 1980s.

5. The aims of this study

The aim of this study is to investigate retrospectively the consequences of carrier testing performed 8–12 years ago in families with X-linked muscular dystrophy (Duchenne and Becker), haemophilia A and B, and chromosomal translocations. Special emphasis is given to investigating the impact of carrier testing in childhood. We have two aims in our study: (i) to discover whether the carrier testing of children has caused any psychosocial changes and (ii) to find out how the children and their families experienced the testing and what they think about it today.

5.1 Subjects

Our study population comprises three groups of children, who have undergone carrier testing in the 1980s in Finland, and their parents.

At the Department of Medical Genetics in the University of Helsinki, carrier testing was performed in 74 families with X-linked muscular dystrophy (Duchenne and Becker) between 1984 and 1988, (Lindlöf et al., 1986). The females tested for carriership were usually sisters or aunts of the affected males. The risk of being a carrier was evaluated in 187 females by direct mutation analyses or by linkage studies. As it was customary at the time to investigate carrier status in young children as well as older females, 38 of those tested were children.

During the period 1984–1988, 302 females in 101 families with haemophilia A or B were tested for carriership (Lehesjoki, 1991). In these families, 46 individuals were tested as children.

At the Department of Medical Genetics, Väestöliitto, genetic counselling and carrier testing have been carried out over the years in five extended families with various chromosomal translocations. The number of children tested in these families is 12. Several additional families tested at the Department of Obstetrics and Gynaecology, Prenatal Genetics, Helsinki University Central Hospital, can also be included in the study. Using this study population, we are able to study the consequences of carrier testing performed in childhood (age 1–17 years when tested, see *Table 1*), in:

- 39 children in 29 families with X-linked muscular dystrophy
- 46 children in 38 families with haemophilia A and B
- 12 children in 5 families affected with various chromosomal translocations.

These three groups of diseases constitute a representative spectrum of genetic conditions that create a burden of varied severity to the patients and their families. Children with unbalanced chromosomal translocations are, as a rule, mentally retarded and have several major and minor malformations. Many of the conditions are lethal early in life. The muscular dystrophies are progressive diseases that are usually considered to be extremely difficult and distressing within the family. The treatment of haemophilia, on the other hand, has improved considerably and in general, the life expectancy of the haemophiliac patients is very good.

5.2 Methods

The families were first contacted by mail to ask whether they wanted to take part in our study. Contact was through the doctor, who was previously responsible for the counselling of the family when the testing was performed. The tested children

Table 1. The number and the testing age of the children of the families with Duchenne or Becker muscular dystrophy, Hemophilia A or B, or chromosomal translocations

Age when tested (years)	Hemophilia A or B	Duchenne /Becker	Chromosomal translocation	Total
1–2	–	1	4	5
3–4	7	8	3	18
5–6	8	3	–	11
7–8	5	3	2	10
9–10	2	6	–	8
11–12	7	4	1	12
13–14	8	3	1	12
15–16	3	7	–	10
17	6	4	1	11
Total	46	39	12	97

(many of them now adults) are contacted only via their parents. Those parents who consent are sent a questionnaire. The tested children, who are at present at least 15 years old, receive their own questionnaire. If the child is younger than 15 years, only the parents fill in the questionnaire.

The questionnaire consisted of about 100 questions. The questionnaire for children differs slightly from the questionnaire of the parents. We used multiple choice questions, which included RAND 36-item Health Survey questions, items on sociodemographic background, life situation, social support and attitudes to illness and anxiety. The RAND 36-Item Health Survey 1,0 questions were used to quantify individuals' functioning and perceived well-being in physical, mental and social domains of life. Its multi-item scales address eight different health concepts: (1) physical functioning; (2) role limitations due to physical health problems; (3) role limitations due to emotional problems; (4) energy/fatigue; (5) emotional well-being; (6) social functioning; (7) bodily pain; (8) general health (Hays *et al.*, 1993).

In addition, we specifically investigated attitudes to genetic testing, including what the family members thought about the testing of children. We asked their opinion on the preferred or ideal age for carrier testing. We would like to know what kind of information or support would be needed before and after the test. We ask whether they have been satisfied with the test and the counselling. We would like to know if the parents have had any problems in telling the test results to their children, and the parents of those tested as children will be asked whether they have passed the information on to their children and whether they considered the situation to be very difficult. We ask whether the tested families still consider carrier testing to be justified and whether they feel that the testing has affected their family planning. The parents are asked whether they would make the same decision again in the same situation. Finally we try to find out whether the information about being or not being a carrier has been correctly understood by the child.

5.3 Preliminary results

We have started to analyse the answers given to the RAND 36-item questions by the families with Duchenne muscular dystrophy and those with haemophilia. According to the RAND 36-item health survey, we have found no differences in the physical, social or mental domains of life in our study population when compared to controls. Most of those, who reported that their test result had identified them as carriers or had given uncertain results, stated that they were worried by this test result. Practically all the tested daughters and parents reported that they were satisfied with the carrier test performed. There were several misconceptions concerning the test results even though each family, although not each individual, had received the results by letter in addition to being given the results in a counselling session or by telephone.

In summary, our preliminary test results indicate that no obvious psychological harm came to those tested in childhood. However, as many of them had not heard the results or had misunderstood them, the aim of the carrier testing had partly failed.

6. Conclusion

The consequences of carrier testing in childhood have never before been systematically studied in the long term. There is no information about how people react if they have grown up knowing of their carrier status from childhood. Through this study are obtaining information about what the effects of testing children could be. However, the reactions of those tested may be very different depending on the cultural background, health care system and the life situation of the family. Thus more research is urgently required in this area.

References

Clinical Genetics Society (1994) The genetic testing of children: Report of a Working Party of the Clinical Genetics Society. *J. Med. Genet.* **31**: 785–797.

Furu, T., Kääriäinen, H., Sankila, E-M. and Norio, R. (1993) Attitudes towards prenatal diagnosis and selective abortion among patients with retinitis pigmentosa or choroideremia as well as their relatives. *Clin. Genet.* **43**: 160–165.

Hays, R.D., Sherbourne, C.D. and Mazel, R.M. (1993) The RAND 36-Item HealthSurvey 1,0. *Health Economics* **2**: 217–227.

Lehesjoki, A-E., Sistonen, V., Rasi, V. and de la Chapelle, A. (1991) Hemophilia A: Genetic prediction and linkage studies in all available families in Finland. *Clin. Genet.* **39**:199–209.

Lindlöf, M., Kääriäinen H., Davies K.E. and de la Chapelle, A. (1986) Carrier detection and prenatal diagnosis in X-linked muscular dystrophy using restrictive fragment length polymorphism. *J. Med. Genet.* **23**: 560–572.

Ranta, S., Peippo, M. and Kääriäinen H. (1994) Haemophilia A; experiences and attitudes of mothers, sisters and daughters. *Ped. Hematol. Oncol.* **11**: 387–397.

Wertz, D.C., Fanos, J.H. and Reilly, R.P. (1994) Genetic testing for children and adolescents. *J. Am. Med. Assoc.* **272**: 875–881.

8

Childhood testing for carrier status: the perspective of the Genetic Interest Group

John Gillott

1. Introduction

The Genetic Interest Group (GIG) is the national alliance of charities, voluntary organizations, and support groups which serve those affected by specific genetic disorders and their families. GIG was officially launched in 1989 by a group of voluntary organizations concerned with genetic disorders who saw the need to coordinate action on the issues the groups have in common. Primarily, these are to improve support and services for people affected by genetic disorders and to advance the knowledge and understanding of human genetics throughout the population. GIG now has a membership of over 100 constituent groups ranging from large, well-established organizations to small support groups for very rare conditions.

GIG believes that it is both possible and necessary to draw up standard, basic, guidelines in relation to childhood genetic testing. GIG also hopes that a wealth of knowledge based on family experience can be passed on from those who have already met the dilemmas of genetic testing in childhood to those who are still awaiting the possibility.

After consultation with its member groups, GIG suggested guidelines it believes ought to frame practice in relation to genetic testing during childhood. These were published in the form of a response to a paper produced by a working party of the Clinical Genetics Society on the same subject (Clarke, 1995; Clinical Genetics Society, 1994; Dalby, 1995). There was much common ground between the two policy papers. Specifically, there was a consensus that testing during childhood was appropriate if it laid the basis for treatment or management of a medical condition; that testing presymptomatically for a childhood onset condition for which medical treatment was not available could also be appropriate as it

---

The Genetic Testing of Children, A.J. Clarke (ed.).

allows parents to plan ahead; and that in general testing during childhood for adult onset conditions was inappropriate. However, there was also disagreement. This chapter principally addresses the issue on which opinion was and remains divided – childhood testing for carrier status. In the next section, an extract from the GIG response to the CGS Working Party Report is reproduced.

2. Extract from the GIG response to the Clinical Genetics Society Report: 'The Genetic Testing of Children'

2.1 Testing for carrier status

Although there are differences to be considered, the arguments in favour of a right to test for childhood-onset conditions also hold good in the case of testing for carrier status. (We note in passing that presymptomatic testing may well reveal a child to be unaffected but a carrier.)

In many, perhaps most, cases, the issue of carrier status will be best dealt with at puberty or when the child becomes sexually active. The child, or young adult, as s/he will then be, would discuss the issues with parents and professionals, and make a decision based upon this. However, in other cases, early knowledge of carrier status could help a child adapt to the consequences of being a carrier over a period of time, rather than having the information presented at puberty, when s/he is going through a time of emotional adjustment and may not best handle the information.

There are additional reasons as to why it might be appropriate to test early for carrier status. Children are often astute, and may well enquire if there is something wrong with themselves or whether they could have an affected child because of the experience of the extended family. For example, a brother may have Duchenne muscular dystrophy, or an elder sister might have given birth to a baby boy with fragile X. Or two cousins may have cystic fibrosis. The CGS certainly advocates openness in answering questions, but seems to prefer answers with a 'worry about it later' slant. This is not unreasonable, and it is important that a child realizes that s/he is not ill and that the issue WILL be addressed later. However, a straight answer, provided that it is well informed, could well relieve pressure in some families. The argument in the [CGS] report, that a child may be treated differently if known to be of carrier status or given erroneous information is unsubstantiated; the vast majority of people are better able to understand the implications than they are often given credit for. Professionals do have a serious responsibility to make sure that the information is understood, and the voluntary organizations can play a large part here. Indeed, our experience is that clinicians often refer families to support groups as they are often best able to provide information in the most appropriate way.

We believe that the interests of the child must be put to the fore, and we appreciate the ethical point of not imposing information on a child, but we believe the seriousness of this information has been exaggerated because it is still relatively new. The child, as an adult, still has the option of whether they want to use it, or not, when they have children of their own. Problems of insurance, employment etc. are problems of public policy and need to be addressed as such.

Overall, we believe that there are distinct advantages for some families who wish to have their children tested for carrier status. Facts of life are best absorbed slowly and when the moment is right rather than during a crisis over a pregnancy. It is distinctly preferable for the child to be involved in the decision to have the test, which could mean delay.

PRINCIPLE: After suitable counselling, parents have the right to make an informed choice about whether or not to have their children tested for carrier status. Ideally, children should only be tested when of an age to be involved in the decision.

2.2 Adult-onset conditions

We believe that adult-onset conditions, for which there are no pre-symptomatic medical treatments, are totally different from childhood ones.

The argument that testing of the child takes away their right to make an informed decision as an adult overrides all other considerations. The low uptake in testing for Huntington's disease shows that many people would prefer not to know that they will be affected at some time in the future.

We also agree with the [CGS] report when it argues that genetic testing as a diagnostic tool for possible childhood onset of a condition which normally affects adults should not be allowed. A positive result does not necessarily confirm the diagnosis – there could be other problems, or the child may be reacting to stress in the family and mimicking symptoms. Time, anyhow, will clarify the situation. We appreciate that some parents are so anxious about their children that they put a great deal of pressure on doctors to have them tested. Parents need support, and again the voluntary organizations can be particularly helpful here, but the rights of the [possibly] affected individual to make an informed choice at a later date have to be held paramount.

PRINCIPLE: Children should not be tested for adult-onset conditions for which there are no pre-symptomatic medical treatments.

3. Clarification

The GIG stance on adult-onset conditions illustrates that GIG is not asking for parents to be allowed to test for anything at all times. Rather, GIG believes that an element of flexibility is needed on the specific issue of childhood testing for carrier status.

GIG shares the geneticists' view that testing for carrier status in childhood should as a normal practice include involving the child in the decision. GIG's experience is that this can meaningfully happen from quite an early age – say 8 years old. However, surveys of GIG member groups and of professional practice indicate that carrier testing before the age of five is not uncommon. GIG's view is that this can be appropriate and that evidence of harm is lacking. Accordingly, GIG would prefer geneticists to take a less proscriptive approach than the one recommended by the CGS report, and leave decisions to the individual family after a suitable exploration of the issues involved.

4. Why testing for carrier status is different from testing for adult-onset conditions

In their expansion on the CGS report, Clarke and Flinter (1996) discuss the issue of carrier testing by analogy with the issue of testing for adult-onset conditions:

'Concerns about testing during childhood arose in the very special case of Huntington's Disease (HD), a particularly devastating condition, but the same principles apply to predictive tests in childhood for other adult-onset conditions, and perhaps also to carrier status tests ... testing children for genetic carrier status flouts their future autonomy and privacy, and is unnecessary and potentially damaging, so that it should usually be deferred at least until adolescence except in very particular circumstances.'

Furthermore, Clarke and Flinter adduce empirical evidence to reinforce their point about the danger of flouting future autonomy and privacy: it is well known that uptake of HD testing is quite low; similarly, they point out, uptake by adults of carrier testing for CF was only about 20% when the test was offered to well-informed medical staff in a medical genetics department in London.

GIG does not believe that the analogy between carrier status testing and testing for adult onset diseases is valid. There is a profound difference between the knowledge that you are a carrier with implications for reproduction, and the knowledge that you are at some point in your adult life going to be personally affected by a genetic disorder which, in the case of HD and many other adult onset conditions, you will be unable to do anything about. The former knowledge is much less serious in character, and the possession of it still leaves open choices about whether to use the knowledge or not.

The statistics presented by Clarke and Flinter are of little use in discussing the issue as the people offered testing for HD come from families with a history of the condition whereas the people offered cystic fibrosis (CF) carrier testing do not. The low uptake of HD testing indicates that the majority of people who know they are at risk make a positive choice not to find out. This is a powerful argument against childhood testing for HD and other adult onset conditions. The valid comparison would be with the uptake of carrier testing in adults with siblings who have CF. Accurate statistics on this would be useful, but anecdotally the uptake of testing by such people is much higher than it is among the general public or the personnel of medical genetics departments. What is more, of those people with a family history of CF who do not get tested for CF carrier status, the reason cited in most cases is oversight or lack of interest rather than a positive decision not to be tested. In relation to the danger of negating adult autonomy, the group that should interest us is those with siblings who have CF who make a conscious decision not to be tested – and anecdotal evidence suggests that this group is very small. (This is the judgement of the Cystic Fibrosis Trust.)

5. Parental autonomy and the danger of harm

Ethicists with an interest in this area commonly emphasize that parents have responsibilities to children not rights over them. The autonomy of the child is emphasized, with the implication that professionals should seek to protect this autonomy where necessary in cases in which parents wish a child to be tested in order simply to satisfy their own concerns.

It is also commonly assumed, however, that parents should be taken to be the best judge of their child's interests and that, unless there is obvious harm, parental and family authority and autonomy should be respected. In their discussion, the American Society of Human Genetics posed the issues in this way:

'Presumption of parental authority is a fundamental principle for families and professionals who are discussing genetic testing for children ... The most compelling justification for parental authority focuses on the well-being of the child and acknowledges that parents are usually in the best position to make such a determination and have the greatest interest in making decisions to promote the well-being of the child. A second justification for parental authority rests on the interests of parents in their own self-determination, including the authority to make decisions on behalf of their children ...

In spite of the presumption of parental prerogative, parental authority can be limited if there are objective reasons to believe that a decision or action has significant potential for an adverse impact on the health or well-being of the child'. (American Society of Human Genetics/American College of Medical Genetics, 1995)

The emphasis on 'objective reasons to believe that a decision or action has significant potential for an adverse impact on the health or well-being of the child' takes us to the heart of the matter. It is the experience of GIG member groups, in which childhood carrier testing is an issue, that allowing parents to make an informed decision on carrier testing does not carry with it the danger of harm. They feel that too much is made of the issue by others, and that allowing parents to judge when it is best to inform their children of health matters that will be important to future reproductive decisions is the natural and sensible way to proceed (see Christine Lavery's piece in this collection). However, Clarke and Flinter do feel that serious harm could follow from allowing parents to exercise authority in this area. They point to social discrimination against carriers and the possibility of harmful family dynamics: 'Particular problems could arise in a family where one child is identified ... as being a carrier whose children may be at risk of a dreaded disease, and another child is found to be unaffected or not a carrier.'

At present, there have been very few studies of the effects of carrier testing during childhood. The assumption of possible harm is just that – an assumption. Some GIG member groups have carried out small-scale studies, including interviews with teenagers who were tested as children. These indicate that some who were tested would have preferred not to have been, while some who were not tested would have preferred to have been. Overall, it does not appear to be possible to apply a general rule.

Michie (1996) calls for more rigorous research. This is certainly a good idea, and GIG would be happy to help work out the best way of developing this work. But in the meantime, GIG sees little reason to adopt a blanket policy against childhood testing for carrier status. Rather, decisions should rest with the individual family.

References

American Society of Human Genetics/American College of Medical Genetics (1995) ASHG/ACMG Report. Points to Consider: ethical, Legal and Psychological Implications of Genetic Testing in Children and Adolescents. *Am. J. Hum. Genet.* **57**: 1233–1241.

Clarke, A. (1995) Reply to Genetic Interest Group response. *J. Med. Genet.* **32**: 492.

Clarke, A. and Flinter, F. (1996) The genetic testing of children: a clinical perspective. In: *The Troubled Helix: social and Psychological Implications of the New Human Genetics* (eds T. Marteau and M. Richards), Cambridge University Press, Cambridge, Chap. 7, pp. 164–176.

Clinical Genetics Society (1994) *The Genetic Testing of Children.* Report of a working party of the Clinical Genetics Society. *J. Med. Genet.* **31**: 785–797.

Dalby, S. (1995) Genetic Interest Groups response to the UK Clinical Genetics Society report 'The genetic testing of children'. *J. Med. Genet.* **32**: 490–491.

Michie, S. (1996) Predictive genetic testing in children: paternalism or empiricism? In: *The Troubled Helix:Social and Psychological Implications of the New Human Genetics* (eds T. Marteau and M. Richards). Cambridge University Press, Cambridge, Chap. 8, pp. 177–183.

9

Telling the children

Heather Skirton

1. Introduction

Birth is an event which marks the realisation of the parental dream of the 'future child'. The natural wish of parents is to ensure the safe and normal development of their offspring. However, in families at risk of genetic disease, the future of a child can be shadowed by the chance that life may be shortened or adversely affected by the condition. Families who seek genetic counselling frequently wish to discuss the issue of telling their child about the condition in the family, and informing the child that they are at personal risk. This issue arises whether or not testing is available. The decision to tell may not be clear-cut, as the desire to inform the individual may be juxtaposed with reluctance to cause anxiety in the child.

Advances in technology have enabled genetic testing to be offered to individuals who are at risk of developing or carrying a genetic condition. However, the possibility of testing has in some ways increased the parental burden, as parents may now also be concerned about whether it is in a child's best interests to be tested. Unless there are medical or management reasons for testing, informed consent is considered to be a necessary component of the process. As the notion of informed consent requires the child to be cognizant of the effects of the condition and the implications of a test result, the issue of telling the child of their risk is particularly relevant to the discussion.

2. Study

A study of the process of disclosure of risk information from parents to their children was undertaken to examine the difficulties surrounding this area. The results of this study will provide the basis of discussion in this chapter. It should be stressed that the term 'children' in the context of this chapter refers to offspring who are in the adolescent or adult age group and are therefore making their own life decisions or preparing to do so. A qualitative study method was chosen so that the reasons for behaviour and the emotional content of the situation could be examined (Pope and Mays, 1995). The data were collected by means of semi-structured interviews with each individual.

The Genetic Testing of Children, A.J. Clarke (ed.).
© 1998 BIOS Scientific Publishers Ltd, Oxford.

The subjects for the study were selected from families affected by Huntington's disease (HD), a dominantly inherited neurodegenerative condition which causes physical, mental, social and psychological impairment. The age of onset is usually around 40 years (Tibben *et al.*, 1993), so parents will have usually had their family before they become affected. The sample of 15 was chosen to include a range of subjects with respect to age, Huntington's status (at risk, affected or spouse of affected), gender and social class. Twelve subjects were parents who had adolescent or young adult children at risk of Huntington's disease. Three subjects were adults at risk of this condition who had a parent in the cohort. *Table 1* contains a brief outline of the subjects interviewed.

Each interview was taped with the subject's consent, and the typed transcripts were used to analyse the data by coding into categories (Miles and Huberman, 1994).

Table 1. Brief profile of subjects

---

(1) Kate is a 44-year-old woman, married with two adult sons. She was unaware of her risk of HD until her mid 30s, when her mother was diagnosed.

(2) Betty is a married woman aged 47 years. She has two adult sons. Betty knew of the presence of the condition in her family from her early adulthood, prior to her marriage. Her mother has been diagnosed with HD and Betty helps care for her.

(3) Joan, aged 50 years, is married to a man currently affected by HD. They have known of his risk for many years. Their three daughters are aged between 18 and 26 years.

(4) Jim's wife died from HD in her 50s. She was affected for many years prior to her death. Jim is aged 66 years and has three sons and a daughter at risk.

(5) Mark is 44 years of age and is married to Carol. He first knew of HD when his mother-in-law was diagnosed several years ago. His children are all in their teens.

(6) Phillip, who is 41 years old, has had HD for several years. He has adult children by a previous relationship. He and his wife Jan have three pre-adolescent daughters.

(7) Malcolm is a 42-year-old man, married to a woman with HD. He cares for her and their young daughter.

(8) Dianne, aged 32 years, has a mother affected with HD. She and her husband Mike have a son aged 8 years. Dianne cared for her mother at home until recently. She is at 50% risk of developing HD herself.

(9) Barbara is aged 46 years and has a father with HD. She and her husband have two young adult daughters. Barbara helps care for her father.

(10) Rose is married to a man at risk of HD, her three children are at 50% risk of the condition. Rose is now 41 years of age and her children are in their teens.

(11) Lyn is the 26-year-old daughter of Joan (Subject 3). Her father is affected, therefore she was born with a 50% of inheriting the gene for HD. She is married and has a baby son.

(12) Pam is affected with HD, and is aged 52 years. She was diagnosed only recently, and had no previous knowledge of the condition because she was fostered from 5 years of age. She has two daughters, one adult and one teenager.

(13) Neal has known of his wife Joy's risk of HD since their courtship. He is 42 years old, and they have three teenage children.

(14) Laura is the 30-year-old daughter of Jim (Subject 4). Her mother had HD throughout her childhood and died recently. Laura is at 50% risk of inheriting HD. She has a long-term partner but no children.

(15) Vincent is 22 years old, and is the son of Kate (Subject 1). He is at 25% risk of the condition. He has a steady partner.

---

3. Setting the scene

3.1 Loss

Before examining disclosure of genetic risk, it is helpful to set the discussion into context by considering the effects of the disease on the life of the family. Not surprisingly, the feelings of those with HD in the family were negative towards the condition, the only remotely positive comments being made by two affected people, who felt it was better than cancer because of the less immediate death. The majority studied felt that they could not envisage anything worse, and felt angry at the 'curse' which had been inflicted upon both themselves and their offspring.

With only one exception, the individuals interviewed had not heard of HD at all prior to a definite diagnosis being made in a close family member. This would appear to indicate that communication in general is poor between family members, as subsequent enquiries within the extended family frequently revealed other affected members. This lack of communication about genetic risk is confirmed by other studies (Ayme *et al.*, 1993). The lack of familiarity with the condition may have contributed partly to the shock reported by families when they heard the news, but whether the diagnosis in a family member was expected or not, it was followed by a period of mourning. The loss attributable to the disease was a theme repeated through each interview, including the loss of quality of life, of personal future and of peace of mind.

Whilst the loss affects every member of a family affected by a genetic condition, the children may suffer the loss of the ability to depend on a parent for warmth, protection and security amounting almost to the loss of childhood itself. A number of people in the study had experienced parental loss, either in real terms through institutionalization or death, or in psychological terms through the loss of skills in an affected parent. Whilst every effort may be made by the unaffected parent to maintain some stability, it seems that this may not always be sufficient. Laura related her feelings of rejection by her mother who appeared uncaring during her childhood:

'My own mother, it was the depression that really affected my life more than anything. She wouldn't even speak to me when I walked through the door when I'd been at school. All these things have a really bad effect I think on a young child. So I think misery has got to be there'.

Laura's obvious relief when she moved to her 'warm' and welcoming grandmother's home at the age of 12 years was palpable. Malcolm's description of his wife's relationship with their daughter seemed eerily similar, and Malcolm quoted his daughter as asking for a different Mummy. Whilst Lyn felt that the knowledge her father was ill may have helped her understand the family situation better, it may be that the needs of a pre-adolescent child for nurturing are so great, that the lack of parenting overwhelms all other considerations that the child would be cognitively capable of understanding.

There was evidence that children were distressed about the suffering they observed in their parents or grandparents in eight different families, and in the opinion of several subjects, this distress was exacerbated by the lack of explanation of the cause of the suffering.

3.2 Coping strategies

In nine families the signs and symptoms of the condition were initially attributed to other causes such as a nervous breakdown, a finding which is certainly consistent with other literature (Phipps and Lazzarini, 1987). Denial or non-acceptance is frequently the first reaction to loss, however, and it may be that the false attribution of symptoms allows the family time to adjust to the news (Kessler and Bloch, 1989). As is true for any loss, individual family members may make the adjustment at different rates, and some may never 'come to terms' with the situation. However, denial may be the only mechanism which allows certain individuals at risk to lead satisfactory lives. Another strategy used for dealing with the potential pain was that of hope; the belief that medical science would find a cure was sustaining at least eight of those interviewed.

> Rose: '... even though they are not at risk now, there is possibilities that by the time they are 40, there could either be a cure or some way of easing the situation'.

When considering the testing of children, it is important to consider whether the test result may interfere with these coping strategies, by removing hope of avoiding the disease or making denial more difficult. There was certainly evidence of intolerance by some family members for the coping styles of others, and a child may certainly experience a lack of support if they chose to act according to their own preference.

4. Issues of disclosure

4.1 Who should tell?

All subjects in the study unequivocally felt that parents should be responsible for discussing genetic risk with their children. However, it was acknowledged that in some cases this might not be feasible because of lack of skill in the parent, and that assistance in this task might be needed. It was stressed that the person who disclosed the news ought to be someone trusted by the young person, a view consistent with those of Jewett (1984) in relation to the breaking of bad news. In most cases the role of the professional is not to break the news, but possibly to act as a support to parents as they fulfil the role themselves. There was no suggestion that lay support groups had a role in this area, other than to provide literature supplying accurate information both to parents and their offspring. The emphasis placed on the support group as a source of companionship, however, is important; just as the burden was lightened in some families by the sharing of the knowledge within the family, the sense of shared difficulties and mutual concern between members of the support group appeared to be constructive in assisting some individuals to deal with the shadow of disease in their lives. This was especially true for two subjects who were actively caring for an affected family member. Throughout the study, family members emphasized that little is known about genetic conditions by the general public or by health professionals and, therefore, the lay support group may be the only forum in which families feel that their predicament is truly understood.

4.2 Intentions

Whether parents intend to tell their offspring about the risk of genetic disease is related not only to their attitude to the condition, but also to their belief in the inherent right of an individual to information which has important personal implications. Of the 12 parents interviewed, only one had formed an intention not to tell her children of their risk. Whilst the 11 others cited were uniformly appalled by the real and potential effects of the genetic disease on the lives of their children, the belief that children should know was obviously a stronger force than the fear of telling. A genuine parallel could be drawn between the parental dilemma and the ethical problems faced in health care, wherein the desire to do no harm is weighted against the pressure to act fairly towards the individual (McLean, 1996). The advantages of disclosure were clearly expressed as offering the opportunity for those at risk to plan and to make informed life decisions and choices. The desire to act honourably and respectfully towards their children appeared to influence some parents. The mother (Joan) who did not intend to tell her daughters of their risk explained that she was motivated by her belief that the news would spoil their happiness:

'Well I suppose basically it depends on your make up and what sort of person you are, but I, perhaps I'm over protective, and I feel that why give them something to worry about which might or might not affect them ? I had it in the back of my mind nearly all my married life and if it wasn't necessary, you know, as long as they needn't know about it, it was the amount of time that I could protect them if you like. Whether or not the ethics of them having the right to know, I didn't think about that . . . I would be hurting them by telling them . . .'.

All the daughters in this family were over 18 years of age, and this approach may reveal an undue feeling of responsibility in this mother for the well-being of her offspring. When interviewed, her daughter Lyn recounted her feelings of frustration concerning her mother's persistence in hiding information which Lyn felt should have been shared. Her anger was centred on Joan's insistence on retaining control over the lives of her children.

4.3 Behaviour

It is clear from the work of Ajzen and Fishbein (1980), however, that intention does not always equate with behaviour. This was demonstrated in the study by parents with an expressed intention to disclose the risk who found reasons for delaying:

'We thought they were too young really, that was our first reaction, keep it from the children . . . you didn't want to tell them nasty home truths . . . we felt that the due time wasn't here and you know, we would tell them eventually'.

The reason for delay quoted by eight parents was to protect the child from anxiety as long as possible. The work by van der Steenstraten (1994) on those at risk of HD, however, indicated that those who were told during adolescence were generally more able to deal with their risk status than subjects who were told in adulthood. There may therefore be a critical age when the disadvantage of delay may outweigh any benefit. It is possible that the delay provides a protective barrier, albeit temporary, for the parent as well as the child. The literature by Claflin and

Barbarin (1991) and Purssell (1994) on this subject also tends to dispute the protective aspects of non-disclosure, as these studies propose that children are aware of concerns, and anxiety is therefore not avoided by silence.

All three of the offspring studied stated categorically that adults ought to be given information which may affect their future, as did all the parents who were affected or at risk. Those who felt that information had been deliberately withheld from them were deeply resentful and expressed anger and feelings of disempowerment as a result. A comment was made in several families that discussion had not taken place because the children had not asked questions, the assumption being that the child did not ask and therefore did not want to know any more. This principle was clearly refuted by the work of Anderson and Clarke (1982) in their study of disabled adolescents, where it was shown that adolescents have concerns about their futures which they do not reveal to their parents. Whilst this might be explained by a natural reticence to discuss personal matters, it may also be a strategy adopted by the adolescent to protect the parent from having to face a painful situation.

Fear for the future was manifested in concern that their child might not find an accepting partner or might be prevented from having children, as noted by Kessler (1993) and this motivated some parents to delay telling, but resentment at not being told was only fuelled by this approach. Those at risk considered that future partners had every right to be aware of the genetic risks before making a long-term commitment.

4.4 Timing

There was no uniform agreement as to the optimal time to break the news, but whilst most parents felt that maturity and ability to comprehend the information was important, opinion as to the age when this occurred varied from 8 to 18 years. A number of subjects ventured the opinion that it is easier for a young person to deal with the news of risk than it would be at an older age. Whilst this was said to be due to the distance of time the young person has between their current age and the age at which they may become affected by the condition, it may also be due to an unconscious awareness of the heightened self-belief of the normal adolescent (Elkind, 1967; Elkind and Bowen, 1979), which acts as a psychological buffer to threatening news.

Further research is needed to clarify this point, as it might also be possible to disrupt to the delicate development of self-esteem by disclosing at the wrong time. In general, parents considered it necessary to inform the young person at risk prior to them having children of their own. To some parents this meant prior to the adolescent becoming sexually active and to others before they formed a stable relationship. The advantages of telling a child gradually as the opportunity presents itself or as questions arise was stressed by a number of respondents, who felt that there were benefits in being able to absorb the information slowly and so gradually adjust to the altered potential of life.

4.5 Benefits of disclosure

Six respondents felt the advantages to offspring of being told were the ability to make informed choices and plan their lives, taking the risk of the disease into account, and this included the possibility of choosing to be tested.

> Betty: 'They should be aware that the test is there and have the facility if they want to get tested'.
> Neal: 'You give those people the right to do something about it . . . that's their decision then'.

As has been shown by the testimony of subjects raised in a family with affected members, children are aware of the differences in both the affected member and the family as a whole, and are therefore aware of secrecy about the cause of the differences. The opinion of several subjects concurs with that in the literature, that children are distressed by the changes in the family (Tyler, 1991), and that explanation of the reasons for the difficulties can help the child or young person to make sense of the situation and begin to adjust. It may be of some reassurance to parents that none of those at risk felt there were any disadvantages to being told the truth.

5. The process of renegotiation

The change of life status which accompanied the realization of personal genetic risk was described by some subjects as an event of calamitous proportions. Life was altered from the moment they became aware of their risk, and the future had to be renegotiated on a different basis. This phenomenon could be likened to the renegotiation process which occurs in adolescence, when life can no longer be faced as a child. In adolescence, the change occurs gradually, and the success of the transition may depend on the support provided to the developing adult during the period of adjustment. Likewise, the adjustment of the individual to being at risk of genetic disease may take place most successfully if the person is given both time and support to achieve the renegotiation process at their own pace.

Andrew Collins (1990) emphasized the role of a loving family in the successful transition during adolescence, and the relevance of support is borne out by Laura and Vincent, who were both told of their risk status during adolescence by caring and supportive parents. Laura particularly commented on the benefit of having time to adjust to her status before adulthood, and compared this to the difficulties faced by a friend who had been given the news in adulthood. Adults who are faced with the news must also try to renegotiate their lives but, as was suggested in a study of subjects considering predictive testing (van der Steenstraten, 1994), the potential for successful renegotiation is possibly greater if a person is made aware during the critical formative stage of adolescence and is able to adapt life plans and expectations accordingly. The accounts of Kate, Betty, Phillip, Diane and Barbara show that a renegotiation of life is possible in the adult years but the cost may be high and the process very painful.

The parents in the study who shared the news immediately with their teenage offspring (Kate, Betty and Phillip) reported that the children appeared to cope well. It may be that in these families, the process of adjustment occurs simultaneously in both parents and children, and mutual support is available.

The findings of this study demonstrate that adjustment of the family is not a function of time, but rather of the ability and willingness of the family members to 'work through' the stages of grief related to the condition and to emerge with a personal model which allows them to invest in the future. Where this has not occurred, other coping mechanisms such as denial must be utilized to allow survival.

Families who considered themselves to be generally 'loving and sharing' felt they utilized these attributes to deal with the issues which arose after the diagnosis was made. Lyn, who felt her family was not 'sharing', seemed wistful at the opportunity which had been lost by the withholding of news. She felt frustrated that she had been denied the opportunity to make informed decisions and to support her parents through a difficult period.

> Lyn stated: 'Families are meant to be sharing . . . something like that when it
> affects everybody, you know, . . . how can people understand what
> the affected person is going through, and make allowances . . .
> when they don't know what it is, or what's wrong'.

The analogy between the change of risk status and adolescence may also be relevant in terms of the shift of power which occurs between parent and child. In adolescence, the child gradually develops the ability to make important decisions and the parental challenge lies in relinquishing control and allowing the young adult to make choices and possibly mistakes. When a parent has knowledge of genetic risk in the family, and is aware of the potential risk to a son or daughter, the parent has the power which accompanies that knowledge. Withholding the knowledge from the child also results in a withholding of the power to make informed choices.

Those parents who have been able to adequately resolve the loss of 'normal', non-risk status and to invest positively in life as an at-risk person will be better equipped to acknowledge their children's risk and support them during the time of adjustment. The means by which the news of genetic risk is relayed cannot be viewed as a discreet entity, but must be seen as part of the continuum of family life, and the passing of information is one component in the gradual devolution of power from one generation to the next. Support for the family as a whole has been identified in previous studies as a helpful way of empowering individuals within the family (Martindale and Bottomley, 1980). It may be that professional support by carers who form a trusting, long-term relationship with the family can help individuals to adapt to the circumstances and enhance their ability to share information necessary for making informed and autonomous decisions.

6. Conclusion

This chapter has attempted to discuss some of the issues raised by parents concerned with telling their children that they are at risk of an inherited condition. It may be that parents who request testing for their children wish either to avoid this task or to prepare their children for the news. Whether parents request a test or the young person does so on their own behalf, it is incumbent upon genetic services to ensure that each individual has the opportunity to adjust to their risk status at his/her own pace and with the support necessary to achieve this.

References

Ajzen, I. and Fishbein, M. (1980) *Understanding Attitudes and Predicting Social Behaviour.* Prentice-Hall, Englewood Cliffs, NJ.

Anderson, E.M. and Clarke, I. (in collaboration with B. Spain) (1982) *Disability in Adolescence.* Methuen and Co. Ltd., London.

Andrew Collins, W. (1990) Parent–child relationships in the transition to adolescence: Continuity and Change in Interaction, affect, and cognition. In: *From Childhood to Adolescence: A Transitional Period?* (eds R. Montemayor, G.R. Adams and T.P. Gullotta). Sage Publications Inc., CA, USA.

Ayme, S., Macquart-Moulin, G., Julian-Reynier, C., Chabal, F. and Giraud F. (1993) Diffusion of information about genetic risk within families. *Neuromuscular Disorders* 3(5–6): 571–574.

Claflin, C.J. and Barbarin, O.A. (1991) Does 'telling' less protect more ? Relationships among age, information disclosure and what children with cancer see and feel. *J. Pediatr. Psychol.* 16(2): 169–191.

Elkind, D. (1967) Egocentrism in adolescence. *Child Dev.* 38: 1025–1034.

Elkin, D. and Bowen, R. (1979) Imaginary audience behaviour in children and adolescents. *Dev. Psychol.* 15: 38–44.

Jewett, C. (1984) *Helping Children Cope with Separation and Loss.* B.T. Batsford Ltd., London.

Kessler, S. and Bloch, M. (1989) Social system responses to Huntington's disease. *Fam. Process.* 28: 59–68.

Kessler, S. (1993) Forgotten person in the Huntington's disease family. *Am. J. Med. Genet.* 48: 145–150.

Martindale, B. and Bottomley, V. (1980) The management of families with Huntington's Chorea; a case study to illustrate some recommendations. *J. Child Psychol. Psychiatry* 21(4): 343–351.

McLean, S.A.M. (1996) The new genetics: a challenge to clinical values. *Proc. R. Coll. Physicians Edinburgh* 26: 41–50.

Miles, M. and Huberman, A. (1994) Qualitative Data Analysis – An Expanded Sourcebook. 2nd Edn. Sage Publications Inc., CA, USA.

Phipps, E.J. and Lazzarini, A. (1987) Fighting dragons: the construction of explanatory systems in genetic disease. *Family Systems Medicine,* 5:304–312.

Pope, C. and Mays, N. (1995) Reaching the parts other methods cannot reach: an introduction to qualitative methods in health and health services research. *BMJ* 311: 42–45.

Purssell, E. (1994) Telling children about their impending death. *BJN* 3(3): 119–120.

van der Steenstraten, I.M., Tibben, A., Roos, R.A., van de Kamp, J.J. and Niemeijer, M.F. (1994) Predictive testing for Huntington's disease: non-participants compared with participants in the Dutch programme. *Am. J. Hum. Genet.* 55(4): 618–625.

Tibben, A., Frets, P.G. and van de Kamp, J.J. *et al.* (1993) Presymptomatic DNA-testing for Huntington's disease: pretest attitudes and expectations of applicants and their partners in the Dutch Programme. *Am. J. Med. Genet. (Neuropsychiatric Genetics)* 48: 10–16.

Tyler, A. (1991) Social and psychological aspects of Huntington's disease. In: *Huntington's Disease* (ed P.S. Harper). W.B. Saunders, London. Chap. 6, pp. 179–203.

10

Family processes in regard to genetic testing

Seymour Kessler

Abstract

Genetic disorders almost invariably have a psychosocial impact on families. For example, Huntington's disease (HD), with onset usually in adult life, has a major disorganizing effect which manifests itself in transgenerational psychiatric dysfunctions. Shame and guilt are promoted. The former is most clearly seen when the disorder is kept secret in the family. Guilt often emerges when individuals struggle to re-establish a sense of control and order in a situation where chance plays a major role.

Family systems dynamics in the context of genetic testing need to be better understood and appreciated. Homeostasis, the tendency for families to maintain or return to the status quo, and role differentiation are important processes to consider in families participating in genetic testing. In particular, the preselected patient phenomenon appears to be a common coping strategy used by family members to deal with adult-onset conditions. The consequences of disconfirming family expectations as a result of genetic testing need to be considered.

The role of the family as a protector of its members is an important consideration. The denial of symptoms in one member of the family by the entire unit may be an expression of the family's need to play a protective role. This role is also discussed in relation to the predictive genetic testing of minors. Questions are raised as to the merit of such testing for disorders for which no treatment or cure currently exists. In the face of the growing reality of genetic technology and the increasing pressures on geneticists to test minors, there is a strong need for professionals, parents, interest groups and support organizations to work together to ensure and promote the psychological interests of children.

1. Introduction

Psychologists attempt to understand human behaviour and focus their attention on what motivates people, the dynamics of their inner lives, their beliefs, values,

The Genetic Testing of Children, A.J. Clarke (ed.).
© 1998 BIOS Scientific Publishers Ltd, Oxford.

relationships with others, and so on. In contrast to ethicists and lawyers, who frame discussions in terms of rights and obligations, clinical psychologists try to integrate cognitive and emotional functioning with behaviour. This is not to say that psychologists are disinterested in ethical issues. On the contrary, their professional work is subject to and guided by ethical beliefs (regarding, *inter alia*, confidentiality, uses of professional power, the best interests of their clients, etc.) that are no less demanding, and in many ways are more rigorous, than those of the medical and other health-related professions.

As a psychologist I have been engaged at the University of California, San Francisco Medical Center (UCSF) in the genetic testing for Huntington's disease (HD). This testing, which began in the late 1980s, is conducted in collaboration with Dr. Charles Epstein, a medical geneticist, Dr. Frank Longo, a neurologist, and Ms. Andrea Zanko, a genetic counsellor.

HD, once called Huntington's chorea, is a progressively neurodegenerative disease marked by cognitive changes and choreic movements (Folstein, 1989; Harper, 1996; Hayden, 1981). The disorder is inherited as an autosomal dominant trait and children born to an affected parent have a 50% risk of having inherited the gene. The usual age of onset is between 35 and 45 years and death often occurs some 15 years after the age of onset. No cure or treatment for the disorder has as yet been found and presently there is no way to slow down the progression of symptoms.

HD is but one of many genetic disorders, almost all of which exert a profound psychosocial impact on individuals and families. When children are born with a genetic disorder, the family unit is frequently subject to enormous stress, sometimes overwhelming the ability of parents to cope effectively with the medical, psychological and interpersonal problems the disorder promotes.

Some disorders, such as HD, appear to have a major disorganizing impact on families (Wexler, 1979). In the HD testing program at UCSF, we found that nearly half of the asymptomatic persons who were tested came from homes in which considerable disruption and disturbances of family functioning were present. The problems were generally multifaceted and included family breakup due to divorce or abandonment by one of the parents. The placement of children into foster care commonly occurred. Whether or not the family remained intact, test participants reported instances of parental drug and/or alcohol abuse, physical, emotional and sexual abuse, major neglect and premature parentification, in which young children had to take over major parental functions such as cooking, care of younger siblings, and other caregiving functions.

2. Shame and guilt

Psychologically, genetic diseases are sources of enormous personal and family shame and guilt (Kessler *et al.*, 1984). Shame, in all its various forms including embarrassment and stigma, generally arises from the narcissistic injury, diminished self-esteem, feelings of deficiency and loss of control following the birth of a child with a genetic disorder. It is a failure to live up to one's inner ideal or image of one's self that comes from producing a perceived less-than-perfect child.

Perhaps shame is seen most clearly among parents who hide their children's disfigurements in public or do not even take their children out of the home for fear of embarrassment from the stares and derisive comments made by others. Most embarrassing and painful are expressions of sympathy on the part of others. Some individuals interpret this as pity, implying that they are pitiable or deviant in the other's eyes. Hiding one's shame to avoid such painful experiences may seem to offer a solution, but this very behaviour reinforces the person's sense of stigma and deviance.

Shame is also seen as an oppositional response in instances where exhibitionistic behaviour occurs as, for example, in the spotlighting – the deliberate calling of excessive attention to a child's problems – in public situations.

The tendency to keep the presence of genetic disease secret from spouses or potential spouses is another example of how genetic problems tend to be laden with an additional burden of shame (Kessler, 1993). Generally, the keeping of such secrets has a ripple effect on a family. I worked with a highly successful professional woman whose husband kept the presence of HD a secret from his wife. They had one child, a boy, who showed multiple behavioural problems from early on, quite likely promoted by the tensions between the parents arising as a result of the husband's secret agenda. By the time the boy was a teenager he had been arrested for minor crimes, drug use and drinking. Meanwhile, the husband began to become symptomatic for HD and had difficulty keeping a job. He ascribed his movement difficulties as being due to job-related anxieties. His wife began to suspect something more was involved and it was only when she confronted her husband's relatives that the truth about HD began to emerge. She was enraged and eventually walked out on the family, largely free of guilt, feeling that her action was justified because she was the one who had been deceived. The son, incidentally, participated in predictive testing and was found not to carry the HD gene.

The ubiquitous experience of guilt associated with the occurrence of genetic disease is also well known. Guilt generally arises when the person believes he or she has done something or failed to do something which might have averted the occurrence of the genetic problem. It emerges invariably as the person struggles with the need to make the disorder conform to the laws of causality, in other words, to find a reason for the occurrence of the disorder and to make sense out of an event which seemingly has no purpose or sense.

The person frequently phrases his/her guilt in terms of questions such as, 'What did I do wrong to cause this to happen?' Obviously, in some instances the person's guilt does have a rational basis; they have contributed to their child's problems as, for example, in fetal alcohol syndrome. But, in many cases, the guilt has no rational basis and professional efforts to help the person feel less guilty are appropriate.

In situations where guilt plays a major role in family dynamics, family members sometimes blame one another or focus their blaming on one member of the family. In a recent study, Chapple et al. (1995) described a family in which two sisters both had sons suspected to have the fragile-X disorder. The family of origin consisted of four highly competitive sisters and a brother. The father and all the sisters were allied in holding the mother responsible for the problems in the family, presumably a pattern that was already in place before the genetic problem was identified. The brother, who was less involved in the intense sibling rivalries, appeared to be more in sympathy with the mother. However, his influence was evidently swamped by the sheer force of numbers, five against two. The family came for genetic counselling, but genetic education did little, if anything, to reduce the mother's guilt. On follow-up, it was clear that she continued to feel stigmatized by the family. It is easy to see that the other family members felt absolved of any responsibility, especially for the genetic condition, so long as she carried the guilt for problems in the family.

3. Family dynamics

Cases like this serve to emphasize the complexity of family systems. Effective counselling or therapy needs to recognize that families are social and psychological entities made up of components whose interactions are governed by implicit and explicit rules. These rules function to maintain the equilibrium and homeostasis of the family unit so that changes in one component or part of the system invariably affect the whole. The tendency in such systems, whenever a change or pertubation occurs, is to attempt to re-establish the previous point of equilibrium.

This simple yet profound finding was discovered in the mid-1950s; in the course of family therapy for schizophrenia, the improvement of the affected individual was often found to lead to the decompensation of another family member (Jackson, 1968).

In regard to genetic testing, the Vancouver, BC group reported on a similar situation where the husband of a woman who tested negatively for the HD gene in presymptomatic testing became clinically depressed after the test results were made known (Huggins et al., 1992). He had recently retired and was planning to devote his retirement years to the care of his wife, who he incorrectly assumed would eventually show symptoms of HD and become an invalid.

Families might also be thought of as ecological systems in which each member strives to find a viable ecological niche or role to play within the greater family. It is rare for two children, even two identical twins, to occupy the same niche within the family.

Distinct niches require distinct roles. Take Elizabth in Jane Austin's *Pride and Prejudice*. She strongly identified with her father rather than her mother and occupied the sensible, rational role among the siblings leaving little if any room for the other siblings to play that specific role. One wonders how the family dynamics would change once the two older sisters were married and were no longer part of the permanent dynamics of the nuclear family. In real life, we frequently see how rapidly ingrained patterns of relating become re-established when family members come together for special occasions, holidays, etc.

In working with families around adult-onset genetic disorders we frequently see how one member of the family takes the lead in obtaining information about the disorder and exploring the possibility of predictive testing. This family member may be the conduit by which the rest of the family stays informed about the disorder; they may be the one who arranges medical appointments, the testing of other family members, and so on. If the leader is an at-risk individual he or she is generally the first to go through a testing protocol.

Information the leader brings to the family is not always welcomed. In fact, many such individuals report that other family members are blasé about things the 'leader' finds exciting in testing or research, and which the others will sometimes ignore, resent or ridicule. Literature given out by the leader will often go unread and will be conveniently lost or misplaced.

What is striking about family systems is the degree to which individual members accept the implicit understandings about one another's roles. The clearest example of this may be the preselected patient phenomenon in which an asymptomatic individual is designated well in advance as the one destined to manifest the hereditary disease segregating in the family (Kessler, 1988). Such designations are often justified on the basis of the flimsiest of evidence – the person's name, gender, physical appearance, etc.

The preselected patient accepts the role and behaves in ways that confirm the designation both to him or herself and to the wider family despite the fact that fulfilling this role may (and generally does) have a severe impact on the person's self-image, self-esteem and overall functioning.

When genetic testing is carried out and the preselected patient is found not to carry the gene in question, it seems difficult for either the individual or the family to believe the results. One such person with whom I worked for several years, after an initial period of relief and elation following presymptomatic testing for HD, became clinically depressed and had to be hospitalized. He has two siblings, neither of whom have been tested. When he told them of his good fortune, one laughed derisively and the other responded as if he had just said he recovered from a cold. In discussions I had with the the siblings a year or so after my client was tested, it was clear that they were surprised that their brother did not have the HD gene and had difficulties accounting for his behaviour.

The tendency of family members to occupy distinctive ecological niches frequently leads to jealousy, resentment, enmity and competition among siblings and parents. These are often mixed with concern and love for one another, but at times antagonism, rather than caring, may dominate the family environment. This is particularly true when the family system is under unusual stress, as for example, when one or more family members come for genetic testing.

For some families, genetic testing is an extremely stressful event, but one in which family members have an opportunity to give each other support and to demonstrate mutual understanding. Testing, however, does not always bring out the best in families. In several instances, the marriages of persons who participated in genetic testing did not survive (Codori and Brandt, 1994).

Before direct testing for the HD mutation became available, testing was possible by linkage analysis. This, however, required the test participant to obtain the cooperation of affected and unaffected relatives in order to obtain tissue samples. For a considerable number of persons the need to negotiate or deal with relatives within an extended family was enough to discourage them from proceeding with testing. Thus, it was not suprising how few at-risk persons availed themselves of linkage testing opportunities (Quaid and Morris, 1993).

Once the direct test for HD was developed, the entry of many at-risk individuals into a testing protocol was facilitated (Huntington's Disease Collaborative

Research Group, 1993). Nonetheless, several at-risk persons told us that they had no intention of letting their family members know that they were being tested. Some did so for reasons of self-protection, whereas others wished to protect their families from finding out that they were HD gene carriers, something they frequently assumed in advance. I have heard of instances where at-risk individuals have made requests to be tested anonymously or have simply lied afterwards to other family members about the outcome of their test.

After a test is carried out, the perceived risk status of untested individuals almost invariably changes. If the test participant is found not to be a gene carrier, untested at-risk persons within the family perceive that their personal risk increases; they feel that, 'If it's not him then it must be me'. Some individuals who turn out to be negative for the HD gene express dismay over the possibility that some other sibling will now be at greater risk for the disorder.

Professionals frequently attempt to correct these misperceptions by emphasizing statistical principles in their contacts with family members, a strategy that sometimes leave the latter feeling that their fears and anxieties have not been understood. This is a common problem in the field of genetic counselling and does little to engender trust in professional opinion on the part of clients.

4. The protective function of families

One of the major expectations of families is that they will protect their members, especially children, the aged and the infirm. This protective function is considered a social norm in human cultures and deviations are subject to legal, social and other sanctions. We hold our children's hands when we cross a busy urban street.

Protection of family members is often seen when genetic disorders such as HD are involved. For example, family members will collude with one another to deny symptoms in one of their members. This serves an adaptive purpose in allowing the family time to adapt to the change in the health status of the affected person. It also gives the affected person time in which to come to accept the changes in their own health status and to develop a safety net of support. Such denial is an expression of a general wish to protect the affected individual from real and imagined hurts and embarrassments. Only when the symptoms become destructive to the affected individual do other family members generally change their defensive denial to a more active form of protection.

The issue of protection comes into play when discussions of the genetic testing of minors arises, especially for disorders such as HD in which no intervention as yet exists to forestall the onset of the disorder or to change its grim prognosis.

We need to keep in mind that children are particularly vulnerable to influences from perceived powerful others; they will tend to acquiesce in, and to endorse, parental desires and wishes even if their own physical and mental well-being is at stake. The recent tragic incident involving Jessica Dubroff, a 7-year-old child, is a case in point. Jessica, under the encouragement and guidance of her parents, attempted to set a record in the Guinness Book of Records for being the youngest person to pilot an airplane across the continental US. The publicity accompanying this feat was intense and each step was carefully stage-managed by her father. Television interviews of the child before the flight suggested that she had been punctiliously prepared to answer the reporters' questions. But, when it came to providing a personal rationale of why she was undertaking the flight she tended to fumble. The flight ended disastrously in a crash in which she died as well as her father and a flight instructor (who it seems was actually piloting the plane).

In retrospect it appears that the narcissistic needs of Jessica's parents, their dreams of fame and possible economic gain that might have resulted from the child's achievments, may have clouded their judgment about Jessica's physical and psychological capacities. This episode illustrates how easy it is to convince children to participate in procedures which, in some cases, may have no short-term and doubtful long-term benefits for them.

Their need for parental approval and affection, combined with their relatively limited life experiences, make children particularly dependent on cues and guidance from adults. This tends to seriously limit their ability to give informed consent for such procedures as genetic testing. Although recent studies (see Michie, 1996 for a review) have not clearly delineated the differences between the cognitive processes of adults and children in their dealings with 'real-world' tasks, this may say more about the research instruments and methodology than about the decision making capabilities of children. The fact remains that the tasks of adulthood and childhood are different and that children are vulnerable to the attitudes, needs, wishes and decisions of adults. Few, if any children make important life decisions independently of, or in isolation from, the influences of significant others.

I believe that none of us would object to genetic testing of minors when the problem in question can be rectified, treated or prevented. When it comes to problems where no cure or treatment is yet possible, however, questions can legitimately be raised as to what purpose the testing of children might serve (Ball et al., 1994). In addition, asking children to make a decision about testing for an adult-onset disorder, which poses no immediate medical or psychosocial threat to the child (and may not even have any concrete relevance for him/her), perhaps is expecting a great deal from them.

One might argue that genetic testing would provide parents and siblings with a sense of reality so that the preselected patient phenomonon might be obviated. It might give parents a chance to plan the child's and the family's economic future more rationally if they knew well in advance what lay in store for a given child. Testing might also reduce uncertainties in the family and may positively improve the family's self-image and relationships (American Society of Human Genetics and American College of Medical Genetics, 1995). There are also precedents one might cite. For many medical conditions the poor prognosis for a child is seldom if ever withheld from parents. Thus, one might ask why shouldn't children be tested and the results given to parents?

Conversely, several studies (Codori and Brandt, 1994; Tibben et al., 1993) show that the major reason people give for participating in genetic testing of adult-onset diseases is the reduction of uncertainty, which is another way of saying the reduction of personal anxiety. However, if parents have anxiety about their child or children, is this a sufficient justification to have the latter tested for an adult-onset disease? Perhaps such parents need to be channelled towards personal counselling to learn more effective ways to contain their anxieties rather than to subject their children to genetic testing. I believe that the position adopted by the Working Party of the Clinical Genetics Society (UK) is an eminently sensible one (Clinical Genetics Society, 1994):

> '[T]esting carried out to satisfy the curiosity of a medical practitioner (except in the context of certain types of research . . .), or to relieve parental anxiety, would not seem to us in general to be appropriate. There may be circumstances where such testing is justified, but we expect these circumstances to be unusual . . . [T]he possible emotional harm to the child, the abrogation of their autonomy and the breaching of their confidentiality, will generally outweigh the possible benefits . . .'.

The experience of genetic testing for HD among adults suggests that a self-selecting process occurs. When pretest counselling is provided, persons tend to delay or not participate in the testing programme when they do not feel ready to hear bad news or when their self-assessment suggests they may have later adjustment difficulties (Codori *et al.*, 1994; Craufurd *et al.*, 1989; Jacopini *et al.*, 1992; Simpson *et al.*, 1992; Tibben *et al.*, 1993). This self-selection seems to promote a relatively low rate of later psychiatric casualties (Kessler, 1994). Clearly, there would be enormous value in providing counselling to parents requesting testing of their children, so that they have a clearer picture and deeper understanding of the implications of their request both for themselves and for the child.

Perhaps in some individual cases the testing of minors may have some value (American Society of Human Genetics and American College of Medical Genetics, 1995). Some children have more mature cognitive, emotional and decision-making skills than others. Also, arguments have been made that some children between 12 and 14 years of age have the ability to evaluate the risks and benefits of certain treatments (Wadlington, 1983; Weithorn, 1983). But, who is to decide which case has merit and which does not? Certainly, geneticists are not in a position to make such determinations, at least not without the input of ethicists, psychologists, laypersons and others. Whoever does make such decisions will need to face the question: who will protect the vulnerable when their customary protectors are not doing so?

5. The psychological needs of children

All knowledge is not necessarily 'good'. In fact, evolution has developed our psychological defences and designed them to limit our access to information and knowledge that threatens our physical and psychological integrity. There are times when denial serves adaptive purposes and contributes to a person's well-being.

Every parent knows that we cannot fully protect our children from harm and the hard knocks of life. Nor should we do so, since that would leave our children ill-prepared to deal with life's unpleasant realities. It is the progressive exposure of children to disappointments, frustrations and setbacks that helps them master their anxieties and develop coping skills. Overprotected children tend to become highly anxious adults because they do not have the inner security that comes from the successful personal mastery of frustrations and defeats earlier in life.

Good psychological defences develop as the child acquires self-knowledge – that is, knowledge of one's capacities, strengths and limitations – as a consequence of one's own active efforts. Knowledge acquired as a result of someone else's agenda does not generally help children since they were not the active agents in acquiring the knowledge. When knowledge overwhelms the child's coping capacities, it erodes their sense of competence and often leaves them less able to deal with life problems later on.

There is a need for professionals, parents, and groups representing parental interests to work together to educate parents and the broader society about the psychological needs of children and the potential for harm which might result from genetic testing. Parents sensitive to such needs are likely to be in a better position to make decisions regarding the genetic testing of their children than those who lack such sensitivity, and, if they choose to pursue testing, may be better equipped to help their children deal with the implications of the test.

References

American Society of Human Genetics/American College of Medical Genetics Report (1995) Points to consider: ethical, legal and psychosocial implications of genetic testing in children and adolescents. *Am. J. Hum. Genet.* 57: 1233–1241.

Ball, D., Tyler, A. and Harper, P. (1994). Predictive testing of adults and children. In: *Genetic Counselling: Practice and Principles* (ed A. Clarke). Routledge, London, pp. 63–94.

Chapple, A., May, C. and Campion, P. (1995) Lay understanding of genetic disease: A British study of families attending a genetic counseling service. *J. Gen. Counsel.* 4: 281–300.

Codori, A.M. and Brandt, J. (1994) Psychological costs and benefits of predictive testing for Huntington's disease. *Am. J. Med. Genet.* 54: 174–184.

Codori, A.M., Hanson, R. and Brandt, J. (1994) Self-selection in predictive testing for Huntington's disease. *Am. J. Med. Genet.* 54: 167–173.

Craufurd, D., Dodge, A., Kerzin-Storrar, L. and Harris, R. (1989) Uptake of presymptomatic predictive testing for Huntington's disease. *Lancet* 2: 603–605.

Folstein, S.E. (1989) *Huntington's Disease.* The Johns Hopkins University Press, Baltimore, MD.

Harper, P.S. (1996) *Huntington's Disease,* 2nd Edn. W.B. Saunders, London.

Hayden, M.R. (1981) *Huntington's Chorea.* Springer-Verlag, Berlin.

Huggins, M., Bloch, M., Wiggins, S., et al. (1992) Predictive testing for Huntington disease in Canada: Adverse effects and unexpected results in those receiving a decreased risk. *Am. J. Med. Genet.* 42: 508–515.

Huntington's Disease Collaborative Research Group (1993) A novel gene containing a trinucleotide repeat that is expanded and unstable on Huntington's disease chromosome. *Cell* 72: 971–983.

Jackson, D. (1968) *Communications, Family and Marriage,* Vol. 1. Science and Behavior Books, Palo Alto, CA.

Jacopini, A.G., D'Amico, R., Frontali, M. and Vivoan, G. (1992) Attitudes of persons at risk and their partners toward genetic testing. *Birth Defects* 28: 113–117.

Kessler, S. (1988) Psychological aspects of genetic counseling. V. Preselection: A family coping strategy in Huntington disease. *Am. J. Med. Genet.* 31: 617–621.

Kessler, S. (1993) The spouse in the Huntington Disease family. *Fam. Sys. Med.* 11: 191–199.

Kessler, S. (1994) Predictive testing for Huntington disease: A psychologist's view. *Am. J. Med. Genet.* 54: 161–166.

Kessler, S., Kessler, H. and Ward, P. (1984) Psychological aspects of genetic counseling. III. Management of guilt and shame. *Am. J. Med. Genet.* 17: 673–697.

Michie, S. (1996) Predictive genetic testing in children: Paternalism or empiricism? In: *The Toubled Helix: Social and Psychological Implications of the New Genetics.* (eds. T. Marteau and M. Richards). Cambridge University Press, Cambridge.

Quaid, K.A. and Morris, M. (1993) Reluctance to undergo predictive testing: The case of Huntington disease. *Am. J. Med. Genet.* **45**: 41–45.

Simpson, S.A., Besson, J., Alexander, D., *et al.* (1992) One hundred requests for predictive testing for Huntington's disease. *Clin. Genet.* **41**: 326–330.

Tibben, A., Frets, P.G., Kamp, J.J.P. *et al.* (1993) Presymptomatic DNA-testing for Huntington disease: Pretest attitudes and expectations of applicants and their partners in the Dutch program. *Am. J. Med. Genet.* **48**: 10–16.

Wadlington, W.J. (1983). Consent to medical care for minors: The legal framework. In: *Children's Competence to Consent* (eds G.B. Melton, G.P. Koocher and M.J. Saks). Plenum Press, New York, NY, pp. 57–74.

Weithorn, L.A. (1983). Involving children in decisions affecting their own welfare: Guidelines for professionals. In: *Children's Competence to Consent* (eds G.B. Melton, G.P. Koocher and M.J. Saks). Plenum Press, New York, NY, pp. 235–260.

Wexler, N.S. (1979) Genetic 'Russian roulette': The experience of being 'at risk' for Huntington's disease. In: S. Kessler (ed.), *Genetic Counseling: Psychological Dimensions* (ed S. Kessler). Academic Press, New York, NY, pp. 199–220.

Working Party of the Clinical Genetics Society (UK) (1994) The genetic testing of children. *J. Med. Genet.* **31**: 785–797.

11

Adolescent requests for predictive genetic testing

Julia Binedell

1. Introduction

The debate concerning genetic testing of minors for adult-onset, untreatable diseases has focussed largely on the testing of young children at third party or parental request. Adolescents who actively request predictive testing of their own accord are seldom considered separately. Yet self-referral by adolescents does occur, albeit infrequently. The precedent set by the Huntington's disease (HD) predictive testing guidelines is that testing should be available only to individuals aged 18 years and over. While provision is made for testing minors who show sufficient maturity to give fully-informed consent, current practice among clinical geneticists reflects adherence to an age-based exclusion criterion.

This chapter presents a case for the assessment of adolescent requests for genetic testing. Where decisions about competence do not depend arbitrarily on age, evaluation of competence raises a number of questions: How should decision-making competence be defined? If competent adults are used as a reference point, how do adults make decisions in the context of genetic risk? How competent are adolescent decision-makers? How should competence to consent be assessed and by whom? These questions will be considered in arriving at a framework for responding to adolescent requests.

2. Current guidelines

The consensus reached by international groups of scientists, ethicists, clinicians and members of lay organisations in drawing up HD predictive testing guidelines is that testing should be available only to individuals who have reached the age of majority (Went, 1990; IHA/WFN, 1994). This exclusion criterion is rationalized on the grounds of (i) the lack of clear benefits and speculated harmful effects of

---

The Genetic Testing of Children, A.J. Clarke (ed.).
© 1998 BIOS Scientific Publishers Ltd, Oxford.

testing children for an adult-onset, untreatable disorder (Harper and Clarke, 1990) and (ii) the loss of autonomy and confidentiality when third parties influence or decide on behalf of children who are unable to give voluntary and informed consent (Bloch and Hayden, 1990).

3. The costs and benefits of childhood genetic testing

The general opinion among professionals is that testing children for adult-onset, untreatable disorders holds more potential for harm than for benefit (Bloch and Hayden, 1990; Clarke and Flinter, 1996). In the absence of evidence or experience of the long-term effects of such testing, policy is guided by the ethical principle of 'first do no harm' (Clinical Genetics Society, 1994). By excluding minors from HD predictive testing on the grounds of the lack of medical treatment, policy makers adopt an essentially medical view of 'benefit'. When possible psychological and social advantages are considered, the balance of costs and benefits is less clear.

While there is no good information concerning the most appropriate age to offer HD predictive testing (Tyler and Morris, 1990), research in the related areas of risk-factor testing and coping with serious illness in childhood allows us to anticipate the likely impact of genetic testing at a young age.

In their review of the psychological impact of childhood genetic testing, Michie and Marteau (1996) observe that adverse consequences are associated primarily with population-based screening, rather than with testing of high-risk groups, and that problems are minimized by the provision of good counselling. Outside of the field of genetic screening and testing, studies of children's adjustment to chronic illness support the benefits of early knowledge (Slavin et al., 1982). Wertz et al. (1994) report that 'those who find out only in adulthood that a serious disorder such as HD exists in the family fare more poorly than those who know earlier.'

In the adult population, there is evidence to suggest that HD predictive testing can be a positive psychological intervention. Follow-up assessment of test applicants in the Vancouver programme indicated that those who received a test result, irrespective of whether it raised or lowered their risk for HD, showed a better psychological outcome than those for whom results were inconclusive (Wiggins et al., 1992). In Belgium, psychometric scores of tested individuals at 1-year follow-up showed that non-carriers were significantly less anxious and depressed than at pre-test assessment, while receiving 'bad news' had no significant effect on anxiety, depression and the personality of carriers (Decruyenaere et al., 1996).

Compared with speculation about the likely impact of giving genetic information, less attention has been paid to the potential effects of withholding testing (Sharpe, 1993). In this regard, Bloch et al. (1992) comment that 'the decision to postpone testing can never be taken lightly. The stress of undergoing testing, receiving a result, and adjusting to the new risk status must be weighed against the stress and uncertainty of living at risk for HD, the blow to the candidate's self-respect by being denied testing and the possible sense of humiliation and helplessness by having one's autonomy undermined.' There may also be regret at time wasted or opportunities missed when testing is postponed until a later age (Clarke and Flinter, 1996).

The need for research to clarify the relative costs and benefits of giving genetic information at a young age versus continued uncertainty has been noted (Marteau, 1994; Michie, 1996). Parents' and children's perspectives on these issues need to be sought, especially given evidence that professional and lay views differ on questions of who should determine the scope of genetic testing (Michie *et al.*, 1995) and whether minors should be tested presymptomatically for HD (Bloch *et al.*, 1989; Tibben, 1993).

4. Self-referral by adolescents

Apart from the concern to protect children from possible harmful effects of genetic testing, another justification for not testing children for adult-onset, untreatable problems is to uphold the child's future right to make an autonomous decision, with confidentiality preserved (Clarke and Flinter, 1996). Some providers make an exception to the exclusion of minors where testing is 'actively sought by a subject who is sufficiently mature to make such a profound decision and is able to freely provide fully informed consent' (Ball and Harper, 1992) and believe that such a request should take precedence over parents' wishes to with-hold genetic information (Wilfond *et al.*, 1995).

It is clear from data from a European Community Collaborative HD Study (1993) that self-referral by adolescents does occur, although rarely (*Table 1*).

In this study, it is likely that the low figures for centres other than Cardiff are as a result of referral of minors being screened out at an early stage. Of the 29 referrals in Cardiff, 12 involved teenagers with a mean age of 15.5 years (Audrey Tyler, personal communication).

5. The legal precedent for assessment of competence

There has been a trend in Britain to award children a growing set of rights, including the right, if competent, to make major personal decisions. Increasingly, decisions about minors' competence to give consent in matters of their health and welfare are being made on the basis of assessment rather than age. 'Sufficient understanding to make an informed decision' (Children Act, 1989) has become the central focus, although no criteria have been established for judging degree of understanding. It has been suggested that legal capacity

Table 1. Referral of minors

Referral	Cardiff	Leiden	Leuven	Rome	Total
Parents	16	0	4	3	23
Professionals[a]	10	0	1	0	11
Self	3	2	0	0	5
Total	29	2	5	3	39

[a] Social worker, 7; psychiatrist/paediatrician, 4.

should be determined in accordance with the nature of the decision (Wertz *et al.*, 1994), but it is still unclear how much an individual must understand to consent to genetic testing (Montgomery, 1994).

The removal of an age-based criterion for determining competence to consent requires the development of strategies for individual assessment.

6. How should decision-making competence be defined?

Apart from the diverse ethical and legal definitions, there is a broad spectrum of psychosocial approaches to defining decision-making competence. These range from a focus on cognitive ability, to social and psychological factors which affect decision making, to a view of competence as a 'way of relating...within a network of relationships and cultural influences' (Alderson, 1992).

Early developmental theorists, such as Piaget, view competence in the context of age-based stages of cognitive development (comparable to stages of physical development) that are seen as basically biological, although facilitated by social and environmental factors (Smith and Cowie, 1995). A distinction is drawn between concrete versus abstract thinking, the latter being a prerequisite for hypothetical-deductive reasoning, taking the future into account, and the ability to understand an issue from another person's point of view (Gordon, 1996). This theory of cognitive development has been challenged by the observation that, with revised research methods, children have been shown to be capable of abstract tasks at a much younger age than was previously recognised (Alderson, 1993). Furthermore, there are wide variations in cognitive ability associated with social, cultural and educational background, and competence may be situation specific (i.e. competence in abstract problem-solving tasks does not necessarily translate to competence in dealing with interpersonal problems in a social setting) (Mann *et al.*, 1989).

There is growing appreciation of factors external to the child that affect the development of competence. In the context of medical decision making, features of the clinical setting, the status of knowledge in the area, and personal characteristics of the health professional will affect the adolescent's ability to give informed consent. The clinician's attitude towards children's rights and abilities, familiarity with the procedure, perceptions of the risks and benefits and tolerance for uncertainty will influence judgements about competence. There is evidence that clinicians' tolerance for ambiguity is significantly related to their practices regarding genetic testing. Health professionals, including medical geneticists, with low ambiguity tolerance are less likely to adopt a new predictive test 'when no others in their specialty offer it' or to respect patient autonomy in clinical decisions (Geller, 1993).

7. What is the evidence concerning competence in adolescents?

The available evidence suggests that many adolescents are competent decision-makers, particularly in terms of cognitive information-processing and problem-solving abilities. Formal operational (abstract) thinking generally develops around the age of 11 (Wertz et al., 1994) and is well-developed by the age of 14 (Nicholson, 1986). In their US study of the competence of 96 children and young adults to consent to hypothetical medical treatment, Weithorn and Campbell (1983) found that 14-year-olds showed comparable levels of competence, as defined by legal standards of consent, to adult study groups. The social and motivational aspects of competent decision making (e.g. independence of choice and commitment to a decision) appear more consistently in later adolescence (aged 15 years and above) (Mann et al., 1989). It is interesting that children's and adolescents' beliefs about the appropriate age at which to be making autonomous health-related decisions (on average 15.1 years) (Taylor et al., 1984, cited in Mann et al., 1989) coincide with evidence for a growth in competence at this age. Korer and Fitzsimmons (1987), in their study of young people in HD families, conclude that 'the young people concerned seemed able to discuss HD and its implications in a rational and thoughtful manner' and that 'there is no evidence to support the supposition that young people are too emotionally volatile to cope with information about HC' (Korer and Fitzsimmons, 1987; note that HD used to be called Huntington's Chorea, HC). Overall, competence in decision making increases towards later adolescence and is greatest where decision making occurs within clearly-defined boundaries or in domains in which the adolescent has experience (Mann et al., 1989; Michie and Marteau, 1996).

While there is evidence that adolescents have the capacity for competent decision making, there are factors that facilitate or inhibit both the development and the use of this competence.

8. Barriers to competent decision making in adolescence

Adolescence is often characterized as a time of rapid change, exploration and experimentation. It is also a time when, whether deliberately or not, decisions are made with important implications for the future. Studies of the effects of peer pressure suggest that, as conformity reaches a peak in early adolescence (Mann et al., 1989), this may threaten autonomy in decision making. Parents generally remain more influential than peer groups in terms of key values and decisions with long-term consequences (Mann et al., 1989). Research among 14- to 15-year-olds, who were asked to make decisions about hypothetical medical dilemmas that included varying degrees of parental influence, suggests that adolescents generally defer to perceived parental wishes (Sherer and Reppucci, 1988). However, the effect of parental influence varies according to the situation, with adolescents being 'more likely to resist parental influence when the consequences or gravity of the decision has serious implications for the adolescent's health' (Sherer and Reppucci, 1988). Adolescents' perceptions of an external locus of control, where other people or events are seen as controlling them, may also threaten their ability to assume control in decision making (Wertz et al., 1994). Evidence suggests that young adolescents are less likely to question the credibility or impartiality of the information they receive (Mann et al., 1989). There is also evidence to suggest that adolescents tend to be present-orientated, attaching too much weight to present as opposed to future concerns and not appreciating that values change over time (Wertz et al., 1994).

Family environment is a major factor in adolescents' experience and confidence in decision making, with decision-making capacities being under-used when adolescents do not perceive themselves as having choices in the first place. Where family dynamics are characterized by low or unequal involvement by parents in family decisions, or where there is conflict or low cohesion among family members, opportunities for decision making will be few (Mann *et al.*, 1989).

9. How do adults make genetic-related decisions?

Assessments of competence in adolescents and children generally use competent adults as a reference point. One approach to establishing appropriate criteria for assessing adolescent competence in decisions about genetic testing is to examine how supposedly competent adults make decisions in the context of genetic risk. Most research in this area has been based on normative models of decision making that specify how individuals ought to make decisions on logical information-processing grounds. The few studies that describe how individuals actually make genetic-related decisions in real life suggest that decision making does not correspond to these models (Lippman-Hand and Fraser, 1979). Normative decision-making models tend to over-estimate people's information-processing abilities and are particularly limited in capturing decision making processes in the context of genetic risk where decisions are subject to the simplified short-cuts and biases typical of judgements under uncertainty (Shiloh, 1996). A better understanding of how adults actually make genetic-related decisions will aid the development of appropriate criteria for assessing competence in adolescents. It may be that, when the actual decision making processes of adults are described, there will be a lesser gap in competence between adults, adolescents and children than current theory suggests.

10. How should competence be assessed, and by whom?

It is relatively easy to measure competence defined in terms of intellectual performance or, as is traditionally used as an indicator of informed consent, the ability to recall and recount medical explanations given (Alderson, 1993). But such an assessment may not be a good indicator of competent decision making in a social context. It excludes less-quantifiable components of decision making such as practical, social and imaginative skills and the way in which information is evaluated, and personal experiences and values are drawn on, in reaching a decision (Alderson, 1993). There are no standardized tools for assessing this broader view of competence.

According to the General Medical Council's guidelines, it is the clinician's duty to make judgements about a child's level of understanding and maturity and to act accordingly (Green and Stewart, 1987). It may, however, be unrealistic to expect clinical geneticists to make judgements about adolescents' competence. A more appropriate way forward may involve assessment using referral to appropriately-trained professionals, such as child/adolescent psychologists or psychiatrists, who have experience in making such assessments in, for example, child custody cases. It may also be useful to gather information from secondary sources, such as school teachers. As with the introduction of HD predictive testing for adults, such assessment should occur within a research protocol which documents both the process and the short-term and long-term outcomes.

11. A framework for assessing competence in adolescence

The following information may be useful in the assessment process:

(i)   The length of time that the adolescent has lived with knowledge of HD in the family and being at risk. An individual will know the most about those risks that he/she has been dealing with the longest (Fischhoff, 1992), therefore the length of time lived with an at-risk status will be a significant factor in the adolescent's ability to evaluate the costs and benefits of living at risk versus receiving a test result.

(ii)   Adaptation to knowledge of HD in the family and being at risk. This has been identified as a good predictor of how an individual will cope with a test result (Wexler, 1979).

(iii) Absence of any controlling influences in the decision to ensure that the adolescent is the primary decision maker. Does the adolescent assume or avoid control in decision making? What is the adolescent's primary reference group with regard to decision making in this context?

(iv)   Adequate information concerning the salient aspects of the test procedure and its implications. Is this knowledge accurate and is it being applied in the decision?

(v)   Appreciation of the broader social and emotional implications of the result for the individual and the family, both in terms of short-term and long-term interests and concerns. What is the nature and number of the costs and benefits considered in reaching a decision?

(vi)   The stability and consistency of the decision and the reasons for the decision over time.

(vii) Assessment of the role of the family context in the request and the function of the request in the family at the time. Does the request stem from a need to confirm beliefs about preselection (Kessler, 1988) or to resolve an anomalous situation where there is clarity about the genetic status of some but not all family members? An unresolved family conflict may underlie a request for testing and signal that intervention is needed at another level.

(viii) Past opportunities for the development of competence in decision-making. What is the pattern of sharing information and making decisions in the family? How and by whom are decisions usually made?

12. Conclusions

The reports dealing with referral of minors for HD predictive testing have concentrated on third party requests and have seldom considered as a separate issue adolescents who request testing of their own accord. The limited research into the consequences of early disclosure of a serious diagnosis or a child's genetic status has yielded equivocal findings (Michie and Marteau, 1996). This makes it inappropriate to enforce rigid criteria for childhood testing (Clarke and Flinter, 1996). The protocol of the UK Huntington's Prediction Consortium acknowledges that 'the exclusion criteria set out in the protocol are not intended to be invariable, and every case should be judged on its merits' (Craufurd and Tyler, 1992).

Some providers do make exceptions to the presumption against predictive testing of minors in the case of a request from a competent adolescent. Variations in maturity and abilities in different domains of decision making require that competence is assessed on an individual and situation-specific basis. In the absence of standardized tools for this purpose, assessment using referral to professionals with expertise in this area, and within a research protocol, seems the appropriate way forward.

It is suggested that adolescents should be encouraged to take extra time to consider all the implications of HD predictive testing (Tyler and Morris, 1990). A similar commitment is needed from clinicians in responding to a request from an adolescent. Apart from the time required in assessing competence, there are several other reasons why exploration of the concerns behind an adolescent's request for testing is warranted. A request for testing may mask a need for intervention at another level. It may reveal misconceptions regarding the genetic status of individuals or unresolved conflicts within the family. Providing an opportunity for discussion may resolve such conflicts or at least provide a space in which they can be acknowledged. As is the case with adult test applicants, many inquiries from minors will not result in a decision to receive a test result. When a decision is made to postpone testing, there may be a need for psychological support for individuals who have to live with continued uncertainty.

Acknowledgements

The author is supported by a grant from the Medical Research Council and the NHS (Wales) Office of Research and Development for Health and Social Care. This chapter is based on a previous paper: Binedell, J., Soldan, J.R., Scourfield, J. and Harper, P.S. (1996) Huntington's disease predictive testing: the case for an assessment approach to requests from adolescents. *J. Med. Genet.* **33**: 912–918. I am grateful to my co-authors and the editor of the *Journal of Medical Genetics* for their permission to draw upon this material.

References

Alderson, P. (1992) In the genes or in the stars? Children's competence to consent. *J. Med. Ethics* 18:119–124.
Alderson, P. (1993) Children's Consent to Surgery. Open University Press, Milton Keynes.
Ball, D.M. and Harper, P.S. (1992) Presymptomatic testing for late-onset disorders: lessons from Huntington's disease. *FASEB J* 6: 2818–2819.
Bloch, M. and Hayden, M.R. (1990) Opinion: Predictive testing for Huntington's disease in childhood: challenges and implications. *Am. J. Hum. Genet.* **46**: 1–4.

Bloch, M., Fahy, M., Fox, S. and Hayden, M.R. (1989) Predictive testing for Huntington disease: II. Demographic characteristics, life-style patterns, attitudes and psychosocial assessments of the first fifty-one test candidates. *Am. J. Med. Genet.* **32**: 217–224.

Bloch, M., Adam, S., Wiggins, S., Huggins, M. and Hayden, M.R. (1992) Predictive testing for Huntington disease in Canada: the experience of those receiving an increased risk. *Am. J. Med. Genet.* **42**: 499–507.

Children Act. (1989) HMSO, London.

Clarke, A. and Flinter, F. (1996) The genetic testing of children: a clinical perspective. In: *The Troubled Helix: Social and Psychological Implications of the New Human Genetics.* (eds T. Marteau, M. Richards) Cambridge University Press, Cambridge, pp. 164–176.

Clinical Genetics Society. (1994) Report of a Working Party on the Genetic Testing of Children. *J. Med. Genet.* **31**: 785–797.

Craufurd, D. and Tyler, A. (1992) Predictive testing for Huntington's disease: protocol of the UK Huntington's Prediction Consortium. *J. Med. Genet.* **29**: 915–918.

Decruyenaere, M., Evers-Kiebooms, G., Boogaerts, A., *et al.* (1996) Prediction of psychological functioning one year after the predictive test for Huntington's disease and the impact of the test result on reproductive decision making. *J. Med. Genet.* **33**:737–743.

European Community Huntington's Disease Collaborative Study Group. (1993) Ethical and social issues in presymptomatic testing for Huntington's disease: a European Community collaborative study. *J. Med. Genet.* **30**: 1028–1035.

Fischhoff B. (1992) Risk taking: a developmental perspective. In: *Risk-taking Behavior.* (ed J.F. Bates) John Wiley and Sons, Chichester, pp. 133–162.

Geller, G., Tambor, ES., Chase, GA. and Holtzman, NA. Measuring physicians' tolerance for ambiguity and its relationship to their reported practices regarding genetic testing. *Med. Care* **31**: 989–1001.

Gordon, C.P. (1996) Adolescent decision making: a broadly based theory and its application to the prevention of early pregnancy. *Adolescence* **31**: 561–584.

Green, J. and Stewart, A. (1987) Ethical issues in child and adolescent psychiatry. *J. Med. Ethics* **13**: 5–11.

Harper, P.S. and Clarke, A. (1990) Should we test children for 'adult' genetic diseases? *Lancet* **335**: 1205–1206.

International Huntington Association and the World Federation of Neurology Research Group on Huntington's Chorea. (1994) Guidelines for the molecular genetics predictive test in Huntington's disease. *J. Med. Genet.* **31**: 555–559.

Kessler, S. (1988) Invited essay on the psychological aspects of genetic counseling. V. Preselection: a family coping strategy in Huntington's Disease. *Am. J. Med. Genet.* **31**: 617–621.

Korer, J.R. and Fitzsimmons, J.S. (1987) Huntington's chorea and the young person at risk. *Br. J. Social Work.* **17**: 521–534.

Lippman-Hand, A. and Fraser, F.C. (1979) Genetic counseling: the postcounseling period: I. Parents' perceptions of uncertainty. *Am. J. Med. Genet.* **4**: 51–71.

Mann, L., Harmoni, R., Power, C. (1989) Adolescent decision-making: the development of competence. *J. Adolescence* **12**: 265–278.

Marteau, T.M. (1994) The genetic testing of children. *J. Med. Genet.* **31**: 743.

Michie, S. (1996) The genetic testing of children: paternalism or empiricism? In: *The Troubled Helix: Social and Psychological Implications of the New Human Genetics.* (eds T. Marteau and M. Richards). Cambridge University Press, Cambridge, pp. 177–183.

Michie, S. and Marteau, T.M. (1996) Predictive genetic testing in children: the need for psychological research. *Br. J. Health Psychol.* **1**: 3–14.

Michie, S., Drake, H., Bobrow, M. and Marteau, T. (1995) A comparison of public and professionals' attitudes towards genetic developments. *Public Understanding of Science* **4**: 243–253.

Montgomery, J. (1994) The rights and interests of children and those with mental handicap. In: *Genetic Counselling: Practice and Principles.* (ed A, Clarke) Routledge, London, pp. 208–222.

Nicholson, R.H. (1986) *Medical research with children: ethics, law and practice.* Oxford University Press, Oxford.

Sharpe, N.F. (1993) Presymptomatic testing for Huntington disease: is there a duty to test those under the age of eighteen years? *Am. J. Med. Genet.* **46**: 250–253.

Sherer, D.G. and Reppucci, N.D. (1988) Adolescents' capacities to provide voluntary informed consent. *Law Hum. Behav.* **12**: 123–141.

Shiloh, S. (1996) Decision-making in the context of genetic risk. In: *The Troubled Helix: Social and Psychological Implications of the New Human Genetics*. (eds T. Marteau and M. Richards) Cambridge University Press, Cambridge pp. 82–103.

Slavin, L.A., O'Malley, M.D., Koocher, G.P. and Foster, D.J. (1982) Communication of the cancer diagnosis to pediatric patients: Impact on longterm adjustment. *Am. J. Psychiatry* **139**: 179–183.

Smith, P.K., and Cowie, H. (1995) *Understanding children's development*, 2nd edition. Blackwell, Oxford.

Taylor, L., Adelman, H.S. and Kaser-Boyd, N. (1984) Attitudes toward involving minors in decisions. *Prof. Psychol. Res. Pract.* **15**: 436–449.

Tibben, A. (1993) What is knowledge but grieving? On psychological effects of presymptomatic DNA-testing for Huntington's disease. *Ph.D. thesis*, Rotterdam University, The Netherlands.

Tyler, A. and Morris, M. (1990) National symposium on problems of presymptomatic testing for Huntington's disease, Cardiff. *J. Med. Ethics* **16**: 41–42.

Weithorn, L.A. and Campbell, S.B. (1983) The competency of children and adolescents to make informed treatment decisions. *Child Dev.* **9**: 285–292.

Went, L. (1990) Ethical issues policy statement on Huntington's disease molecular genetics predictive test. *J. Med. Genet.* **27**: 34–38.

Wertz, D.C., Fanos, J.H. and Reilly, P.R. (1994) Genetic testing for children and adolescents: who decides? *JAMA* **272**: 875–881.

Wexler, N.S. (1979) Genetic 'Russian Roulette': The experience of being 'at risk' for Huntington's disease. In: *Genetic Counselling: Psychological Dimensions*. (ed S. Kessler) Academic Press, New York, NY pp. 199–220.

Wiggins, S., Whyte, P., Huggins, M. *et al.*, (1992) The psychological consequences of predictive testing for Huntington's disease. *N. Engl J. Med.* **327**: 1401–1405.

Wilfond, B.S., Pelias, M.Z., Knoppers, B.M. *et al.* (1995) Points to consider: ethical, legal, and psychosocial implications of genetic testing in children and adolescents. *Am. J. Hum. Genet.* **57**: 1233–1241.

12

Moving away from the Huntington's disease paradigm in the predictive genetic testing of children

Cynthia B. Cohen

Abstract

Health care professionals in genetics are usually unwilling to offer predictive genetic testing for children. They are concerned that this might have harmful psychological and social effects with few offsetting benefits. They have reached this conclusion on the basis of the devastating results for some adults of testing for Huntington's disease (HD), a debilitating and fatal late-onset condition. However, it seem inaccurate to generalize from testing for this unique condition to testing for all late-onset conditions. Each adult-onset condition bears somewhat features and testing for each might therefore involve different sorts of harms and benefits.

A whole complex of factors needs to be considered in weighing the harms and benefits of providing predictive testing for a specific condition to a specific asymptomatic child. These factors include the emotional effects of testing, impact on family dynamics, effect on ability to plan for the future, import for ethical principles, social and economic effects, remoteness of time of onset of condition, degree of probability of incurring the condition, degree of seriousness of the condition, possible future medical benefits of testing, level of maturity of the child, and availability of genetic counselling.

There is tremendous uncertainty about the benefits and harms of providing predictive testing for children, for few empirical studies of this question have been done. In such situations of uncertainty, in which value judgments play an especially significant role, there is a presumption that the decisions of parents should prevail. Parents care most about the welfare of their children, are responsible for

The Genetic Testing of Children, A.J. Clarke (ed.).
© 1998 BIOS Scientific Publishers Ltd, Oxford.

maintaining their daily lives, and impart values to them. Yet health care professionals have a responsibility to protect children should their parents take inappropriate decisions not in the interests of their children.

Further, we must begin to consider the impact of allowing predictive testing on our underlying social values and cohesiveness as a community. We owe it to our children to promote social policies that provide support and care for those who are vulnerable to future illness.

1. Introduction

Genetic testing allows a diagnosis of an increased risk of adult-onset conditions at an earlier stage of life than ever before. Yet health care professionals in genetics have been reluctant to accede to parental requests to have asymptomatic children tested for such conditions. Because most adult-onset conditions have no way of being prevented, no predictable time of appearance, and no effective treatment, many professionals consider it harmful to provide parents and children with such potent and potentially devastating information. They have reached this conclusion largely on the basis of the results of presymptomatic testing for Huntington's disease (HD), which reveal that some adults tested for this debilitating, fatal disorder experience serious psychosocial burdens. Children, they believe, should not similarly have to live with a genetic 'sword of Damocles' hanging over their heads.

Guidelines for testing for HD have provided the model for testing children for all other adult-onset conditions. The HD guidelines recommend against presymptomatic testing of children (Huntington's Disease Society of America, 1989; World Federation of Neurology, 1990). Yet HD is a distinctive condition, terrible in its manifestations, progression, and fatal outcome. Therefore, it is inaccurate to generalize from testing for genes for this disease to testing for all other adult-onset conditions. Each of these conditions bears somewhat different features; testing for each might therefore involve different sorts of harms and benefits.

A whole complex of factors needs to be considered before providing predictive testing for a specific condition to specific children. Many of these factors will vary according to the kind of condition at issue, the child to be tested, and the personal and social values at stake. We should not rule out *a priori* all predictive genetic testing of children on the basis of the manifestations of one truly terrible condition.

2. Disadvantages of testing children for adult-onset conditions

We have little information about the effects on children of predictive genetic testing. We must instead rely on studies in which children have been tested for serious conditions without a genetic component. These, however, are also limited in number and may not be relevantly similar to adult-onset genetic conditions. Hence, the conclusions that we draw about the harms of genetic testing of children are necessarily very tentative.

2.1 Creates emotional burdens

Adults tested for HD, including those with positive and those with negative results, have been left with a heavy emotional burden (Bloch *et al.*, 1992; Huggins *et al.*, 1992). It seems reasonable to assume that children would experience even greater emotional distress upon receiving positive or negative results, since they would have a longer period in which to live with this knowledge and fewer capabilities for dealing with it. Moreover, professionals maintain that the sense of self-esteem and worth of children who test positive for a serious adult-onset condition might be gravely damaged (Bloch and Hayden, 1990).

2.2 Impacts negatively on family

Predictive genetic testing also bears the risk of changing family dynamics in ways that could be damaging. Children who test negative while siblings test positive may be plagued by 'survivor guilt' (Tibben *et al.*, 1992). Those who test positive may inadvertently be stigmatized by their parents and made to feel that they are sick and damaged (Institute of Medicine 1994; Wertz *et al.*, 1994). Some parents may develop negative expectations about the future development of affected children, while others may reject them entirely (American Society of Human Genetics and American College of Medical Genetics, 1995).

2.3 Impairs future planning

Positive test results might provoke a feeling of fatalism in children and lead them to believe that they are powerless to change the inevitable (Monmaney, 1995). Families may shift resources away from children who receive a positive test result and fail to prepare them for future roles. Planning for the future of their children in some families, Wertz and colleagues suggest, may, in effect, become restricting of their future (Wertz *et al.*, 1994).

2.4 Disregards principle of autonomy

A distinctive ethical harm of predictive testing of children that has been raised is to the principle of autonomy. This principle, some suggest, requires that children should have the freedom to decide whether or not to be tested when they become adults; this freedom should not be co-opted by parents during childhood (Working Party of the Clinical Genetics Society, 1994). We know that about 85% of adults at risk of HD choose not to undergo testing (Bloch *et al.*, 1992; Quaid, 1991), yet children who are tested before they become adults are denied this choice in violation of their autonomy as adults.

2.5 Leads to discrimination and stigmatization

Discrimination against children in education, employment, and insurance can follow from predictive testing of children (Billings *et al.*, 1992). Social stigmatization is also a serious risk of testing, for others may consider children who test positive to be 'diseased' even though they are not ill.

These possible harms seem sufficiently serious to lead many in genetics to refuse to test children for adult-onset conditions. Yet we have no evidence whether it may also be harmful not to meet requests of parents to have their children tested (Marteau, 1994). Moreover, we must balance these possible risks of testing children for adult-onset conditions against what little we know about its possible benefits (Harper and Clarke, 1990).

3. Benefits of testing children for adult-onset conditions

We are learning that despite its potential harms, testing for HD carries benefits for some adults. Children may also gain some benefits from being tested for this condition, as well as for other adult-onset conditions. The following possible benefits of testing take into account many of the same factors considered in enumerating its possible harms and therefore parallel those harms.

3.1 Provides emotional benefits

Children in families with a history of an adult-onset condition such as breast cancer or HD can become aware that they might be at risk for the condition, even though they live in households in which this is not explicitly discussed. It is difficult to keep these sorts of family secrets from them (Baran and Pannor, 1989). The most obvious benefit of testing such children for adult-onset conditions is that this may reduce their anxiety and uncertainty and that of their parents. There is some evidence that adults at risk of HD who are tested are relieved to know whether or not they will have it (Brandt et al., 1989; Kessler, 1994; Meissen et al., 1988). Indeed, in one study those who learned that they carried the gene for HD, adjusted better than did those who chose not to be tested or whose test results were inconclusive (Wiggins et al., 1992). Children who are tested for an adult-onset condition after counselling and support may also experience relief from the inchoate uncertainty that has hovered over them (American Society of Human Genetics and American College of Medical Genetics, 1995; Koocher 1986).

3.2 Enhances family dynamics

Predictive testing of children may bolster, rather than impair family relationships. The fact that members share the same vulnerability and the same information, even though some test negative and others positive, can lead families openly to air their concerns and options with one another as appropriate to their age. Parents may develop supportive and sympathetic attitudes toward their children, both affected and unaffected, that serve to draw them closer together (Petersen and Boyd, 1995). Indeed, it has been argued that waiting until children are adults to test them for a serious condition may lead to anger and resentfulness within families (Cotton, 1995).

3.3 Aids future planning

Children who learn they have a gene associated with an adult-onset condition can take advantage of early preventive, ameliorative, and therapeutic measures as they become available (Li *et al.*, 1992). Moreover, those who receive a negative test result are saved the discomfort and cost of screening procedures and therapies as these are developed. In addition, obtaining genetic information through testing can allow parents and children to plan for the future of the children in terms of educational goals, occupational choices, and specific career plans (American Society of Human Genetics and American College of Medical Genetics, 1995; Hayden *et al.*, 1995).

3.4 Respects the principle of beneficence

It is not necessarily unethical for parents to make choices about their children's welfare that disallow them such choices when they reach adulthood. Parents do this often, as when they choose to have them baptized, circumcised, or tested for various intellectual, vocational or other capabilities. To promote their children's well-being, as required by the principle of beneficence, parents need to address the current and future needs and concerns of their offspring. Some will find that they cannot put off predictive testing because their children are aware of their family history and its implications for them. Current beneficence in such situations can ethically outweigh future autonomy.

3.5 Lessening of discrimination

The possibility of discrimination against children who test positive for an adult-onset condition is a real and pressing problem. There are some signs of change in the United States due to new state and federal laws that, in combination, would bar employers and health care insurers from refusing jobs and insurance to those who have been genetically tested in many instances (Lord, 1996). Other countries are also taking measures against genetic discrimination. Until public policy across the nations bans such discrimination entirely, however, the developing benefit of lessened discrimination will be limited. Although predictive testing may provide some benefits to some children, many health care professionals in genetics believe that its harms outweigh these benefits. Because many base their assessment of harms on testing for HD, which represents a 'worst case' scenario, they may overstate the possible harms of testing children for other adult-onset conditions and overlook the benefits.

HD, the early signs of which generally begin between ages 35 and 45, involves abnormal involuntary movements, severe personality changes, cognitive deficits and an inevitable death after 10–15 years of unremitting degeneration. Another serious adult-onset condition, breast cancer, does not bear these disturbing physical and psychological features and does not necessarily lead to death. It would seem, therefore, that the uncertainty about whether one will incur breast cancer if one tests positive for the altered *BRCA1* gene is not as awful to bear as is the certainty that one will be stricken with HD should one test positive for the gene. The psychological harm of learning that one is at increased risk of Alzheimer's disease might be greater than that of learning that one is at risk of HD because one has not had lifelong preparation for the former diagnosis (Burgess, 1994). The harms of testing for different adult-onset conditions, therefore, may well differ, depending on the condition and the person undergoing testing.

4. Other factors relevant to a decision

In assessing the harms and benefits of predictive testing of children, a variety of factors need to be taken into consideration related to the distinctive features of various adult-onset conditions and the particular situations of children and their families. These factors include not only those cited above – emotional effects, impact on family dynamics, effect on ability to plan for the future, import for ethical principles, and social and economic effects – but such additional ones as the following.

4.1 Remoteness of time of onset of condition

How far into the future will or might the condition occur? Full-blown Alzheimer's disease would occur at such a distant point in time that it would not affect the child's medical status, emotional development, or understanding of the future for many years. Testing children for susceptibility to this condition therefore would not benefit them. However, testing children of an age close to the time when a condition might occur could redound to their welfare. Testing an adolescent for a gene associated with a 100% risk of a form of colon cancer that would appear in early adulthood can provide information relevant to the child's health (Petersen *et al.*, 1993), and follow-up and preventive measures can be considered. This factor would also weigh the balance in favour of testing a mature 16-year-old girl who is aware of her family history of breast cancer for the associated genes. Indeed, in the first tests for the mutated *BRCA1* gene, a 16-year-old girl was allowed to undergo testing despite a policy of waiting until minors reach age eighteen (Biesecker, personal communication).

4.2 Degree of probability of incurring the condition

When the likelihood of having an adult-onset condition is small, that weighs against predictive testing of a child. There is no readily discernible reason, for instance, to test an asymptomatic child with no family history of HD for that condition. When a condition runs in a family, however, there is greater reason to test a child. Yet even this statement must be qualified. Being at risk of HD is part of the self-identity of a child from a family with a history of this condition. In contrast, a child in a family with a history of Alzheimer's is not likely to have developed an identity related to that condition because its genetic associations have only recently been recognized (Parker, 1995). Therefore, the fact that the condition runs in the family would not necessarily provide reason to test the latter child.

Moreover, we cannot derive a generally applicable rule stating that the more certain it is that a condition will occur, the more beneficial it is to test for it. We have learned from testing adults at risk for HD that the certainty that this condition will strike can be a deterrent to being tested to some and a reason for being tested to others. In the case of testing for the altered *BRCA1* gene – in which those who test positive have an 80% chance of breast cancer (Biesecker *et al.*, 1993) – it is the lack of certainty about whether the condition will appear that leads some to decide against testing and others to request it.

4.3 Degree of seriousness of the condition

This is a difficult factor to measure objectively, for it will vary depending on the perceptions of the family and child involved (Wilfond and Nolan, 1993). Some would rate Alzheimer's, which affects mental status, as more serious because of this than heart disease, which is physically disabling and threatens death. Others would judge the reverse. Even so, life-threatening conditions, such as HD, are generally rated as more serious than chronic conditions, such as diabetes. Furthermore, conditions that involve major debilitation, pain, and anguish are to be considered more serious than those which involve almost imperceptible deterioration.

The UK Clinical Genetics Society report maintains that the more serious the condition, the stronger the arguments in favour of testing would need to be (Working Party of the Clinical Genetics Society, 1994). It can also be argued, however, that the more serious the condition, the stronger the arguments against testing would need to be. Many parents and children might have greater reason to want test results when the condition at issue is very serious so that they can keep abreast of medical developments and prepare in other ways. This seems to be a factor that will vary depending on the family and child involved.

4.4 Possible future medical benefits

Although there are currently few measures available to alleviate or cure most adult-onset conditions, research may be underway that promises to lead to prevention or therapy for a specific condition. The decision to provide predictive testing to children will be affected by whether surveillance would be beneficial and whether the condition might be amenable to treatment in the near future.

4.5 Level of maturity of child

The degree of emotional, social, and intellectual maturity of the child should also figure in the decision. Professional groups in the United States have been willing to provide genetic testing for such conditions as HD or breast cancer only for those who are 18 years old or above (Biesecker *et al.*, 1993). This age limit does not take into account the capacities for decision making of younger children. By contrast, only a few respondents queried in the United Kingdom had a firm age below which they would not test the genetic status of a child – and that age varied widely (Working Party of the Clinical Genetics Society, 1994). The maturity of the individual child and his or her family situation may have been considered relevant to reaching a decision about testing.

Adolescents of 16 or 17 years of age may have mature decision-making capacities (Wertz *et al.*, 1994). Younger adolescents, while not able to make independent choices, may have substantial abilities to make decisions (Melton *et al.*, 1983).

Some of these adolescents could reasonably be given a voice in the decision whether to be tested. The capacities needed to understand the causes of disease and to recognize that one is making a choice between alternatives develop substantially between the ages of 8 and 11, and are often well-developed by ages 12–14 (Melton *et al.*, 1983). Furthermore, children at the ages of 12 or 13 are beginning to develop the capacity to anticipate the future. The ability of children to understand and choose will vary depending on the child and the testing choice at issue (Buchanan and Brock, 1989).

These observations suggest that it would be wise to forgo adult-onset testing for children below the age of 12, for they do not tend to understand the causes of disease or recognize that they are making a choice between alternatives. Moreover, they do not generally have the capacity to anticipate the future. Some children between the ages of 12 and 18, however, do have these abilities. Depending on their maturity and circumstances, they might derive benefits from predictive genetic testing.

4.6 Availability of genetic counselling

A significant factor to take into account is whether genetic counselling is available for children and their parents. A major disadvantage of testing children, some note, is that they cannot be given counselling from a child's perspective before or in parallel with the testing process (Working Party of the Clinical Genetics Society, 1994). Yet counsellors can offer age-appropriate counselling to children. They can also assist their parents to understand the issues children face and the options open to them. Predictive genetic testing need not be separated from counselling, either for parents or for children. The major difficulty that the need for genetic counselling creates is that there are insufficient numbers of trained counselors who can provide this important service to them.

5. The roles of parents and professionals in the decision

If it were clear that predictive genetic testing creates overriding harm for children, a firm rule against such testing would be justified. But there is tremendous uncertainty about the benefits and harms of testing children for various sorts of adult-onset conditions. Recommendations made for testing children, Marteau points out, are not based on empirical evidence, for 'there is, as yet, next to none' (Marteau, 1994). To refuse to test children for adult-onset conditions is to assume without confirmation that the harms of testing outweigh its benefits.

Value judgements are at the heart of decisions about testing children for adult-onset conditions. The different sorts of risks and benefits ascribed to the predictive testing of children cannot be totted up as though they are on an evaluative par without regard to their disparate medical, psychological, and moral weight. There will be differences in the way that different people in different circumstances evaluate these in coming to a decision about predictive testing for a child, depending on their judgement about where the good for the child lies. This sort of judgment should be made by those who usually care most about the welfare of these children, are responsible for maintaining their daily lives, and impart values to them – their parents.

A presidential commission in the United States recommended that when there is ambiguity or uncertainty about the probable harms and benefits to children of both treatment and nontreatment, the decisions of parents should prevail (President's Commission, 1983). The commission recognized that such decisions involve value considerations and therefore maintained that they should be made in 'a way that appropriately protects and encourages the exercise of parental judgment.' Yet the commission also emphasized that parents, as their children's surrogates, need the close cooperation of physicians, for the latter can provide them with most recent and complete information. Further, the relationship between professional genetic counsellors and their clients or surrogates reflects a shared decision making process which guarantees to clients or surrogates the authority to make choices reflecting their values (Clarke, 1991).

Physicians and genetic counsellors, however, have a responsibility to protect children should parents take 'inappropriate decisions' in particular cases, the commission stated. This does not require that parents make the best possible decision for their children, but one that supports their basic interests. Buchanan and Brock point out that a failure by parents to optimize the interests of their children:

> 'is not sufficient to trigger justified interventions by third parties, or even a challenge to the parent's decision-making authority. . . . the best interest principle is to serve only as a regulative ideal, not as a strict and literal requirement . . .'

(Buchanan and Brock, 1989).

Many of the factors discussed above should be taken into account by professionals when they consider whether to overrule parental decisions that they believe might seriously harm a child. As a general presumption, though, in situations of uncertainty about the harms and benefits of predictive testing of children, the informed decisions of parents should prevail.

6. Social import of predictive testing of children

It is not enough to assess the harms and benefits of predictive testing for certain conditions to specific children when developing policy. We must also consider the potential of this profoundly powerful technology for changing our underlying social values and our cohesiveness as a community. Would the increased use of genetic testing for adult-onset conditions in children lead us to expand genetic discrimination to childhood? Would the development of search-and-destroy missions against certain adult-onset conditions fuel our drive toward perfectionism and tempt us to disvalue children who are 'merely normal'? Or would greater use of predictive testing in children give us increased impetus to address the needs of children susceptible to adult-onset genetic conditions? Would it render us more accepting of such conditions as part and parcel of our endowment as human beings? The value of predictive genetic testing, not only for our children, but for the society in which our children are nurtured and brought to adulthood, will depend on whether we promote social policies to provide support and care for those who are vulnerable to future illness, accepting them as essential members of the human community.

References

American Society of Human Genetics and American College of Medical Genetics. (1995) Points to consider: ethical, legal, and psychosocial implications of genetic testing in children and adolescents. *Am. J. Hum. Genet.* 57: 1233–1241.

Baran, A. and Pannor, R. (1989) *Lethal Secrets*. Warner, New York, NY.

Biesecker, B.B., Boehnke, M., Calzone, K., Markel, D.S., Garber, J.E., Collins, F.S., Weber, B.L., (1993) Genetic counseling for families with inherited susceptibility to breast and ovarian cancer. *J. Am. Med. Assoc.* 269: 1970–1974.

Billings, P.R., Kohn, M.A., deCuevas, M., Beckwith, J., Alper, J.S. and Natowicz, M.R. (1992) Discrimination as a consequence of genetic testing. Am. J. Hum. Genet. 50: 476–482.

Bloch, M., Adams, S., Wiggins, S., Huggins, M., and Hayden, M.R. (1992) Predictive testing for Huntington disease in Canada: the experience of those receiving an increased risk. Am. J. Med. Genet. 42: 499–507.

Bloch, M. and Hayden, M.R. (1990) Opinion. Predictive testing for Huntington disease in childhood: challenges and implications. *Am. J. Hum. Genet.* 46: 1–4.

Brandt, J., Quaid, K.A., Folstein, S.E., *et al.*, (1989) Presymptomatic diagnosis of delayed-onset disease with linked DNA markers: the experience in Huntington's Disease. J. Am. Med. Assoc. 261: 3108–3114.

Buchanan, A.E. and Brock, D.W. (1989) *Deciding for Others: The Ethics of Surrogate Decision-Making.* Cambridge University Press, Cambridge.

Burgess, M.M. (1994) Ethical Issues in Genetic Testing for Alzheimer's Disease: Lessons from Huntington's Disease. *Alzheimer Dis. Assoc. Disorders* 8: 71–78.

Clarke, A. (1991) Is non-directive genetic counselling possible? *Lancet* 338: 998–1001.

Cotton, P. (1995) Prognosis, diagnosis, or who knows? Time to learn what gene tests mean. *J. Am. Med. Assoc.* 273: 93–95.

Harper, P.S. and Clarke, A. (1990) Should we test children for 'adult' genetic diseases? *Lancet* 335: 1205–1206.

Hayden, M.R., Bloch, M. and Wiggins, S. (1995) Psychological effects of predictive testing for Huntington's disease. In: *Behavioural Neurology of Movement Disorders*, (eds W.J. Weiner and A.E. Long). Raven Press, New York, pp. 201–210.

Huggins, M., Bloch, M., Wiggins, S., *et al.*, (1992) Predictive testing for Huntington disease in Canada: adverse effects and unexpected results in those receiving a decreased risk. *Am. J. Med. Genet.* 42: 508–515.

Huntington's Disease Society of America. (1989) *Guidelines for Predictive Testing for Huntington's Disease.* The Huntington's Disease Society of America, New York, NY.

Institute of Medicine. (1994) *Assessing Genetic Risks: Implications for Health and Social Policy.* (eds L.B. Andrews, J. E. Fullarton, N.A. Holtzman and A.G. Motulsky). National Academy Press, Washington, DC.

Kessler, S. (1994) Predictive testing for Huntington's disease: a psychologist's view. *Am. J. Med. Genet.* 54: 161–166.

Koocher, B.P. (1986) Psychological issues during the acute treatment of pediatric cancer. *Cancer* 58: 468–472.

Li, F.P., Garber, J.E., Friend, S.H., *et al.*, (1992) Recommendations on predictive testing for germ line p53 mutations among cancer-prone individuals. J. Am. Cancer Inst. 84: 1156–1160.

Lord, D. (1996) Something in the genes. *Am. Bar Assoc. J.* 86: 82–86.

Marteau, T.M. (1994) The genetic testing of children, *J. Med. Genet.* 31: 743.

Meissen, G.J., Myers, R.J. and Mastromauro, C.A. (1988) Predictive testing for Huntington's disease with use of a linked DNA marker. *N. Engl. J. Med.* 318: 535–542.

Melton, G.B., Koocher, G.P. and Saks, M.J. (1983) *Children's Competence to Consent.* Plenum Press, New York, NY.

Monmaney, T. (1995) Genetic testing: kids' latest rite of passage. *Health* 46–48.

Parker L. (1995) Ethical concerns in the research and treatment of complex disease. *Trend. Genet.* **11**: 520–523.

Petersen, B.M. and Boyd, P.A. (1995) Gene Tests and Counseling for Colorectal Cancer Risk: Lessons from Familial Polyposis. *J. Nat. Cancer Inst.* **87**: 67–71.

Petersen, G.P., Francomano, C., Kinzler, K. and Nakamura, Y. (1993) Presymptomatic direct detection of adenomatous polyposis coli (APC) gene mutations in familial polyposis coli. *Hum. Genet.* **91**: 307–311.

President's Commission for the Study of Ethical Problems in Medicine and Biomedical and Behavioral Research. (1983) *Deciding to Forego Life-Sustaining Treatment*. US Government Printing Office, Washington, DC, p. 218, Table 1.

Quaid, K.A. (1991) Predictive testing for HD: maximizing patient autonomy. *J. Clin. Ethics* **2**: 238–240.

Tibben, A., Vegter-van der Vlis, M., Skraastad, M.I. *et al.*, (1992) DNA-Testing for Huntington's disease in the Netherlands: a retrospective study on psychosocial effects. *Am. J. Med. Genet.* **44**: 94–99.

Wertz, D.C., Fanos, J.H. and Reilly, P.R. (1994) Genetic testing for children and adolescents. *J. Am. Med. Assoc.* **272**: 875–881.

Wiggins, S., Whyte, P., Huggins, M., Ada, S., Theilmann, J. and Bloch, M. (1992) The psychological consequences of predictive testing for Huntington's disease. *N. Engl. J. Med.* **327**: 1401–1405.

Wilfond, B.S., Nolan, K. (1993) National policy development for the clinical application of genetic diagnostic techniques. Lessons from cystic fibrosis. *J. Am. Med. Assoc.* **270**: 2948–2953.

Working Party of the Clinical Genetics Society (UK) (1994) The genetic testing of children. *J. Med. Genet.* **31**: 785-797.

World Federation of Neurology. (1990) Research Committee Research Group. Ethical Issues Policy Statement on Huntington's Disease Molecular Genetics Predictive Test. *J. Med. Genet.* **27**: 34–38.

13

Cancer susceptibility testing: risks, benefits and personal beliefs

Andrea Farkas Patenaude

'The doctor could never persuade Vinod to believe in the genetic reasons for either his or his son's dwarfism. That Vinod came from normal parents and was nonetheless a dwarf was not in Vinod's view the result of a mutation. The dwarf believed his mother's story: that, the morning after she conceived, she looked out the window and the first living thing she saw was a dwarf. That Vinod's wife, Deepa, was a normal woman...didn't prevent Vinod's son Shivaji, from being a dwarf. However, in Vinod's view, this was not the result of a dominant gene, but rather the misfortune of Deepa forgetting what Vinod had told her. The morning after Deepa conceived, the first living thing she looked at was Vinod and that was why Shivaji was also a dwarf. Vinod had told Deepa not to look at him in the morning, but she forgot.'

A Son of the Circus
John Irving
©1994 Random House, New York

1. Introduction

Medicine has entered a new era in which genetics offers the possibility of improved approaches to cancer prevention, treatment and cure. Genetic testing brings to the individual level the assessment of lifetime cancer risk. Research investigations of the psychosocial impact of genetic testing for breast, ovarian, and colon cancers in high-risk populations are underway. Commercial testing for breast cancer susceptibility and other cancer genes has also become available, despite professional controversy about the ethics and utilization of testing members of the general population without identified risk factors (Olopade, 1996).

This chapter aims to provide a brief overview of the current state of genetic testing for cancer genes and to describe some of the ethical and psychological issues which may affect the impact of genetic testing for cancer susceptibility in paediatrics.

The Genetic Testing of Children, A.J. Clarke (ed.).
© 1998 BIOS Scientific Publishers Ltd, Oxford.

2. Which children are potential testing candidates?

There are a variety of circumstances under which children may be considered for cancer genetic testing. One of the most frequent motivations offered by adults seeking genetic testing is the wish to know more about their children's risk of developing cancer. The natural extension of this wish is the testing of the children themselves when genetic identification of a parent as a carrier of a cancer susceptibility gene has been made or when other circumstances suggest that the child may be likely to be at increased risk of cancer. Gene testing may also be sought by adults who had cancer as a child or the parents of minors who are paediatric cancer survivors to ascertain if genetic markers are present which signify increased risk for second malignancies. The risk of a second cancer in survivors of some paediatric malignancies may be as high as 50%, depending on the individual's genetic status (Desjardins *et al.*, 1991). Survivors may also want testing to determine the genetic risks of cancer to their offspring or to other relatives (Strong, 1993). Scientists may, in addition, wish to identify children who are at increased risk of primary or secondary malignancies in order to enter them onto protocols aimed at testing environmental or genetic interventions to prevent the development of malignancies (Pollock and Knudsen, 1996).

The testing of children for cancer susceptibility genes raises even more complex ethical and medical issues than testing adults (Wertz *et al.*, 1994). Many feel that children should not be tested unless medical benefit in childhood is likely to result (Clarke and Flinter, 1996). Others feel this is too restrictive, that predictions of harm are not based on empirical data, and that prohibitions against testing children may limit advances in paediatric cancer treatment (Michie, 1996). As the number of identified cancer genes increases and as new approaches to the prevention and cure of cancer based on genetic knowledge become possible, the forces pro and con the genetic testing of children are likely to become more vocal.

3. Complexity of cancer risk identification

Cancer is a multi-factorial disorder (Vogelstein and Kinzler, 1993). This is contrasted to a circumstance like Huntington's disease (HD), where a fully penetrant gene predicts virtual certainty of developing the disease (Huntington's Disease Collaborative Research Group, 1993). Inherited susceptibility is responsible for only a small percentage (about 5%) (Schimke, 1992) of most cancers and many other agents in our modern existence are known or suspected to have links to cancer aetiology. Genetic risk is likely to be only one factor in determining whether a gene carrier develops cancer. Smoking has been definitively linked to lung cancer (Valanis, 1996), radiation has long been known to cause cancers (Doll and Peto, 1981), and studies are underway to evaluate the relationship between cancer causation and exposure to electromagnetic fields (Meinert and Michaelis, 1996) occupational toxins (Harty *et al.*, 1996), and dietary factors such as fat, fibre (Hunter and Willett, 1996), and beta-carotene (Challem, 1996). The multi-factorial nature of cancer aetiology colours the meaning of genetic risk status in several ways:

(i) It raises the possibility that at-risk individuals could alter the likelihood of getting cancer through changes in diet, exercise, occupational exposure, etc. However, currently the effectiveness of such changes in genetically predisposed individuals is unknown.

(ii) The presence of potentially mediating factors helps some gene carriers maintain hope that they will never develop cancer. However, the possibility that cancer may not develop may lead others to disregard or downplay their added risk and to ignore recommendations for screening or changes in health behaviors.

The evaluation of the roles of genetic and other factors in cancer aetiology increases the importance attributed to identification of cancer susceptibility genes. The hope is that by observing which susceptible individuals do and do not develop cancer, much can be learned about cancer prevention and, possibly even, cancer cure.

The multi-factorial nature of cancer also leaves considerable room for the development of personal theories about cancer aetiology. These personal beliefs often guide much of an individual's and a family's health behaviours, especially in families where cancer occurs with greater than normal frequency (Petersen and Boyd, 1995; Richards, 1996). Such personal beliefs may explain some of the lack of congruity between knowledge of cancer risk and adherence to recommendations for cancer screening. Examples of the lack of congruity are the fact that women identified by family history as being at increased risk for breast cancer often do not have any higher rates of adherence to recommendations for mammography or clinical breast examination than women at average risk of breast cancer (Kash et al., 1992). These beliefs may underlie the finding that survivors of childhood cancer smoke at similar rates as other young people (Hollen and Hobbie, 1996). They could also help explain why some individuals who seem eager to be tested for cancer susceptibility genes may have no identified physicians and may never have sought regular medical check-ups.

4. Testing for which genes? Under what conditions? With what safeguards?

Some existing guidelines for genetic testing recommend that genetic testing of children should only occur if direct, timely medical benefit is an expected outcome of the testing (American Society of Human Genetics Board of Directors and the American College of Medical Genetics, 1995; American Medical Association Council on Ethical and Judicial Affairs, 1995). This would be likely to include, for example, the testing of children for familial adenomatous polyposis (FAP). This is an inherited condition in which a high number of precancerous polyps develop in the colon. If more than 100 polyps are found, the individual undergoes removal of sections of the colon (Rhodes and Bradburn, 1992). Although the surgery usually occurs after age 16, screening with annual sigmoidoscopies beginning around age 10 allows for early removal of small numbers of polyps and careful follow-up of at-risk individuals (Petersen, 1994). Genetic testing makes identification of the 50% of children in these families who are not at increased risk, and thus who can be relieved of the burden of annual colonoscopies, possible. It also provides information to aid in long-term planning for the care of children with increased genetic risk. Genetic testing of children at risk for FAP is underway at a few centres in the US and Britain (Codori et al., 1996; Marteau et al., 1996) and assessment of the impact of testing is simultaneously under study. Difficulties in successful translation of cancer genetic testing results conducted outside research programmes is illustrated by a recent study which identified that over 30% of general practitioners scheduled to provide FAP genetic testing results to patients or parents had an incorrect or incomplete understanding of the meaning of the test result. In addition only 20% of the patients who had been tested had undergone genetic counselling or had given written informed consent (Giardiello et al., 1997).

Genetic testing of paediatric cancer survivors to determine their likelihood of developing a second malignancy could potentially result in medical benefit if genetic identification improved the survival rates of patients who developed secondary malignancies. An example of a group whose increased risk for second malignancy could be identified from genetic analysis, but not from the physical characteristics of their primary disease, are patients who have unilateral retinoblastoma. If these patients carry an altered *RB1* gene, they have a 50% chance of developing a secondary cancer (Desjardins et al.,1991; Eng et al., 1993), considerably above the risk they carry without an altered *RB1* gene. In the future, survivors of leukaemia, Hodgkin's disease or other malignancies of childhood may also be able to utilize genetic testing to determine their secondary cancer risk.

However, the wish to appear 'normal' (Rait et al., 1992) and to put one's cancer experience out of mind (Last, 1992) may lead some cancer survivors to shun genetic risk identification correlated with the risk of a second primary cancer. Information about genetic testing specially geared to paediatric cancer survivors and their needs may be helpful in allowing survivors to determine the personal utility of testing and to explore the options for surveillance which do not require testing.

There are other conditions where the testing of children for cancer susceptibility is currently possible, but the advisability of testing is more controversial. Li-Fraumeni syndrome (LFS) is a rare familial cancer syndrome which predisposes individuals to both adult and paediatric cancers (leukaemia, brain tumours, sarcomas, breast cancer, adrenocortical carcinoma, etc.)(Li et al., 1991). For individuals within these families, the risk of developing cancer if one inherits an altered *p53* gene is 40% by the age of 20 and 90% by the age of 60 (Williams and Strong, 1985). The risk of inheriting the dominant gene from a parent who carries it is 50%. Predisposition testing of adult members of these families is currently underway in the US and UK. Testing of children with cancer for *p53* has occurred at a number of centres; predisposition testing of healthy children is not currently occurring, but is under consideration at a number of centres. One significant issue is that there is no screening to recommend for children who are found to be at increased genetic risk due to an alteration in their *p53* gene. Screening of a child known or suspected to carry a *p53* mutation is limited to regular physical examination and immediate follow-up of any symptoms, behaviours which do not require genetic testing. However, knowing a child does not carry an altered *p53* gene could lift the sword of Damocles from the 50% of the children in LFS families who don't carry the mutation. The impact of such knowledge for both gene-positive and gene-negative children on family relationships is as yet unknown.

This raises the larger question of whether short-term medical benefit ought to be the prime determinant of the advisability of testing children. More controversial issues arise when considering the testing of children for genes which predispose only to adult-onset cancer, such as breast or ovarian cancer (Harper and Clarke, 1990). Many argue that such testing is inadvisable since there are typically no medical screening measures for children who are gene carriers and, hence no immediate medical benefit (Clinical Genetics Society, 1994). Incorporation of a positive test result in the child's medical record could result in later insurance or employment discrimination (Strong, 1993).

Additionally, the impact on family equilibrium of genetic testing in childhood, with potential differential findings for siblings leading potentially to differential parental response to the children, is unknown. Parents' decision to test their children takes away the child's right to decide as an adult about genetic testing. These factors strongly suggest to many that testing for adult-onset malignancies ought to be postponed for minors until they can legally act independently. Others disagree with these cautions and suggest that there are potential medical and psychological benefits of testing for adult onset-disorders in at least some at-risk children (Fryer, 1995). They believe that broad prohibitions against the testing of children result in a kind of paternalistic neglect of children through the inhibition of research which offers potential, although currently undefined, benefit. As more genes are identified for lung and other cancers, arguments are likely to be advanced that early preventive interventions, either biological or behavioural, could be aided by genetic identification of teenagers or children at increased genetic risk for adult-onset cancers. Some researchers believe that the testing of children in high-risk cancer families could allow for targeted dietary interventions which could help clarify the gene–diet interaction in these diseases and could reduce the likelihood that children at increased risk would go on to develop cancers (Pollock and Knudsen, 1996). Proponents of testing in childhood also worry that, if genetic testing of minors is discouraged, no one will later accurately inform young adults from high-risk families of their increased risk or the possibility of testing (Pelias, 1991).

One question which remains controversial is whether reduction of a parent's anxiety about a child's genetic status by itself justifies testing when there is no clear or immediate medical benefit. This question is particularly controversial with regard to pre-natal diagnosis. The question arises whether pre-natal genetic diagnosis for adult-onset disorders, like breast cancer, is unethical and the start of a slippery slope towards eugenics. Or is it a blessing, a technique which could free some individuals who have grown up in families heavily burdened with cancer from the family's genetic 'curse'? (Andrews et al., 1994).

A related question concerns the rights of a child who experiences uncertainty about his/her genetic status as particularly distressing and who wishes to be tested prior to age 18. Many of the adults from high-risk families coming forward for genetic testing report that their worries about getting cancer developed early in childhood as they became aware of the illness and the deaths around them as being attributable to cancer. Should a child have to wait until age 18 to learn if his/her fears about cancer risk are justified ?

5. Risks of genetic testing of children for cancer susceptibility

The generic risks of genetic testing have been described or estimated by many authors (Holtzman, 1996; Hubbard and Lewontin, 1996). The possibility of being rendered permanently uninsurable before adulthood, when individuals typically purchase health or life insurance, is particularly worrisome. Such might be the case for a child whose insurer became aware that he/she had been found by genetic analysis to carry a highly increased lifetime risk of developing cancer. As insurers become more sophisticated, they are more likely to ask either about genetic testing of the individual or about the prior testing of any family members. Employers or insurers could even require testing for individuals who have a family history suggestive of inherited cancer susceptibility (Hudson et al., 1995). Tests for cancer genes are likely to receive more attention than genetic tests for less common diseases, so insurers may be particularly likely to increase premiums or exclude individuals with known genetic susceptibility to cancer. Employment discrimination, especially exclusion from jobs where occupational exposure is possible, is particularly likely when the inherited susceptibility is to cancer.

Finally and perhaps most importantly, what are the unknown risks of awareness of genetic susceptibility on family dynamics and on the development of self-esteem in children found to be or not to be carriers of genetic alterations? What is the relative impact of living with a fear of developing cancer based on a family history replete with cases of cancer versus individual identification of which family members do and do not carry altered genes? The answers are not clear and it is not even clear by which parameters the empirical study of these issues could be carried out. There is considerable hysteria in the discussion about genetic testing of children for cancer susceptibility. To answer these questions, empirical research will be needed (Michie, 1996), but other questions must first be answered.

6. Will parents want to test children?

Do professionals and parents share similar views on the advisablilty and pitfalls of testing? Current research suggests that parents may have high levels of interest in testing children for genetic susceptibility. Wertz (unpublished) found that 530 of the 1000 US adults she questioned believed that children should be tested for conditions where there was neither cure nor treatment. In families where one child has been diagnosed with cancer, parents are almost universally concerned about the cancer risk to their other currently healthy children. Patenaude et al. (1996a,b) found that, if testing were available, and even if no medical benefit were predicted, about half of the mothers of paediatric oncology patients interviewed would want to test their healthy children to determine if they carry a cancer susceptibility gene. If there were to be potential medical benefit to participants, 92% of the mothers reported that they would test their healthy children.

Several recent reports of on-going genetic testing of adults reveal lower than expected rates of acceptance of testing when offered (Craufurd et al., 1989; Lerman et al., 1996; Patenaude et al., 1996b). These reports suggest that estimates of interest in testing made prior to the actual availability of testing may over-estimate the level of interest, at least when respondents are asked about their interest in testing for themselves. The question of whether the hesitancy shown by at least 50% of those adults offered genetic testing for HD (Quaid and Morris, 1993), *p53*, and *BRCA1* would also affect parents and lead to lower than estimated rates of testing children can only be answered as the testing of children becomes an acceptable option, at least under research conditions. The question of whether surrogate decision-making is significantly different than deciding for oneself is an important one in policy decisions about the genetic testing of children (Buchanan and Brock, 1989) and is an area where further research is needed.

7. The impact of personal beliefs on the utilization of genetic testing

The ultimate utilization of genetic testing depends not only on the accuracy of the predictive information provided but also on the skill of the translators of the information to those at risk. Physicians, genetic counsellors, nurses, etc., who convey genetic test results, are charged with teaching individuals seeking testing (or their parents) the meaning and limitations of the results and the value, if any, of altering health behaviours on the basis of genetic status. Conveying the meaning and implications of a genetic finding necessitates clear, careful communication of often complex scientific material. Information about the predictive limitations of the information often strains the understanding of risk and probability held by most members of the general public (Parsons and Atkinson, 1992).

As the John Irving quotation at the beginning of this chapter illustrates, personal and family beliefs about events with mysterious aetiology are often given more weight and power than medical explanations. A significant and largely unexplored factor in communication to at-risk individuals and their family members about aetiology, risk, and prevention is the personal belief system into which the genetic information is incorporated. Richards and co-workers have illustrated the impact of kinship beliefs on individuals' understanding of genetics (Richards, 1996). Beliefs about the aetiology of a child's cancer are likely to be of particular importance in the parent's assimilation of genetic information.

A parent's ability to hold both the medical explanations they receive from their child's doctor and their own personal aetiological beliefs is illustrated by research conducted on 47 mothers of paediatric oncology patients (Patenaude, 1996a). The majority of mothers recalled being told that the origins of childhood cancers are largely unknown, with only a small percentage of cases thought to be genetic in origin. This mirrors the prevailing explanation offered by paediatric oncologists. However, when mothers of paediatric oncology patients were asked about their personal beliefs about the origins of their children's cancer, mothers easily volunteered a wealth of deeply held convictions (see *Table 1*). These beliefs reflected incorrect genetic and immunological understandings, guilt for personal or spouse behaviours, connections drawn between the child's past illnesses and experiences and the development of cancer, and a wide variety of links to possible environmental factors. The depth of the beliefs was illustrated not only by the language

Table 1. Examples of mothers' personal theories of cancer aetiology

---

Environmental

'I questioned the microwave. I got rid of it'
'The house across the street was a landfill, then a junk yard'
'They dumped carpet cleaner. I wondered if it got into the ground water'
'Is there benzene in the walls?'
'I question what's in the lake. It's a mill town'

Genetic

'I think it goes back genetically to the chromosomes. The predisposition
   virus attacked her body and activated it'
'I thought cancer is associated with NF'
'Genetic thing. Something in the family'
'My dad served in Africa. Maybe there, his genes mutated'

Child's Prior Illness or Injury

'I knew it had something to do with mono two years ago'
'She had X-rays at birth'
'She fell in the tub and got a concussion'
'She was attacked by a cat. Did the adrenaline cause it?'

Parental Past Behaviour

'I smoked during pregnancy'
'My husband chews tobacco. Is that related?'
'I worry about everything I ever did – smoking pot in college...'
'Carnal guilt. Something I did while pregnant?'

Parental Illness or Injury

'I had chicken pox when I was 7 months pregnant and wondered if it caused it'
'I had a bad delivery, I got a haematoma, and my AFP was positive when
   they checked for Down's Syndrome'
'When I was pregnant I had a skin cancer on the back of my leg'
'I thought maybe when I was pregnant it was something that happened with
   her'

Psychological Stress

'A lot of stress in the household'
'I had problems. I was very pregnant and stressed and so upset. Did it cause
   cancer?'
'She had an affair with a boy, but the relationship ended. Emotionally, she
   was weakened.'
'I read psychological trauma can lower the immune systems of everyone. We
   had psychological trauma 2 years before and we moved to Boston 7
   months before her diagnosis. I'm not convinced that wasn't part of it. I'm
   haunted by it'

---

used by the parents to describe their beliefs, but also by the fact that they reported making personal and family decisions on the basis of the beliefs. If they attributed the child's illness to emissions of a microwave oven or to the use of electric blankets, then they removed those articles from their environment. Old houses with lead or new houses with radon were sold in some cases and, in others, parents felt guilty for not leaving what they thought to be harmful environments for their children. Bacteria in the backyard pool, chicken pox during pregnancy, steroids a husband took in college, or mothers' own sexual adventures all carried different implications for blame and different soil for the planting of the genetic interpretation. Information about a genetic alteration might relieve the burden from some parents. The genetic information might allow parents tò put less blame on themselves or their spouse. But it is also possible that the finding of a genetic alteration in a child might increase some parents' guilt about past behaviors they connected to their child's disease. In turn, the parent's willingness to follow screening recommendations resulting from identification of a genetic mutation or the lack of a mutation may vary considerably depending on the personal meanings the information takes on. Discussion of personal beliefs regarding cancer aetiology and prevention should routinely be included in discussion with parents of children considered for cancer susceptibility testing. Such discussions should include not only provision of medical risks, benefits, and limitations to the parents, but should also include a discussion of the personal meanings which parents have made of the information provided.

8. Professional guidelines and a need for public education

The ethical, legal, and social complexities of genetic testing of children for cancer susceptibility creates a need for specialized professional education and the setting of professional standards. Certification of those offering genetic counselling and testing of children for cancer susceptibility genes involves ascertainment that those offering testing appreciate and communicate the ethical concerns about confidentiality which are especially important in this area. It is also crucial that those offering counselling regarding the testing of children understand the complex psychological and legal issues involving surrogate decision-making and the growth of competence. They should take into consideration both medical risk, benefit, and limitations of testing, as well as the psychological and social factors involved. Multidisciplinary professional groups should be involved in the setting of such standards.

Public education must also be offered to advance lay understanding of the new areas of medical genetics and to aid parents in thinking about whether testing of themselves and/or their children is advisable. Cancer genetics is a topic of considerable social interest and updating public understanding in this rapidly changing arena will be particularly challenging. Recognition of the complexity of decision-making around testing and the individual, familial, and social impacts will help establish the proper role of genetic information in the puzzle of cancer aetiology and will aid in maximizing the utilization of information gained by genetic studies.

9. Summary

The question remains, how will these controversial issues be decided – on a case-by-case basis by professionals close to the circumstance but often with limited awareness of the risks or limitations of genetic testing, or through the setting of guidelines or recommendations by professional organizations? Ideally a means will be found to involve some combination of professional recommendation and individual evaluation. This would allow for review of the complex ethical, developmental, and societal issues, but would also allow for flexibility based on an individualized assessment of the child's and the family's needs, strengths and vulnerabilities. The coming decade is likely to be one of controversy and experimentation as these topics are addressed.

In conclusion, children with cancer, their siblings, and the offspring of individuals who are carriers of cancer susceptibility genes are all potentially subjects for genetic cancer susceptibility testing. More data are needed to ascertain whether the risks outweigh the benefits of such testing and who should control the advisability of testing. Improved empirical data regarding testing by professional and consumer groups would help determine whether more or less paternalistic approaches to the genetic testing of children for cancer susceptibility are preferable.

Acknowledgement

I would like to acknowledge the sensitive, persistent work of Dr. Laura Basili who conducted the parental interviews quoted in Table 1.

References

Andrews, L.B., Fullarton, J.E., Holtzman, N.A. and Motulsky, A.G. (1994) *Assessing Genetic Risks: Implications for Health and Social Policy.* National Academy Press, Washington, D.C, pp. 167–170.

American Medical Association Council on Ethical and Judicial Affairs (1995) *Testing children for genetic status.* Code of Medical Ethics, report 66.

American Society of Human Genetics Board of Directors, the American College of Medical Genetics (1995) ASHG/ASMG Report: Points to consider: ethical, legal, and psychosocial implications of genetic testing in children and adolescents. *Am J. Hum. Genet.* 57: 1233–1241.

Buchanan, A.E. and Brock, D.W. (1989) *Deciding for Others: The Ethics of Surrogate Decision-making.* Cambridge University Press, Cambridge.

Challem, J.J. (1996) Risk factors for lung cancer and for intervention effects in CARET, the Beta-Carotene and Retinol Efficacy trial. *J. Nat. Cancer Inst.* 89: 325–326.

Clarke, A. and Flinter, F. (1996) The genetic testing of children: a clinical perspective. In: *The Troubled Helix: Social and Psychological Implications of the New Human Genetics* (eds T. Marteau and M. Richards) Cambridge University Press, Cambridge, pp. 164–176.

Clinical Genetics Society (1994) *The Genetic Testing of Children* Report of a working party of the Clinical Genetics Society. Clinical Genetics Society, Birmingham.

Codori, A.M., Petersen, G.M., Boyd, P.A., Brandt, J. and Giardiello, F.M. (1996) Genetic testing for cancer in children: Short-term psychological effect. *Arch. Pediatr. Adol. Med.* 150: 1131–1138.

Craufurd, D., Dodge, A. and Kerzin-Storrar, L., (1989) Uptake of presymptomatic testing for Huntington's disease. *Lancet* 2: 603–605.

Desjardins, L., Haye, C., Schlienger P., Laurent, M., Zucker, J.M and Bouguila H. (1991) Second non-ocular tumours in survivors of bilateral retinoblastoma. *Opthal. Paediatr. Genet.* 12:145–148.

Doll, R. and Peto, R. (1981) *The Causes of Cancer: Quantitative Estimates of Avoidable Risks of Cancer in the United States Today.* Oxford University Press, Oxford.

Eng, C., Li, F.P., Abramson, D.H., Ellsworth, R.M., Wong, F.L., Goldman, M.B., Seddon, J., Tarbell, N. and Boice, J.D. (1993) Mortality from second-tumors among long-term survivors of retinoblastoma. *J. Natl Cancer Inst.* 85: 1121–1128.

Fryer, A. (1995) Genetic testing of children. *Arch. Dis. Childhood* 73: 97–99.

Giardiello, F.M., Brensinger, J.D., Petersen, G.M., Luce, M.C., Hylind, L.M., Bacon, J.A., Booker, S.V., Parker, R.D. and Hamilton S.R. (1997) The use and interpretation of commercial APC gene testing for familial adenomatous polyposis. *N. Engl. J. Med.* 336: 823–827.

Harty, L.C., Guinee Jr. D.G., Travis, W.D. *et al.*, (1996) p53 mutations and occupational exposures in a surgical series of lung cancers. *Cancer Epidem., Biomarkers Prevention* 5: 997–1003.

Harper, P.S. and Clarke, A. (1990) Should we test children for 'adult' genetic diseases. *Lancet* 335: 1205–1206.

Hollen, P.J. and Hobbie, W.L. (1996) Decision-making and risk behaviors of cancer-surviving adolescents and their peers. *J. Pediatr. Oncol. Nursing* 13: 121–134.

Holtzman, N.A. (1996) Are we ready to screen for inherited susceptibility to cancer? *Oncology* 10: 57–64.

Hubbard, R. and Lewontin, R.C. (1996) Pitfalls of genetic testing. *N. Engl. J. Med.* 334: 1192–1194.

Hudson, K., Rothenberg, K. and Andrews, L. (1995) Genetic discrimination and health insurance: an urgent need for reform. *Science* 270: 391–393.

Hunter, D.J. and Willett, W.C. (1996) Nutrition and breast cancer. *Cancer Causes and Control* 7: 56–68.

Huntington's Disease Collaborative Research Group (1993) A novel gene containing a trinucleotide repeat that is expanded and unstable on Huntington's disease chromosomes. *Cell* 72: 971–983.

Kash, K.M, Holland, J.C., Halper M.S. and Miller, D.G. (1992) Psychosocial distress and surveillance behaviors of women with a family history of breast cancer. *J. Natl. Cancer Inst.* 84: 24–30.

Last, B.F. (1992) The phenomenon of double protection. In: *Developments in Pediatric Psychosocial Oncology* (eds. B.F. Last and A.M. van Veldhuizen) Swets and Zeitlinger, Amsterdam, pp. 39–51.

Lerman, C., Narod, S., Schulman, K. *et al.* (1996) *BRCA1* Testing in families with hereditary breast-ovarian cancer: A prospective study of patient decision-making and outcomes. *JAMA* 275: 1885–1892.

Li, F.P., Correa, P. and Fraumeni Jr. J.F. (1991) Testing for germ line p53 mutations in cancer families. *Cancer Epid. Biomarkers Prev.* 1: 91–94.

Marteau, T., Michie, S., Smith, D., and Bobrow, M. (1996) Genetic testing: The example of familial adenomatous polyposis. Abstract presented at the *28th Meeting of the European Society of Human Genetics*, London, 1996.

Meinert, R. and Michaelis, J. (1996) Meta-analyses of studies on the association between electromagnetic fields and childhood cancer. *Radiat. Environ. Biophys.* 35: 11–18.

Michie, S. (1996) Predictive genetic testing in children: paternalism or empiricism. In: *The Troubled Helix: Social and Psychological Implications of the New Human Genetics* (eds T. Marteau and M. Richards) Cambridge University Press, Cambridge, pp. 177–183.

Olopade, O.I. (1996) Genetics in clinical cancer care. *N. Engl. J. Med.* 335: 1455–1456.

Parsons, E.P. and Atkinson, P. (1992) Lay constructions of genetic risk. *Sociology of Health and Illness* 14: 437–455.

Patenaude, A.F., Basili, L., Fairclough, D.L. and Li, F.P. (1996a) Attitudes of 47 mothers of pediatric oncology patients towards genetic testing for cancer predisposition. *J. Clin. Oncol.* 14: 415–421.

Patenaude, A.F., Schneider, K.A, Kieffer, S.A., Calzone, K.A., Stopfer, J.E., Basili, L.A., Weber, B.L., and Garber, J.E. (1996b) Acceptance of invitations for p53 and BRCA1 predisposition testing: Factors influencing potential utilization of cancer genetic testing. *Psycho-Oncol.* 5: 241–250.

Pelias, M.Z. (1991) Duty to disclose in medical genetics: a legal perspective. *Am. J. Med. Genet.* 39: 347–354.

Petersen, G.M. (1994) Knowledge of the adenomatous polyposis coli gene and its clinical application. *Ann. Med.* 26: 205–208.

Petersen, G.M. and Boyd, P.A.(1995) Gene tests and counseling for colorectal cancer risk: lessons from familial polyposis. *Nat. Cancer Inst. Monogr.* 17: 67–71.

Pollock, B.H. and Knudson Jr, A.G. (1996) Preventing Cancer in Adulthood: Advice for the pediatrician. In: *Principles and Practice of Pediatric Oncology*, Third Edition. Chap. 59, (eds P.A Pizzo and D.G. Poplack). Lippincott-Raven, Philadelphia, PA.

Quaid, K.A. and Morris, M. (1993) Reluctance to undergo predictive testing: The case of Huntington's disease. *Am. J. Med. Genet.*, 45: 41–45.

Rait, D.S., Ostroff, J.S. and Smith, K. (1992) Lives in a balance: Perceived family functioning and the psychological adjustment of adolescent cancer survivors. *Fam. Proc.* 31: 383–397.

Richards, M. (1996) Lay and professional knowledge of genetics and inheritance. *Public Understand. Sci.* 5: 1–14.

Rhodes M. and Bradburn, D.M. (1992) Overview of screening and management of familial adenomatous polyposis. *Gut* 33:125–131.

Schimke, R. (1992) Cancer in families. In: *The Genetic Basis of Common Diseases* (ed R. King) Oxford University Press, New York, NY.

Strong, L.C. (1993) Genetic implications for long-term survivors of childhood cancer. *Cancer Suppl.* 71: 3435–3440.

Valanis, B.G. (1996) Epidemiology of lung cancer: a worldwide epidemic. *Sem. Oncol. Nursing* 12: 251–259.

Vogelstein, B. and Kinzler, K.W. (1993) The multistep nature of cancer. *Trends Genet.* 9: 138–141.

Wertz, D.C., Fanos, J.H. and Reilly, P.R. (1994) Genetic testing for children and adolescents. *J. Am. Med. Assoc.* 272: 875–881.

Williams W.R. and Strong, L.C. (1985). Genetic epidemiology of soft tissue sarcomas in children. In: *Familial Cancer* (eds H. Mueller and W. Weber). Karger, Basel.

14

The genetic testing of children: adult attitudes and children's understanding

Martin Richards

1. Introduction

In recent years there has been growing discussion about the genetic testing of children. There have been reviews of ethical (Chapple *et al.*, 1996; Wertz *et al.*, 1994), legal (McLean, 1995) and research issues (Michie and Marteau, 1996). In this chapter, I will briefly introduce the issues that have dominated this debate and then discuss two questions. The first concerns children's understanding of inheritance and genetics. If children are to be involved in decisions about genetic testing and are to understand the implications of genetic tests, they need to understand at least the basic principles of inheritance. A child's conceptualization of the process may be very different from that of an adult, a point with obvious implications for anyone who may discuss these issues with a child. The second question concerns public attitudes towards the testing of children for adult-onset genetic diseases. While there has been much speculation about attitudes, I am unaware of any previous research-based information about public attitudes.

Before discussing these two questions, the general issues of the genetic testing of children will be introduced.

2. Diagnostic genetic testing

Diagnostic testing is the least controversial of the areas of genetic testing in childhood. For disorders that develop in childhood, genetic testing may be employed as one of the diagnostic procedures used to ensure that those affected receive appropriate care and treatment. A good example is familial adenomatous polyposis (FAP). This is a dominantly inherited condition in which polyps may begin to appear in the colon in late childhood and may in time become malignant. However, their presence can be detected by colonoscopy and they can be removed surgically. Children of parents with the condition have a 50:50 chance of inheriting the gene mutation linked to polyp formation. Genetic testing can be used to identify those children who carry the mutation so that appropriate screening of their colons can be started, while those in the family who are found not to have the mutation can be spared these unpleasant and invasive procedures.

With a condition like FAP, which produces symptoms in childhood and where effective treatment procedures are available, the issues in genetic testing are not, in principle, different from those related to any kind of medical treatment. This will involve questions of consent (Alderson, 1990, 1992; King and Cross, 1989) and understanding, as discussed further below. However, if effective treatment is not available for the condition, the matter becomes much more complex. An example of such a programme is the screening of newborns in Wales for Duchenne's muscular dystrophy. Newborn boys are screened with blood spot enzyme test which is then confirmed with a DNA test (Bradley et al., 1993). The earlier diagnosis that this screening programme may make possible does not have any direct benefits in terms of treatment for an affected child. The programme is justified on the grounds that it can give parents information which they can use to avoid births of further affected sons and it may prevent the sometimes long drawn out and potentially stressful traditional process of recognition and diagnosis (Parsons and Bradley, 1994). Of the nine families reported who have had sons diagnosed on the Welsh programme, eight have been generally in favour of the programme, but one has been reported to be traumatised and generally hostile (Bowman, 1993), wishing that their son had not been tested and feeling that they and their son have been robbed of the normal childhood that they would have experienced until the time symptoms of the condition would have become apparent.

3. Predictive testing

Professional opinion among clinical geneticists has been against the genetic testing of children for adult-onset conditions (Clarke et al., 1994) and lawyers have questioned its legality (McLean, 1995). It is argued that such testing has no benefit for the individual during childhood and denies their autonomy and their chance of making their own choices as an adult. Furthermore, the identification of those carrying gene mutations could lead to discrimination within the family and more widely, and may lead to difficulties in such things as obtaining insurance (Harper, 1993; Nys, 1993). An opposing view has come from some parents and patient groups who have argued that parents have a right to knowledge of the genome of their children. Requests for such testing have come from parents who are at risk of having children with Huntington's disease (Bloch and Hayden, 1990), arguing that they would rather know than have to live with the uncertainty of not knowing if their children had inherited the disorder. It has also been suggested that parents may want to use testing to ensure that they have at least some children who are free of the disease. Given that the great majority of adults at risk choose not to be tested, it would seem very hard to justify parents' requests to have their children tested. According to published accounts, such requests have been denied. However, even if the present professional position is maintained, it may be very difficult to police testing as it becomes commercially available through the post. The code of practice for genetic testing offered commercially direct to the public (Advisory Committee on Genetic Testing, 1997) suggests that testing should not be offered to those under the age of 16 and that persons under the age of 16 should not be tested presymptomatically for adult-onset conditions for which there are no clinical treatments. However, it is not clear how a commercial company would determine whether a sample had in fact come from someone aged 16 or over.

Somewhat different issues are raised by adolescents who actively seek predictive testing. Here, in the case of Huntington's disease, clinicians have argued for a case-by-case consideration of requests by minors, involving the concept of Gillick competence (Binedell *et al.*, 1998) (see Chapter 11). Assuming that a young person makes the request without pressure from others and fully understands the consequences and implications of testing, it is very hard to see on what grounds such a request from a Gillick-competent minor could be denied. In this context it is relevant to note the evidence that there is no difference between the competence to consent of 14-year-olds and 18- and 21-year-olds, according to an American study (Weithorn and Campbell, 1982).

4. Carrier detection in recessively-inherited disorders

Requests for carrier detection tests for young children often come when a sibling is diagnosed with a recessively-inherited disorder or when it is known that both parents are carriers. Indeed, some paediatricians and other clinicians have felt it to be appropriate to carry out such tests routinely on all family members when a diagnosis such as that of cystic fibrosis is made. Arguments for and against carrier testing are broadly the same as those made about predictive testing for adult conditions, but, additionally, it is sometimes suggested that an early carrier test is helpful to children as they can have the implications explained progressively and as appropriate through childhood, so that they are well prepared before they need to make any choices about partners or reproduction.

However, I am unaware of any evidence to support this view and clearly it denies the child the possibility of making his or her own decision about whether or not to be tested in adulthood. In some communities, the uptake of testing for carrier status by young people is relatively high (at least, a great deal higher than predictive testing for, say, Huntington's disease), so this argument against testing may not be as powerful as it is in the case of predictive testing for adult-onset disease. Nevertheless, carrier testing of a child at an age when that child is unable to give fully informed consent denies that child autonomy and removes the possibility of choices at a later age.

5. Children's understanding of inheritance

It is widely accepted that clinicians have a duty to maximize children's involvement in discussions related to their health care, so that children together with their parents, may be progressively involved in decision making as they mature (King and Cross, 1989). To be involved, a child must understand at least the fundamentals of what the procedures involve and their likely consequences. However, while there are many discussions of children's abilities to consent to medical interventions, including genetic testing, and to research (Alderson, 1995; Binedell *et al.*, 1998; Fanos and Reilly, 1994; King and Cross, 1989; Morrow and Richards, 1996), I have yet to find a discussion which considers the specific question of children's understanding of inheritance. Given the difficulties many adults have in understanding Mendelian genetics (Richards, 1996a), what is known of children's understanding?

There is a second aspect of consent which may raise specific and particular issues in the case of genetic testing. This concerns implications of a test result for other family members, with the attendant implications for family relations, confidentiality issues and other matters related to family dynamics. Family implications of genetic testing, and genetic disease more generally, are discussed elsewhere (Richards, 1996b, 1998). Here, the focus will be on children and their understanding of inheritance and family relations.

There has been a long tradition of research which has set out to assess children's understanding of genetics against the standard of the Mendelian genetics that is taught in the school curriculum, presented in popular accounts of science or in genetic clinics. This finds children's knowledge, and that of young adults, to be 'deficient' (Clough and Wood-Robinson, 1985; Kargbo *et al.*, 1980; Lewis *et al.*, 1997; Longden, 1982; Wood-Robinson, 1994; Wood-Robinson *et al.*, 1996). These studies show that many children and young people (like many adults) do not have a concept of a genetic material that is passed between generations and is independent of phenotypic characters, and that often their model is one of a mixing or blending of parental characteristics. There is widespread adherence to a Lamarkian model in which characteristics acquired in a lifetime may be passed on to children, and to beliefs that the inheritance from one or other parent may be particularly powerful (or 'dominant') and account for a greater proportion of a child's endowment. Such ideas may be expressed as a child 'taking after' one parent or family member.

I have argued elsewhere that there is a widespread lay knowledge of, and indeed interest in, inheritance which is learnt in the context of the family and departs in important respects from the scientific Mendelian account (Richards, 1996a, b; Richards and Ponder, 1996a). I have suggested that this pre-existing lay knowledge which conflicts with the Mendelian account may inhibit the assimilations of the latter in schools, genetic clinics and in public education. I have further argued that the lay knowledge is derived from and grounded in social concepts of kinship which are themselves sustained by everyday social activities and relationships which may make them particularly resistant to change.

This perspective focuses attention on children's acquisition of concepts of family and kinship and its relationship to their notions of inheritance. As it happens, there is a rich source of largely American research in the area. This work has been driven, not by an interest in genetics and its understanding, but by ideas about cognitive development in children and the possibility of domains of knowledge that are privileged in development, or may even have 'innate' roots. Be that as it may, research in the field has produced a large body of empirical data about children's understanding of inheritance and family which are highly relevant to my present discussion.

Let us first consider family and kinship. Hirschfeld (1989, 1994) argues that young children (pre-school) conceptualize people as living in highly salient groups, namely families. He suggests, drawing on much earlier work by Jean Piaget, that a young child's notion of family is their attempt to capture the sense of relatedness (and, as we shall see, of resemblance) reflected in co-residence and which sees the family as internally differentiated in terms of roles, gender and relative age (mother, father, sister, brother). The last point is well illustrated by the common difficulty young children have in recognizing that grown-ups may have brothers and sisters. Interestingly, these notions of family seem to result not simply from what Piaget (1928) called 'direct or spontaneous experience' and 'observations of the external world' but from the child's interpretation of the verbal concept used by adults. While the more anthropologically inclined research has stressed relatedness as being a salient feature of children's notion of family, other research shows that the quality of relationship is very important to children. Gilby and Pederson (1982) used a series of questions about living arrangements, relatedness and quality of relationship to assess children's concepts of family. These researchers individually presented children with cardboard figures and then asked questions such as: 'Here are Mr and Mrs Brown. They are married. They live together. They have no children. Are they a family?' Groups of six-, eight- and ten-year-old children and university students were questioned (*Table 1*). Co-residence proved to be very important, as was contact between people and the quality of their relationship, but in the older age groups there was an increasing emphasis on blood ties and legal relationship – essentially, the older children approach adult concepts. These results are paralleled in other studies using different methodologies (Morrow, 1998).

There is abundant evidence that young children expect that children will resemble their parents. They expect blue-eyed parents to have blue-eyed children, dogs to have puppies, not kittens and so on (Carey and Spelke, 1994). Springer (1992), for example, told four- to eight-year-olds that a pictured animal had an unusual property (e.g. this horse has hair inside its ears) and then asked about a physically similar horse who is either a friend of the first horse or its baby. At all ages children were more likely to say the character was present in the baby and not the friend. This, together with other similar experiments with both human and animal examples, establishes that family resemblance and the inheritance of a physical character are linked by young children.

But do these young children have a model of biological inheritance? A number of studies suggest not. Solomon and his coworkers (Solomon *et al.*, 1996) carried out a series of experiments which set out to test the idea. In one of their experiments, four- to seven-year-old children were told a story in which a boy was born to one man and adopted by another. The biological father was described as having one set of features (e.g. green eyes) and the adoptive father another (e.g. brown eyes). The

Table 1. Percentages stating that each group was a family (adapted from Gilby and Pederson, 1982)

Group	Age group			
	6 years	8 years	10 years	Students
Mr and Mrs Brown	45	45	70	85
Mr and Mrs Brown and their son, Billy	90	95	100	100
Mrs Brown and Billy	55	40	80	95
Billy's father lives in a different house. Is he Billy's family?	5	35	50	85
Billy's grandparents	60	55	85	100
Billy's grandparents. What if they live in another city and he never sees them?	0	35	85	85
Mr and Mrs Brown and Billy, plus Miss Jones who helps take care of Billy. Is Miss Jones in Billy's family?	90	20	15	40
Mr and Mrs Brown and Billy, plus Billy's friend Joe. Is Joe in Billy's Family?	70	20	20	45
Mr and Mrs Brown and Billy. They all live together but they don't love each other. Are they a family?	15	55	55	70
Here are two very good friends, Miss Black and Miss Smith. They live together. Are they a family?	70	25	5	25

children were asked who the adoptive child would resemble when they grew up. (The experiment was also carried out with the example of mothers.) The young children showed little understanding of the processes which link the features to the parent. It was not until age seven that a substantial proportion of the children associated the boy with his birth parents' physical features and the adoptive parent with his beliefs. Young children seem to understand that the birth mother determines the kind, or species, of animal – cats have kittens and dogs have puppies – but not that individual features are derived from a parent (Johnson and Solomon, 1997). So young children may well know where babies come from, but this does not mean that they understand inheritance, as the following conversation between the investigator (I) in an American study and a preschool child (C) serves to illustrate:

I   'How did the baby happen to be in your mommy's tummy?'
C   'It just grows inside.'
I   'How did it get there?'
C   'It's there all the time. Mommy doesn't have to do anything. She waits until she feels it.'

I 'But you said the baby wasn't in there when you were there.'
C 'Yeah, then he was in the other place. In . . . in America.'
I 'In America?'
C 'Yeah, in somebody else's tummy.'
I 'In somebody else's tummy?'
C 'Yeah, and then he went through somebody's vagina, then he went in, um, in my Mommy's tummy.'
I 'In whose tummy was he before?'
C 'Um. I don't know, who his, her name is. It's a her.'

(Bernstein and Cowan, 1975)

So young children may understand that babies are created inside mothers (and something about gender) but it may take them much longer to construct an explanatory theory of biological origins. So while knowledge of the facts of birth may be one of the cornerstones on which children construct a theory of biology (Springer, 1994), it does not seem to be until about the age of seven that children put together information about birth, family relationships and the potential contribution of both parents to the development of individual characteristics. But, typically, the way these are put together does not accord with a scientific Mendelian account of inheritance. For many individuals, the childhood notion is more akin to Darwin's idea of blended inheritance (also involving the transmission of acquired characteristics) rather than a post-1900 scientific conception based on an equal contribution of genetic material from the two parents and segregating Mendelian characters (Richards, 1996a).

What are the implications of the widespread and non-Mendelian lay concept of inheritance and the manner in which this develops through early childhood for the genetic testing of children, and for genetic medicine more generally? First, to repeat the basic point that developmental psychologists have long insisted on – children cannot be regarded as either incomplete or ignorant adults, they are quite simply different. They see and conceptualize the world in ways that are unlike those of adults. As with the changing concept of family, children may use the same term as adults, but construe it in a different way. Of course, not all adults use words in the same way. As many observers have noted, terms such as gene, chromosome and DNA are being used in different ways by the lay population and among developmental biologists and geneticists. So when we hear a phrase like 'it's in the genes' it may be that its meaning is a reference to a characteristic that appears to run in a family, rather than any more precise claims about the mode of inheritance. Here, the misleading oversimplifications, with implications of genetic determinism, that are used by some scientists and their popularizers in phrases like 'the breast cancer gene' or the 'gene for aggression' may encourage scientific terminology to be taken up and attached to lay concepts of inheritance (Richards, 1998). What children are doing may be seen in a similar way. Words used by adults are taken up and used by children but as a label for the way a child defines and understands a concept. So to a child a family may be the small group who live in the same household and love each other. Not surprisingly, therefore, when asked to list their family it is common for small children to include pets, as well as unrelated adults who may be part of the household.

Clearly, if you are trying to explain to a child what the implication of a genetic test may be, you cannot assume that a child's apparent familiarity with the terms you are using means that you are discussing the same concepts. One has to explore the child's meaning.

The second important implication of my argument is that we should not assume that the child has no understanding of inheritance. They will have a concept of family resemblances and how these may come about, but this is unlikely to be compatible with the Mendelian account that a genetic counsellor may be trying to explain. As some of those who have worked on children's understanding of inheritance in the classroom have argued (Clough and Wood-Robinson, 1985) and has been more generally suggested in science education (Driver *et al.*, 1994), one should begin with the child's concepts and build on these (Martins and Ogborn, 1997). The approach must be bottom-up. Top-down approaches, which regard children as blank slates on which the scientific explanation is to be drawn in simple terms, are not effective. In the clinic it is very important to be clear what are the points that need to be understood for a child to have a view on what may take place. Let us suppose the issue is a genetic test for a mutation in a family that carries FAP. What is crucial for someone to understand is that they may or may not have inherited the gene mutation associated with the disorder. If they have inherited the gene mutation, they are very likely to develop polyps and so need screening in order for the polyps to be identified and removed to avoid the possibility of cancer. They need to know what the screening entails and what may happen if they are not screened. On the other hand, if they are tested and found not to have the gene mutation, then the growth of polyps is very unlikely and screening is unnecessary. All of this needs to be explained in an age-appropriate way using terminology and concepts that are part of the child's world, or at least are related to the children's concepts. What the child does not require, and what is likely to be very unhelpful, is the kind of description of Mendelian genetics that would be found in a GCSE level textbook.

I have presented evidence that there is a close connection between a child's developing concepts of kinship and family and their notion of inheritance, and I have argued that their knowledge of inheritance grows from their concept of kinship. A test of this hypothesis was carried out using the difference in English kinship between the very close kinship connection between a parent and child and the less close link between siblings. As predicted, respondents believed the genetic connection between a parent and child, or child and parents to be closer than that between siblings (Richards and Ponder, 1996a). It is characteristic of English kinship ties that there is a strong line of descent from grandparent to parent to child and less strong social links and obligations with lateral kin (Craig, 1979). If the idea that lay knowledge of inheritance is derived from kinship knowledge is correct, and in general it is correct, then we should expect that lay knowledge of inheritance will vary across different kinship systems. We would predict that where kinship systems put great emphasis on lateral kin – as in many systems in the Indian sub-continent – there will be larger gaps between Mendelian and lay accounts than would be the case for kinship systems that emphasise descent. Could this be the reason why many of those from the Indian sub-continent ethnic minority communities find it particularly difficult to understand what is said in genetic clinics?

6. What are public attitudes to the genetic testing of children for late-onset genetic disorders?

As I have already mentioned, the predominant professional view is that children should not be given predictive genetic tests for conditions that develop in adulthood. However, some parents take the opposite view that parents have a right to know. What does the public think about this question?

As part of a study of public knowledge of genetics and attitudes toward the deployment of new genetic technologies, we carried out a study of Women's Institute (WI) members and first-year social science university students (Richards and Ponder, 1996b; Rosena and Richards, personal communication). We asked questions about the genetic testing of children. The WI (a national voluntary adult educational organization) held a three-day national residential conference on the New Genetics in 1995, which was sponsored by the Medical Research Council, the Wellcome Trust and the Biotechnology and Biological Science Research Council. We sent a wide-ranging postal questionnaire to participants before and after the conference. Each attender ($N = 64$) was also asked to pass on questionnaires to two members of their local WI branch of a similar age and with a similar number of children. These 'friends' gave us a further 128 respondents. A very similar questionnaire was also completed by 73 first-year university social science students. The average age of the WI women was 55 and the students 19 years. Seventy-five percent of the WI sample were mothers and just over 40% were grandmothers. One of the students was a parent and 82% said that they would like to have children in the future. Compared with population norms, the WI samples were slightly better educated on average. Given their age and background they might well have been mothers of our student sample.

The questionnaire used included a number of statements and respondents were asked whether or not they agreed with them. One of these was 'Parents have a right to ask for their child to be tested for genetic disorders that develop in adulthood'. The responses are shown in *Table 2*. Overall, respondents were more or less equally divided in their opinions with a hint that students are less happy for parents to have a right to ask for their children to be tested than were the WI respondents.

It is interesting to look at these opinions on the testing of children alongside another question about the testing of respondents themselves for late-onset

Table 2. Parents have a right to ask for their child to be tested for genetic disorders that develop in adulthood (values given as percentages)

	Agree completely	Agree somewhat	Disagree somewhat	Disagree completely	No opinion
WI conference $N = 64$	21	31	18	26	3
WI friends $N = 100$	22	38	14	21	5
Students $N = 73$	18	29	30	15	5

Table 3. If tests were available, would you want to take them to discover which inherited diseases you are likely to suffer from later in life? (Values given as percentages.)

	Yes	No	Don't Know
WI conference women N = 64	25	46	29
WI friends N = 128	26	57	17
Students N = 75	19	62	19

disorders. We asked 'If the tests were available, would you want to take them to discover which inherited diseases you are likely to suffer from in later life?' As *Table 3* shows, only a small minority were interested in taking such tests – many fewer than the proportion who thought that parents should have the right to have their children tested. Of course, it is quite consistent to believe that such a right should be generally available but not choose to exercise that right in the case of your own family or yourself. After the conference we asked attenders a similar question about three conditions, Alzheimer's disease, breast cancer and heart disease, all of which had been discussed at the conference (*Table 4*). Here, women were more interested in testing for breast cancer and heart disease. This is probably because these are both disorders for which the women believe they can do something.

Overall, a bare majority of middle aged women and marginally fewer students thought that parents should have a right to ask for their children to be tested for adult-onset genetic disorders, while about a quarter of the sample said that they were interested in taking such tests for themselves. Among the group of women who had attended the conference (where there was a good deal of discussion about genetic testing) the percentage saying they would be interested in a predictive test for Alzheimer's disease remained much the same proportion as for the response to the general question before the conference, but figures for breast cancer and heart disease were much higher (67% in each case).

Our results suggest that a majority of women would grant parents the right to have predictive genetic tests for their children. Of course, we can argue that they may not have been fully aware of the implications of testing children and perhaps

Table 4. If such tests were available, would you be interested in taking them to discover whether you are likely to suffer from the following diseases later in life?[a]

Disease	Yes	No	Don't Know
Alzheimer's disease	25	59	16
Breast cancer	67	20	13
Heart disease	67	21	11

[a] The results were taken from post-conference WI members (N = 73)

saw the issue as more of a general one about the position of parents in relation to their children than a balancing of the pros and cons of genetic testing from both the parents' and the child's perspective.

I suspect that our respondents who support parents' right to have their children tested will have history on their side. While I would argue strongly that children should be protected from unwanted genetic testing and that their autonomy must be protected, I suspect that as the range of tests that are commercially available continues to grow, within a global economy where information is freely available to those with access to the Web, it is going to be increasingly difficult to prevent anyone from having tests performed on any DNA sample, whether it comes from an adult or a child.

References

Advisory Committee on Genetic Testing (1997) *Code of practice and guidance on human genetic testing services supplied direct to the public.* Health Departments of the United Kingdom, London.

Alderson, P. (1990) *Choosing for Children: Parents' Consent to Surgery.* Oxford University Press, Oxford.

Alderson, P. (1995) In the genes or in the stars? Children's competence to consent. *J. Med. Ethics* **18**: 119–124.

Bernstein, A. and Cowan, P. (1975) Children's concepts of how people get babies. *Child Developm.* **46**: 77–91.

Binedell, J., Soldan, J.R., Scourfield, J., Harper, P.S. (1988) Huntington's disease predictive testing: the case for an assessment approach to request from adolescents. *J. Med. Genet.* **33**: 912–918.

Bloch, M. and Hayden, M.R. (1990) Opinion: predictive testing for Huntington's disease in childhood: challenges and implications. *Am. J. Hum. Genet.* **46**: 1–4.

Bowman, J.E. (1993) Screening newborn infants for Duchenne muscular dystrophy. *Br. Med. J.* **306**: 349.

Bradley, D.M., Parsons, E.P. and Clarke, A.J. (1993) Experience with screening newborns for Duchenne muscular dystrophy in Wales. *Br. Med. J.* **306**: 357–360.

Carey, S. and Spelke, E. (1994) Domain-specific knowledge and conceptual change. In: *Mapping the Mind* (eds L.A. Hirschfeld and S.A. Gelman). Cambridge University Press, New York.

Chapple, A., May, C. and Campion, P. (1996) Predictive and carrier testing of children: professional dilemmas for clinical geneticists. *Eur. J. Genet. Soc.* **2**: 28–38.

Clarke, A., Fielding, D., Kerzin-Storrar, L., Middleton-Price, H., Montgomery, J., Payne, H., Simonoff, E. and Tyler, A. (1994) *The Genetic Testing of Children.* Report of a Working Party of the Clinical Genetics Society, London.

Clough, E.E. and Wood-Robinson, C. (1985) Children's understanding of inheritance. *J. Biol. Educat.* **19**: 304–310.

Clough, E.E. and Wood-Robinson, C. (1994) Young people's ideas about inheritance and evolution. *Stud. Sci. Educat.* **24**: 29–47.

Craig, D. (1979) Immortality through kinship: the vertical transmission of substance and symbolic estate. *Am. Anthropol.* **81**: 94–96.

Driver, R., Asoko, H., Leach, J., Mortimer, E. and Scott, P. (1994) Constructing scientific knowledge in the classroom. *Educational Research* 5–12.

Gilby, R.L. and Pederson, D.R. (1982) The development of the child's concept of the family. *Can. J. Behav. Sci.* **14**: 110–121.

Harper, P.S. (1993) Insurance and genetic testing. *Lancet* **341**: 224–227.

Hirschfeld, L.A. (1989) Rethinking the acquisition of kinship terms. *Int. J. Behav. Developm.* **12**: 541–548.

Hirschfeld, L.A. (1994) Is the acquisition of social categories based on domain-specific competence or on knowledge transfer? In: *Mapping the Mind* (eds L.A. Hirschfeld and S.A. Gelman). Cambridge University Press, New York.

Johnson, S.C. and Soloman, G.E.A. (1997) Why dogs have puppies and cats have kittens: the role of birth in young children's understanding of biological origins. *Child Developm.* **68**: 404–419.

Kargbo, B., Hobbs, E.D., and Erickson, G.G. (1980) Children's beliefs about inherited characteristics. *J. Biol. Educat.* **14**: 137–146.

King, N.M.P. and Cross, A.W. (1989) Children as decision makers: Guideline for paediatricians. *J. Pediat.* **115**: 10–16.

Lewis, J., Driver, R., Leach, J. and Wood-Robinson, C. (1997) Understanding of basic genetics and DNA technology. Learning in Science Research Group, University of Leeds. Working Paper 2.

Longden, B. (1982) Genetics – are there inherent learning difficulties? *J. Biol. Educat.* **16**: 135–140.

Martins, I. and Ogborn, J. (1997) Metaphorical reasoning about genetics. *Int. J. Sci. Educat.* **19**: 47–63.

McLean, S.A.M. (1995) Genetic screening of children: the UK position. *J. Contemp. Health Law and Policy* **20**: 113–130.

Michie, S. and Marteau, T.M. (1996) Predictive genetic testing in children: the need for psychological research. *Br. J. Health Psychol.* **1**: 3–14.

Morrow, V. (1998) Unpublished Report. Centre for Family Research, University of Cambridge.

Morrow, V. and Richards, M.P.M. (1996) The ethics of social research with children: an overview. *Children and Society* **10**: 90–105.

Nys, H. (1993) *Predictive Genetic Information and Life Insurance: Legal Aspects – Towards European Community Policy?* Maastricht, Rijksuniversiteit Limburg, Vakgroep Gezondheidsrecht.

Parsons, E. and Bradley, D. (1994) Ethical issues in newborn screening for Duchenne muscular dystrophy. The question of informed consent. In: *Genetic Counselling. Practice and Principles* (ed. A. Clarke) Routledge, London.

Piaget, J. (1928) *Judgement and Reasoning in the Child.* Routledge, London.

Richards, M.P.M. (1996a) Lay and professional knowledge of genetics and inheritance. *Public Understanding of Science* **5**: 217–230.

Richards, M.P.M. (1996b) Families, kinship and genetics. In: *The Troubled Helix: Social and Psychological Implications of the New Human Genetics* (eds T. Marteau and M.P.M. Richards). Cambridge University Press, Cambridge.

Richards, M.P.M. (1998) Annotation: genetic research, family life and clinical practice. *J. Child Psychol. Psychiat.* **39**: 291–306.

Richards, M.P.M. and Ponder, M. (1996a) Lay understanding of genetics a test of an hypothesis. *J. Med. Genet.* **33**: 1032–1036.

Richards, M.P.M. and Ponder, M. (1996b) Public understanding of genetics: a study of adult women and students. I Knowledge of inheritance. II Attitudes towards the new genetic technologies. Unpublished Report, Centre for Family Research, University of Cambridge.

Solomon, G.E.A., Johnson, S.C., Zaitchik, D. and Carey, S. (1996) Like father, like son: young children's understanding of how and why offspring resemble their parents. *Child Developm.* **67**: 151–171.

Springer, K. (1992) Children's awareness of the biological implications of kinship. *Child Developm.* **63**: 950–959.

Springer, K. (1994) Young children's understanding of a biological basis for parent–offspring relations. *Child Developm.* **67**: 2841–2856.

Weithorn, L. and Campbell, S. (1982) The competency of children and adolescents to make informed treatment decisions. *Child Developm.* **53**: 1589–1578.

Wertz, D., Fanos, J. and Reilly, P. (1994) Genetic testing for children and adolescents. Who decides? *J. Am. Med. Assoc.* **272**: 878–881.

Wood-Robinson, C. (1994) Young people's ideas about inheritance and evolution. *Stud. Sci. Edu.* **24**: 29–47.

Wood-Robinson, C., Lewis, J., Driver, R. and Leach, J. (1996) Young people's understanding of 'The New Genetics'. Working Paper No. 1. Centre for Studies in Science and Mathematics Education, University of Leeds.

15

Predictive genetic testing in children: the need for psychological research*

Susan Michie and Theresa M. Marteau

Abstract

Recent advances in molecular genetics have led to the ability to test individuals for an increasing range of genetic diseases and conditions that may not become apparent until later life. There are a variety of views amongst health professionals about the desirability of offering predictive genetic testing to children, as reflected in a variety of clinical practices. Little is known about the views of parents and children on this matter. Similarly, little is known about the ability of children to make informed decisions about such tests, and the psychological consequences of testing for children and their families. This article reviews research that is relevant to these questions and highlights the areas in which studies are urgently needed.

1. Introduction

Recent advances in molecular genetics have led to the ability to test individuals for an increasing range of genetic diseases and conditions that may not become apparent until later life. Tests fall broadly into two categories: the prediction of disease risk for late-onset dominantly inherited conditions, where having the gene leads to the disease in 100% of cases (e.g. Huntington's chorea, myotonic

---

* Reprinted with permission from the *British Journal of Health Psychology* 1996, vol 1, pp. 3–14. Punctuation has been modified and the text has been altered for consistency with the rest of this volume in relation to the use of the terms 'screening' and 'testing'.

---

dystrophy, familial adenomatous polyposis); and testing for risk factors for multi-factorial diseases (e.g. hypertension, diabetes, cancer). These conditions can be further subdivided into those for which treatment is available, and those for which no such treatment is available.

Genetic tests exist for late-onset genetic diseases for which there is no treatment (e.g. myotonic dystrophy), and for which there is no treatment available in child-hood (e.g. autosomal dominant polycystic kidney disease). Such tests are routinely available for adults, but not for those under 16 years of age. Although some parents are asking for their children to be tested (Morris *et al.*, 1989), the response to such requests varies according to the views of the geneticists consulted.

The question of whether children should be offered testing has led to much debate which, at present, is mainly restricted to health professionals. As there have been few studies of children and families undergoing predictive testing, this has tended to be a debate informed more by anecdotal than empirical evidence. This situation raises the questions of who should decide whether tests are offered to children, and on what basis the decision should be made. Debates about the basis for the decision centre around children's ability to make informed decisions, and the psychological consequences for the child and the family of knowing in childhood that as an adult the child will develop a particular disease.

This article reviews studies in three relevant areas: views of health profession-als, parents and children on predictive testing in childhood; children's ability to make informed decisions and the impact of predictive testing in childhood. Studies were identified from a computerized search of Psychinfo and Medline as well as informal sources.

2. Views of users and providers

Amongst health professionals, there is agreement about the value of predictive testing of children where there is clear medical benefit. This is true where symp-toms and complications can be screened for and disease can be prevented or treated, as in familial adenomatous polyposis, a familial form of bowel cancer (Clinical Genetics Society, 1994). Disagreement exists where there is no such medical benefit, and where possible advantages and disadvantages are either psy-chological or social.

The majority of views expressed by geneticists about testing in children herald caution:

'To test for progressive and currently untreatable disorders whose onset is gen-erally in adult life, is clearly unwise'. (Harper and Clarke, 1990);

'Children clearly cannot make an informed decision about whether to partici-pate in predictive testing ... we oppose the testing of children'. (Bloch and Hayden, 1990);

As there have been no formal studies of predictive testing in children, the extent to which these statements are valid is unknown.

An opposite point of view has been put forward by Sharpe (1993). He suggests that the policy of not testing children is the result of geneticists imposing their own values and beliefs about what is best for the child and the family. He goes on to argue that parents are in a better position to weigh the social, familial, emotional and economic factors:

'The issue is whether the decision not to test effectively abrogates what it seeks to protect, the child's personal rights and dignity subordinated to, if not replaced by, the objectives, values and rationality of the geneticist'. (Sharpe, 1993).

We know very little about the views of parents and children. The surveys of parents' attitudes that have been carried out suggest that parents have more favourable attitudes than health professionals. In the Netherlands, 41% of 70 individuals at risk for Huntington's disease thought that children under 18 years of age should be allowed to have the test if they so chose (Tibbens, 1993).

A UK study examined parental attitudes towards testing children for carrier status amongst families carrying inherited translocations. It was found that most parents favoured testing (Barnes, 1994). Of 211 children studied, 75 had not been tested. Only eight children had parents who were against testing and 16 had parents who were undecided about testing.

If future research is to address issues of importance to the consumers of health care, as well as issues seen as important by the providers and purchasers, we need more information about parents' and children's views.

3. Decision making in children

Children's competence to make decisions about medical treatment tends to be discussed in terms of the child's cognitive ability or maturity. The child is considered to be able to provide effective consent if he or she is capable of appreciating the nature, extent, and probable consequences of the proposed treatments or procedures. The question of who defines competence, or on what basis, is usually avoided.

Developmental psychologists have suggested that it is necessary for children to demonstrate formal operational (abstract) thought if they are to appreciate the consequences of the procedures and alternatives, to reason rationally or meaningfully about these alternatives, and to reach a reasonable decision (e.g. D'Zurilla and Goldfried, 1971). Such formal operational thinking has been considered to begin at about the age of 11 years in Western culture and to be developed by 14 years old (Inhelder and Piaget, 1958). However, cognitive development is not synonymous with Piaget's stage model and few psychologists now accept this simplistic approach. The ages at which children reach different levels of cognitive development have been lowered as advances are made in the research methods used to study children (Bryant, 1974; Donaldson, 1978). A wide range of individual differences in children's levels of understanding has also been found (Eiser, 1989).

However, the ability to think abstractly may not be crucial in taking concrete decisions in a social context. The question of informed consent in adults has usually been addressed in terms of how much of the information given has been recalled. Neither abstract thought, nor recall of information, are sufficient cognitive variables in considering competence to make decisions or give informed consent. The questions of how the information is analysed and interpreted, how the information is integrated with values and beliefs, and how decisions are taken, need to be addressed. There is little research about these processes, and their variation, in adults who are assumed to be giving informed consent. This is needed, as well as research about the developmental progression of these cognitive processes.

Areas in which there have been empirical studies of the ability of children to understand and give information, and to make decisions are: coping with chronic illness; consent to medical treatment and participation in research.

3.1 Coping with chronic illness

Until the 1970s, adults attempted to protect children from the knowledge that they were confronting a life-threatening illness, on the assumption that children neither understand nor should be worried about death and illness. Research has shown, however, that these patients are aware of the seriousness of their illnesses and experience a great deal more anxiety than other chronically ill or healthy children (Spinetta, 1974); they sense the distress of their parents. Several authors have commented upon the frightening, guilty fantasies, mistrust, and isolation experienced by sick children when their diagnoses and prognoses are treated as taboo subjects (see Slavin *et al.*, 1982).

Slavin *et al.*, (1982) hypothesized that survivors of childhood cancer who had learned of their diagnosis at an early stage would tend to be better adjusted at follow-up than those who learned they had cancer long after the diagnosis was made. They studied 116 survivors on average 12 years after diagnosis and found that good psychological adjustment was associated with patients' early knowledge of the diagnosis. A high proportion of the survivors, their parents, and siblings felt that the cancer diagnosis should be shared with the child early on. However, the limitation of retrospective studies should be noted. It may be, for example, that families who had the best adjusted children before the diagnosis tended also to use open communication styles both before and during the illness.

3.2 Consent to medical treatment

Few issues affecting the therapeutic professions are as much discussed and as little understood as 'informed consent' (Appelbaum *et al.*, 1987). Informed consent can be seen as having three parts: (a) the extent to which the information presented is understood and retained; (b) the process of decision-making, for example, how the information is processed and what factors influence this and (c) the extent to which the presented information influences the decision.

Competence to give informed consent depends, therefore, on both inner qualities (abilities, memory, confidence) and on outer influences (nature and circumstances of the decision, its salience to the individual, others' expectations and views and available support).

These qualities and influences may not develop evenly and may vary widely across individuals and families. Concern has been expressed that parental pressure upon children may be stronger than pressures upon adults, thus limiting their autonomy, their 'freedom to choose'. Defining and measuring these concepts – informed consent, autonomy, parental coercion – are key to developing empirical work in this area. The literature search upon which this review is based revealed gaps that are in urgent need of being filled.

The social climate of the last three decades has resulted in an increased interest in allowing children to participate in their own health care. The presumption of the child's incapacity to make decisions is weakening and the possibility of children giving valid consent is being entertained (King and Cross, 1989). Increased understanding of the cognitive development of children and of the requirements of informed consent have made it possible to make case-by-case determinations of an individual's capacity to consent to treatment (Gaylin, 1982). In addition, the patients' rights movement has fostered the growing belief that fairness requires that children assume an increased role in decisions that affect them, including health care decisions.

There is no agreed standard of competence for informed consent, even for adults. King and Cross (1989) suggest that four sets of interrelated factors should be considered in assessing capacity: (1) reasoning; (2) understanding; (3) voluntariness and (4) the nature of the decision to be made.

In a study in the United States, Weithorn and Campbell (1982) examined the competence of 96 children and young people (9, 14, 18 and 21 years of age) to consent to treatment. They were administered a test based on four legal standards of consent (evidence of choice, reasonable outcome, rational reasons and understanding). The test consisted of four hypothetical treatment dilemmas and a structured interview. Overall, the 14-year-olds did not differ from the adults, but the 9-year-olds did. The 9-year-olds were less able than adults to understand and reason about the information provided but could express their preferences as well as adults.

A milestone in children's decisions about their use of medical technology and health services came from the British Parliament's House of Lords (Gillick, 1986). This ruling was that, once a girl had sufficient maturity and understanding to weigh the issues, she could seek and be given contraceptive advice by a health professional without securing her parents' permission. Children's consent to a wide range of issues is now expected to follow the 'Gillick principles' by which children under 16 years of age have the right to give or withhold consent, without their parents' knowledge or agreement, if they have sufficient understanding to grasp the implications of the decision they are being asked to make. The questions of who decides what 'sufficient understanding' is, and how this decision is to be reached, are not included.

A central principle running through the 1989 Children Act is that a child's perspective must be given due weight in decisions about their care. The Act states that the wishes and feelings of children must be taken into account according to their level of understanding. According to the Act, children under 16 may refuse or consent to medical treatment 'if the child is of sufficient understanding to make an informed decision'. The criteria for judging sufficient understanding are absent.

Another principle of the Children's Act is that both parents and children should be kept informed about what happens to children and should participate in any decisions about their future. Promoting children's participation does not mean simply handing over responsibility to them. It does, however, imply offering more information, consultation and choice in relation to the child's particular needs. Research has long since confirmed the common-sense belief that children need to exercise choice in order to acquire self-respect and a sense of responsibility (Clarke and Clarke, 1976).

Research has found children to be quite competent to learn decision-making skills and to take control over decision-making about their own health, especially when the scope of decision-making is clearly defined. In an empirical study of decision-making about experimental chemotherapy for childhood cancer, children were considered to be able to provide effective consent if they were capable of appreciating the nature, extent, and probable consequences of the proposed treatments or procedures (Kamps et al., 1987). Amongst relatively young children (8–15 years), as many as 13 out of 120 interviewed were found to be the main decision-maker about proposed surgery (Alderson, 1993). The main resistance to involving children in decision-making comes from doctors, with some parents also feeling threatened by this process (Lewis and Lewis, 1990).

In the USA, most states allow 13–18 year olds to make their own decisions about contraception, preferences in child custody hearings, substance abuse treatment, or other types of psychological and medical treatment (Weithorn, 1983). It has been suggested that the decision to withhold information regarding predictive testing may constitute a breach of the doctrine of informed consent in America and Canada (Sharpe, 1993).

3.3 Children's participation in research

The ethical problems presented by child participants stem from the perception that they are more vulnerable to stress, and from their reduced knowledge and experience compared with adults (Ethical Standards for Research with Children, Society for Research in Child Development). A cardinal principle of conducting research with adults and children states that they must consent to being studied with full knowledge of what their participation entails. Consent is seen as needing three elements: the provision of adequate information; the capacity to understand it and the voluntariness of any decision taken (Medical Research Council (MRC), UK, 1993). The MRC guidelines on the ethical conduct of research in children suggest that whether a person has the capacity to understand the information depends on the ability to comprehend the nature and purpose of any course of action and the short-term and long-term risks and benefits of what is proposed. Situational differences are taken into account: 'An individual may be in a position to consent to take part in some studies, but not others; capacity to consent will depend not only on the nature of the research itself but also on the nature of the explanation'.

One study of young people aged from 7 to 20 years found that age was unrelated to their knowledge about the elements of informed consent to participation in biomedical research (Susman et al., 1992). The role of experience rather than age is stressed by the British Paediatric Association (1992) in considering the ethical conduct of research with children: 'Children's ability to consent develops as they learn to make increasingly complex and serious decisions. Ability may relate to experience rather than to age, and even very young children appear to understand complex issues'.

It has also been found that many children with learning and behaviour problems have the competence necessary for understanding their psychological and educational problems and for effectively participating in treatment decision making. In studies of a group of 10–19 year olds which measured the accuracy of their self-estimates of skill, self-explanations of behaviour, and on-task functioning, a relatively high level of ability to understand, evaluate, communicate and perform, was found (Adelman *et al.*, 1985).

4. Psychological impact of predictive testing

The ability to make an informed decision about genetic testing, while necessary, is not considered a sufficient criterion for determining the availability of such tests (the Nuffield Council on Bioethics, 1993; the Royal College of Physicians, 1989). It is also necessary to establish that undergoing a test confers more benefit than harm. One of the main factors that should influence approaches to the early genetic detection of risks for adult-onset diseases is how children and their families respond to such information. There have been very few studies about the short-term effects of testing children for adult-onset diseases, and none assessing the long-term effects. These few studies have been rather limited in their methods. In the absence of much evidence, several possible consequences have been suggested, as listed in *Table 1*.

Given technical feasibility, there are two key reasons for offering testing for children. First, knowledge of risks may give the chance to alter health outcome. An example of this would be testing for hypercholesterolaemia, very high levels of cholesterol, in newborns or children. Second, knowing whether a child will or will not develop a condition later in life may facilitate social and psychological adjustment. In the shorter term, testing may reduce the aversiveness of uncertainty and may also allow more specific planning for the future. For example, a child who carries a gene for the dominantly inherited myotonic dystrophy could be counselled away from electronic engineering towards teaching.

The extent to which testing achieves any of these objectives is largely unknown, and has been considered mainly from the perspective of the first objective, that of altering health outcome.

4.1 Testing to reduce risk

We know from many studies of adults' responses to behavioural advice that non-adherence is a common response (Meichenbaum and Turk, 1987). Within families, responses show more variation. There is some evidence that testing followed by dietary advice can reduce the risk of heart disease in children. In one of the largest studies of children with hypercholesterolaemia (73 children with a mean age of 5.7 years), plasma cholesterol was reduced by 9.6% by diet alone, and by a

Table 1. Suggested advantages and disadvantages of predictive genetic testing in childhood

Result of test	Possible disadvantage	Possible advantage
Faulty gene not present	Rejection by family for being unaffected, for example, where many members are affected. False reassurance about health status.	Avoids clinical monitoring for disease signs. Emotional relief. Ability to plan life with this knowledge. Avoids the documented adverse effects of later disclosure (Block *et al.,* 1992). Avoids 'pre-selection' as the 'sick' individual inheriting the gene. Relieves anxiety about possible early signs of the disorder.
Faulty gene present	The child's self-esteem and long-term adjustment may be impaired. The family's perception and treatment of the child may be adversely affected. May lead to discrimination in education, employment, insurance, mortgage. May adversely affect relationships with future partners. May not have wanted testing if asked when older. May generate unwarranted anxiety about possible early signs of the disorder, before any genuine manifestation of the disorder.	Child has time to adjust and to avoid emotional problems associated with a later disclosure. Enables parents to help child prepare psychologically for the future. Allows child and parents to prepare practically for the future, for example, education, career, housing and family finances. Allows the child to take informed decisions from an early age, for example, education choices. Avoids problems associated with 'family secrets'. Child doesn't miss the opportunity of testing. Relieves child's anxiety/uncertainty about his/her future (child may know more than others realise). Relieves parental anxiety/uncertainty. 16 years may not be a good age to be tested, given the problems of adolescence associated with personal identity, sexuality and family relationships.

further 12.5% by medication (Glueck *et al.,* 1980). However, in another study, parents 'over-adhered' to dietary advice, and eight out of 40 children were made to follow diets that were so restrictive that growth failure ensued (Lifshitz and Moses, 1989). These children were more likely to come from families with a member affected by acute heart disease.

While this study demonstrated 'over-adherence', another study illustrates what could be called 'paradoxical adherence', where the outcome was the precise opposite of what was advised. This occurred in a Swedish programme screening neonates for alpha 1-antitrypsin deficiency. The main purpose of the screening programme was to document the natural history of this recessive condition, and to provide an opportunity to protect the children from concentrated air pollutants, primarily tobacco smoke, to prevent or postpone the lung disease associated with this deficiency in adults. At follow-up one or both parents smoked in more than half of these families. Mothers of affected children smoked as much as controls, while fathers smoked twice as frequently as control fathers. Parental reports suggested that their views of themselves as parents, and their attitudes towards their children were unaffected by learning of the deficiency in their children. Parents of children with the deficiency did, however, view their children as more sickly than did parents of children without the deficiency. This was not reflected in the children's medical records. Children with the deficiency were observed to show more problematic behaviour when interacting with their mothers in the context of a structured test (McNeil et al., 1988). The screening programme was abandoned because of these psychological effects.

Most commonly studies that have assessed the psychological effects of testing have sought to determine whether there are any adverse emotional or cognitive consequences. The results show a mixed pattern.

The emotional and cognitive effects of screening for a predictor of diabetes were studied in 15 children and 21 adults (Johnson, 1991). Children, like adults, exhibited clinically significant anxiety when they were first told that they were at risk. This returned to normal levels within 2 to 3 months. The majority of those found to be at risk did not believe that such a status resulted in diabetes, in contrast with the beliefs of relatives. This minimization of a health threat is well documented amongst adults undergoing risk factor testing (Croyle and Barger, 1993), however this pattern is not always seen; sometimes people perceive the health threat as greater than it is. This was documented in several screening programmes for sickle cell anaemia in the early 1970s in the United States of America. In the process of detecting those at risk for the disease, carriers of the trait were also detected. For some, this caused much concern, in part arising from confusion over what carrying a trait actually meant. In one study, 50% of parents saw sickle trait as a disease, requiring restriction of play, dietary supplementation and rest (Hampton et al., 1974). Parents also reported more symptoms in their children and attributed these to carrying the trait. These findings suggest that information about risk of disease affects how people think about their health and how they interpret symptoms.

This phenomenon of over-attributing symptoms as a result of a perceived risk status has been described in other areas attempting to explain how people respond to symptoms or risk information:

'Given symptoms, an individual will seek a diagnostic label, and given a label, he or she will seek symptoms' (Leventhal et al., 1980).

The third main area where the effects of testing in childhood have been studied concerns children's actual behaviour. The behavioural effects of testing at-risk children for hypercholesterolaemia have been studied in 4–17 year olds by a Canadian group (Rosenberg et al., 1992). No differences were found in anxiety, depression, social competence or behavioural adjustment between the 34 who tested positive and the 12 who tested negative. There were, however, more child behaviour problems reported by the mothers of children who were found to have hypercholesterolaemia. High levels of behaviour problems 1 month after diagnosis declined by 12 months and were at levels found in chronically ill children by 2–5 years after diagnosis. What is unclear is the extent to which this reflects problem behaviour in the children or an increased sensitivity on the part of the parents, to possible problems following detection of a risk factor for heart disease.

More recently, a genetic screening programme has been set up to detect a disease, Duchenne muscular dystrophy, for which no treatment is available. The purpose of such screening includes giving parents information about risks to future pregnancies, and because symptoms are evident. Thus far, twelve boys have been found to carry the gene, and initial responses of parents have been favourable. Mindful of the potential adverse effects of such screening, the programme has been set up to ensure good support from primary and community services (Bradley *et al.*, 1993).

In summary, the results from these studies show that problems do not always occur for children or parents following childhood predictive testing. At present, we do not know the circumstances in which they are more likely to occur. The types of problems most likely to be reported include an increase in parental reports of illness or of problem behaviour. We do not know how long these problems last. Finally, we know little about possible advantages to testing or screening.

As this review shows, the effects of predictive testing cannot be predicted by knowing about the test. Future research needs to be guided by questions that will examine the influence of the processes of testing or screening and of individual differences upon responses to testing (see Marteau, 1993). One aspect of the testing process that is likely to be important in determining the effects of testing is communication of the test results. Evidence to support this comes from a study in which the records of school children were searched to find those for whom there was some mention of a heart problem (Bergman and Stamm, 1967). Amongst 93 found, 75 were examined by paediatricians who found no evidence of heart disease. Amongst these children, 30 were restricted in their activities by their parents, some severely so. The main discriminators concerned parental recall of advice given at the time of detection of a heart murmur or rheumatic fever. Parents who restricted their children's activities were less likely to recall being told that the problem was benign, more likely to recall being told to restrict the child's activities, and more likely to report feeling confused. Some of the individual differences likely to predict outcomes of testing include family functioning, social support, and the extent to which test results differ from those that had been expected.

5. Conclusion

Clinicians and researchers are understandably cautious about offering predictive testing for children. There is, however, mounting uncertainty about the extent to which this view conflicts with the views of families and the extent to which fears of psychological disadvantage are justified. This review highlights the complexity of addressing these questions.

Research into several areas is needed. First, we need to know more about the attitudes of children, their parents and of health professionals. The second key area for research concerns informed consent. Some of the debate about predictive testing in childhood centres on the questions of whether children are competent to give informed consent. This raises the questions of what the criteria for competence are, how they should be measured, and who should be the judge. Informed consent in adults has been approached in terms of how much information is recalled (Alderson, 1993). This should be broadened to include how that information is analysed and interpreted, how values and beliefs are integrated and what processes are used in decision-making. We have little information about the variation of these processes in adults, let alone their variation with age.

The third area for research concerns the psychological consequences of predictive testing in childhood. Several approaches are appropriate. First, studies are needed to determine the psychological effects upon children of being brought up in families at risk for adult-onset diseases. This information is needed as a background against which to judge the effects of testing children in these families. Controlled studies of the short-term and long-term effects of testing, and of the mediators of those effects are also needed.

As the UK (Nuffield report, 1993 and Clinical Genetics Society report, 1994) and North America (Institute of Medicine of the National Academy report, 1993) move towards policy and legislation in the area of genetic testing, psychological research will help ensure that these guidelines and regulations are informed by evidence rather than by prejudice.

Several themes emerge from the published research. First, population-based screening is associated with more problems than testing of high-risk groups. Second, provision of good counselling is associated with fewer problems. Problems are not inevitable, and not testing may cause as many problems as testing. There is sufficient information to proceed cautiously with testing if requested, provided it is done within the context of good counselling and informed decision-making. Such testing, however, should only be performed as part of a research protocol (Marteau, 1994). In this way we will accumulate systematic information on the effects of testing. Provided no major problems are evident, it may be appropriate to move towards formal trials of offering testing.

Acknowledgements

This paper was prepared while both the authors were supported by a programme grant from the Wellcome Trust, entitled 'Psychological and social aspects of the New Genetics'. Our thanks to Dr. Hilton Davies, UMDS, for his helpful comments on an earlier draft.

References

Adelman, H.S., Lusk, R., Alvarez, V. and Acosta, N.K. (1985) Competence of minors to understand, evaluate and communicate about their psycho-educational problems. *Prof. Psychol. Res. Practice* **16**: 426–434.

Alderson, P. (1993) *Children's consent to surgery.* Open University Press, Milton Keynes.

Appelbaum, P.S., Lidz, C.W. and Meisel, A. (1987) *Informed Consent: Legal Theory and Clinical Practice.* Oxford University Press, Oxford.

Barnes, C. (1994) Testing children for balanced chromosome translocations: a study of parental views and experiences. *Medizinische Genetik* **3**: 345.

Bergman, A.B. and Stamm, S.J. (1967) The morbidity of cardiac nondisease in schoolchildren. *N. Engl. J. Med.* **276**: 1008–1013.

Bloch, M., Adam, S., Huggins, M. and Hayden, M.R. (1992) Predictive testing for Huntington's disease in Canada: the experience of those receiving an increased risk. *Am. J. Med. Genet.* **42**: 499–507.

Bloch, M. and Haydon, M.R. (1990) Predictive testing for Huntington's disease in childhood: challenges and implications. *Am. J. Hum. Genet.* **46**: 1–4.

Bradley, D.M., Parsons, E.P. and Clarke, A.J. (1993) Experience with screening newborns for Duchenne muscular dystrophy in Wales. *BMJ* **306**: 357–360.

British Paediatric Association. (1992) *Guidelines for the ethical conduct of medical research involving children.*

Bryant, P. (1974) *Perception and understanding in young children.* Methuen, London.

Clarke, A.M. and Clarke A.D.B. (1976) *Early experience: myth and evidence.* Open Book, London.

Clinical Genetics Society (1994). *The Genetic Testing of Children.* Report of a Working Party of the Clinical Genetics Society. Clinical Genetics Society, Birmingham.

Croyle, R.T. and Barger, S.D. (1993) Illness cognition. In: *International Review of Health Psychology* (eds S. Maes, H. Leventhal and M. Johnston), John Wiley & Sons, Chichester.

Donaldson, M. (1978) *Children's Minds.* Fontana, London.

D'Zurilla, T. and Goldfried, M. (1971) Problem solving and behaviour modification. *J. Abnorm. Psychol.* **78**: 107–126.

Eiser, C. (1989) Children's concepts of illness: towards an alternative to the 'stage' approach. *Psychol. Health* **3**: 93–101.

Gaylin, W. (1982) Competence: no longer all or none. In: *Who speaks for the child: the problems of proxy consent.* (eds W. Gaylin and R. Macklin). Plenum Press, New York, NY.

Gillick (1986) *Gillick v West Norfolk and Wisbech Area Health Authority,* 1 FLR 224.

Glueck, C.J., Tsang, R.C. and Mellies, M.J. (1980) Long-term (2–6 years) therapy of familial hypercholesterolaemia and hypertriglyceridemia in childhood. In *Childhood Prevention of Atherosclerosis and Hypertension.* (eds R.M. Lauer and R.B. Shekelle). Raven Press, New York, NY.

Hampton, M.L., Anderson, J., Lavizzo, B.S. and Bergman, A.B. (1974) Sickle cell 'nondisease': A potentially serious public health problem. *Am. J. Dis. Child.* **128**: 58–61.

Harper, P.S. and Clarke, A. (1990) Should we test children for 'adult' genetic diseases? *Lancet* **335**: 1205–1206.

Huggins, M., Bloch, M., Wiggins, S. *et al.* (1992) Predictive testing for Huntington's Disease in Canada: Adverse effects and unexpected results in those receiving a decreased risk. *Am. J. Med. Genet.* **42**: 508–515.

Inhelder, B. and Piaget, P. (1958) *The growth of logical thinking.* Basic, New York, NY.

Institute of Medicine of the National Academy (1993). *Assessing Genetic Risks: Implications for Health and Social Policy.*

Johnson S.B. (1991) The psychological impact or risk screening for chronic and life-threatening childhood illnesses. In: *Advances in Child Health Psychology* (eds J.H. Johnson and S.B. Johnson, University of Florida Press, Gainesville, FL.

Kamps, W.A., Akkerboom, J.C., Nitschke, R. and Kingma, A. (1987) Altruism and informed consent in chemotherapy trials of childhood cancer. *Loss, Grief and Care* **1**: 93–110.

King, N. and Cross, A. (1989) Children as decision makers: Guidelines for paediatricians. *J. Pediatr.* **115**: 1–16.

Leventhal, H., Meyer, D. and Nerenz, D. (1980) The common sense representation of illness danger. In: *Contributions to Medical Psychology,* Volume 2 (ed S. Rachman), Pergamon Press, Oxford, pp. 7–30.

Lewis, M.A. and Lewis, C.E. (1990) Consequences of empowering children to care for themselves. *Pediatrician* **17**: 63–67.

Lifshitz, F. and Moses, N. (1989) Growth failure. A complication of dietary treatment of hupercholoesterolemia. *AJCD* **143**: 37–542.

Marteau, T.M. (1993) Towards an understanding of the psychological consequences of screening. In: *Psychosocial Effects of Screening for Disease Prevention and Detection* (ed T.R. Croyle). Oxford University Press, New York, NY.

Marteau, T.M. (1994) The genetic testing of children. *J. Med. Genet.* **31**: 743.

McNeil, T.F., Sveger, T. and Thelin, T. (1988) Psychosocial effects of screening for somatic risk: the Swedish alpha-1-antitripsin experience. *Thorax* **43**: 505–507.

Meichenbaum, D. and Turk, D.C. (1987) *Facilitating treatment adherence.* Plenum Press, New York, NY.

Morris, M.J., Tyler, A., Lazarou, L.P., Meredith, L. and Harper, P.S. (1989) Problems in genetic prediction for Huntington's disease. *Lancet* **2**: 601–603.

Nuffield Council on Bioethics (1993) Genetic screening. *Ethical issues.* Nuffield Council on Bioethics, London.

Rosenberg, E., Lamping, D.L., Joseph, L. Blaichman, S., Pless, B. and Lambert, M. (1992) *Cholesterol screening of high-risk children: Psychological and behavioural effects.* Report to the Fonds de la Recherche en Sante du Quebec (FRSQ).

Royal College of Physicians (1989) *Prenatal Diagnosis and Genetic Screening: community and service implications.* The Royal College of Physicians, London.

Sharpe, N.F. (1993) Presymptomatic testing for Huntington's Disease: Is there a duty to test those under the age of eighteen years? *Am. J. Med. Genet.* **46**: 250–253.

Slavin, L.A., O'Malley, M.D., Koocher, G.P. and Foster, D.J. (1982) Communication of the cancer diagnosis to pediatric patients: impact on long-term adjustment. *Am. J. Psychiatry* **139**: 179–183.

Spinetta, J.J. (1974) The dying child's awareness of death: a review. *Psychol. Bull.* **81**: 256–260.

Susman, E.J., Dorn, L.D. and Fletcher, J.C. (1992) Participation in biomedical research: the consent process as viewed by children, adolescents, young adults, and physicians. *J. Pediatr.* **121**: 547–552.

Tibbens, A. (1993) On psychological effects of presymptomatic DNA-testing for Huntington's disease. *Ph.D thesis,* Rotterdam University, The Netherlands.

Weithorn, L. (1983). Children's capacities to decide about participation in research. *IRB* **5**: 1–5.

Weithorn, L. and Campbell, S. (1982).The competency of children and adolescents to make informed treatment decisions. *Child Dev.* **53**: 1589–1598.

Wiggins, S., Whyte, P., Huggins, M., Adam, S., Theilmann, J., Sheps, S.B., Schechter, M.T. and Hayden, M.R. (1992) The psychological consequences of predictive testing for Huntington's disease. *N. Engl. J. Med.* **327**: 1401–1105.

16

Exploring the approach of psychology as a discipline to the childhood testing debate: issues of theory, empiricism and power

Lucy Brindle

1. Introduction

In response to calls for debate to be more strongly grounded upon objective empirical findings arising out of psychological research, this chapter sets out to examine the theoretical assumptions underpinning mainstream or cognitive psychology as an approach to informing the childhood testing debate. Arguments have been made for basing clinical practice on objective empiricism rather than medical opinion and prejudice (Marteau, 1994; Michie, 1996). From a social constructionist perspective, however, it can be argued that it would be problematic to base a critique of medical power on calls for objectivism. An alternative position can be taken from which the activities of both psychological research and clinical practice are understood to be culturally and historically located. The construction of the informed, consenting subject within mainstream psychology and clinical genetics is explored. The implications of this theorization of the subject are discussed in relation to research into informed decision making and the debate about genetic testing in childhood.

2. Empiricism as the key to objectivity

The generation of empirical findings has been called for by those arguing for more objective means of approaching the issues involved in debating the practice

The Genetic Testing of Children, A.J. Clarke (ed.).
© 1998 BIOS Scientific Publishers Ltd, Oxford.

of genetic testing in childhood. The Clinical Genetics Society report (1994) claims to base its responses on, among other things, 'a concern to maintain the generally accepted principles of medical ethics', but also suggests that it is preferable to base policies on firm evidence in the statement that, 'Unfortunately, there is a dearth of evidence on which to base policies. Various blends of ethics, anecdote and prejudice are likely to shape people's attitudes, including our own.'

Arguments have also been made for basing policy and practice on empirical findings rather than prejudice or paternalism. Michie (1996) has argued for empiricism rather than paternalism and Marteau (1994) 'for policy to be built, not upon the shifting sands of 'conjecture, anecdote, and prejudice', but upon the more solid rocks of empirical evidence'.

What constitutes an appropriate knowledge base has been apparent as an area of potential dispute. For example, Michie (1996) has questioned the utility of the information arising from a follow up of families undergoing testing as suggested by Clarke and Flinter (1996) and has identified a need for a 'well designed research protocol' that would allow comparative assessments of the consequences of predictive testing in childhood. Even though the legitimate methods of investigation and types of knowledge are sometimes stated explicitly, the theoretical assumptions underlying these alternative methodologies and their products are rarely discussed.

3. Formulating objects of debate

It could be argued that the application of empirical evidence has served to expand debate rather than to resolve areas of difference. When empirical evidence is drawn upon, it can be used to support opposing arguments and can be interpreted in a number of ways. For example, the low uptake of presymptomatic testing for Huntington's Disease (HD) has been used as evidence to support the reluctance to test children for the condition. It has been suggested that, as adults, they would perhaps be unlikely to request testing and therefore their adult autonomy would be infringed by testing in childhood (Clarke and Flinter, 1996). Alternatively, in response to this argument, the claim has been made that the low uptake of testing does not suggest that the consequences of testing are harmful to children (Michie, 1996) and therefore should not necessarily be used as an argument against testing. The presentation of ethical consequences as factual objects within the debate can also be deconstructed. The argument, that testing in childhood removes the future adult's capacity to make their own autonomous decision, can be opposed by the claim that 'not testing' is not only a case of maintaining choices for the future adults but also limits the options available to them, that is, of having known the test result since childhood (Michie, 1996).

4. Accounting for inconsistency

Research within health psychology, adopting a positivistic methodology, has generated a 'proliferation of statistically significant but inconsistent findings'

(Yardley, 1996). Differences can be approached by attempting to establish the factors responsible for variation. This attempt to explain variation assumes underlying rules or a theory which is able to predict and therefore explain behaviour in an 'a–social' way. In accounting for differences, the positions and findings of others are also criticized on methodological grounds. This is akin to establishing the constructed nature of the other's account while still presenting one's own as less problematically representing a pre-existing reality. Edwards *et al.*, (1995) have described the questioning of an opponent's method, assumptions, or rhetoric in the debate of opposing factual or empirical evidence, as the selective application of "the tools of relativist–constructionist analysis.'

5. An alternative paradigm

Taking a social constructionist position with respect to childhood testing research is not to argue against the value of empirical research. What is at issue is a form of empiricism, an epistemology which assumes the possibility of a single accurate representation of the world, whether in relation to explanations of parents' and children's involvement in the practice of predictive testing or of the consequences of childhood testing. The possibility of alternative representations makes apparent the values inherent in the knowledge produced by any research endeavour.

A rejection of the scientific goal of producing one accurate representation of a pre-existing reality entails the adoption of theories which view the research procedure as a constructive activity. Therefore it would not be possible to regard responses collected by one method as more validly or accurately accessing the 'external or internal reality of the respondent' than those collected by another (Yardley, 1996). Even within a methodological framework that supposedly allows a less limited range of responses from the participant, the interpretation or representation of the research interaction still occurs through the application of theories. When attempting to document families' experiences of having a Down's Syndrome child, Costigan and Fidler (1995) commented on the greater variety of responses which emerged when multiple methods were used. The differing interpretations allowed by an adaptation of the theoretical approach at a later stage of the research were also described. Taking a personal constructivist and phenomenological approach to the interpretation of narratives of experience, the researcher claimed that 'These [biological, social and historical] circumstances constrained, but did not necessarily determine, the scope of the families' constructions of the impact their Down's Syndrome children made on their lives'. When the responses from interview data were incorporated and a social constructionist approach was taken, families were described as powerfully influenced by social attitudes and the provision or withholding of support (Costigan and Fidler, 1995).

Research attempting to ascertain the psychological impact of childhood testing is underway and will provide comparative assessments of psychological states which may be drawn upon in the construction of 'understandings' of childhood testing. For example, preliminary findings from a study assessing the psychological impact of testing pre-school children for an adult onset colon cancer have suggested that 'whilst adults are more anxious after positive than after negative results, such differences are not evident for children' (Michie *et al.*, 1996). The relevant importance assigned to these scores and their interpretation will require much discursive work. The importance of the lack of a detectable negative consequence of testing in the form of an anxiety score will have to be debated with respect to other implications of testing.

Stainton Rogers and Stainton Rogers (1992) have discussed the multiplicity of positions available to those involved in the cultural practice of debating and constituting areas of child concern, while acknowledging the likelihood of consensus over extreme cases. They state that, 'Whatever we think and say and do about children could have some potential benefits for some, and do some potential harms to some others (depending also on shifting definitions of harm and benefit, child protection and child mistreatment could be two different readings of the same treatment)'. The Clinical Genetics Society report on the genetic testing of children also commented on the lack of any agreement concerning the time scale that would be appropriate in considering the implications of genetic testing for a family and the means of assessing harm or benefit to families (Clinical Genetics Society, 1994). The epistemological framework within which outcomes of the research process are interpreted will also determine the possible application of the finding. Whether a questionnaire is understood as accessing some pre-existing state such as anxiety, or as involved in its construction, has implications for the ways in which this finding could be used as a discursive resource within the debate.

This is not to suggest that those arguing for a 'scientific' basis for the debate assume that research findings will unproblematically represent reality. From within a modernistic paradigm, the limitations of representations can be viewed as problems with research methods and with gaining access to specific truths. The presentation of a version of scientific enquiry which does not acknowledge the constructive nature of research conceals the values and theories inherent in any application or interpretation of research findings. Through adopting a constructionist epistemology, it becomes possible to examine psychological theory as constituting rather than simply describing the human subject and to formulate or draw upon alternative theories. Such an alternative theory may itself dissolve the objects or psychological phenomena of attitudes and beliefs about which the researcher establishes truths and the debater constructs arguments.

6. Psychology as contemporary theory

Rather than allowing the creation of knowledge free of anecdotes, prejudice or value-laden stories, psychological theory can be understood as being historically and culturally located. Burman and Parker (1993) have discussed the influence of culturally available ideologies on psychological theory in terms of 'culture contain[ing] particular distinct types of psychology which seep into and mould the discipline of psychology.' The theory constituting the subject within mainstream psychology can be examined as existing within and contributing to contemporary culture.

Marteau and Johnstone (1989) have described a psychological explanation as starting with a 'view of the patient as an active processor of information whose behaviour is influenced by a rationally based decision'. In accounting for 'thoughts, feelings and actions', a cognitive or mainstream psychology considers how individuals represent and interpret information (Eiser, 1995). Therefore in explaining an individual's interactions within an environment, the cognitive approach attempts to identify phenomena residing within the individual. An interest in explicating that which underlies human action or speech, or phenomena which can be separated from that which is social and therefore variable, is compatible with ideas concerning rationality (Shweder, 1991) and with the aims of science as described by Woolgar (1988), in which knowledge only becomes scientific through the exclusion of social factors. Cognitive theory also implies rationality in terms of there being rational principles which can allow the production of accurate internal representations of the world (Gillet, 1995). Shweder discusses rationality in terms of it being based upon the idea that there is some underlying phenomenon that can be separated from people's everyday existence.

> 'The idea that existence is a negation of pure being can be traced in the West from Plato through Descartes to various contemporary "structuralisms" which aim to recover the abstract forms, universal grammar or pure being hidden beneath the superficialities of any particular person's mental functioning or any particular people's social life' (Shweder, 1991).

The dominance of a cognitive theorization of the subject can also be explained in terms of this subject being compatible with a widely held statistical epistemology, rather than being a product of contemporary theory (Spicer, 1995). The representation of the individual as internally compartmentalized allows the 'discovery' of cognitive categories such as cognitive schemas and beliefs, which can function as researchable variables. In contrast, 'theories which emphasize the situated meaning of actions and social practices and processes cannot be mapped easily, if at all, onto the dominant research discourse'. (Spicer, 1995). Therefore Spicer has argued that the methodological discourse is determining theory within psychology. But the emphasis on methodology within psychology can also be explained as an attempt to assert a scientific status for a discipline in which theories were always unlikely to obtain a fully rational status (Danziger, 1990).

7. Constructing the subject of genetic testing

In arguing for practice to be based on psychological evidence, it has been proposed that attitudes to predictive testing in childhood and the informed decision-making process are both objects requiring research (Michie and Marteau, 1996a). Knowledge, attitudes, beliefs and decisions are parts of the psychological possessions and activities of individuals which, within cognitive theory, are positioned between a technological development making genetic testing a possibility and the individual being provided with information regarding their genetic status. The psychological phenomena of attitudes and the process of decision-making are applied to the understanding of interest in and uptake of various genetic tests. The genetic counselling process has been described as providing individuals with appropriate information which they use vigilantly, 'to choose among the alternatives made available by the newly developed genetic technologies' (Shiloh, 1996). Although the options are provided by others, the act of deciding or directing action is then presented as an individual process. Kessler described genetic counselling as shifting towards a psychological paradigm, in line with the move away from eugenics (Kessler, 1980). The use of the concept of autonomy has allowed the presentation of genetic counselling as an activity undirected by the goals, beliefs and values of other individuals and groups. A non-directive genetic counselling aims to maintain client autonomy within an interaction in which information giving is an important component. A cognitive psychological paradigm could be understood as being compatible with an emphasis on autonomy, by constructing an individual whose internal processes and representations operate as causal mechanisms in the determination of action.

The conceptualization of a subject able to maintain an independence from and not be altered by the information which it is processing, implies the existence of a context- and information-independent processing mechanism. Shweder has claimed that such a processing device underlies the dominant model of the human subject within mainstream psychology:

'The main force in general psychology is the idea of that central processing device. The processor, it is imagined, stands over and above, or transcends, all the stuff upon which it operates. It engages all the stuff of culture, context, task, and stimulus material as its content. Given that image the central processor itself must be independent of both context and content. That means, in effect, that the processor must be describable in terms of properties that are either free of context/content (abstract, formal, structural properties) or general to all contexts/contents (invariant, universal properties).' (Shweder, 1991, cited in Harre and Stearns, 1995)

The social environment is not totally absent from this account of the subject of genetic testing. Difficulties in working through options for which no social guidelines exist has been attributed to difficulties in decision-making in contexts of genetic risk (Shiloh, 1996). Knowledge acquisition in counselling situations has been described as a dynamic process taking place within a social context (Michie and Marteau 1996b). But despite the recognition of the social provision of options and interactions with the environment, the way that the individual is being theorized within these models is Cartesian and formulates an intact individual separable from a social context. The isolation of the psychological subject from social context can be demonstrated by contrasting this conception of the subject with one in which the subject and conceptual variables of knowledge, beliefs, desires or intentions are constituted through discourse.

8. The formulation of informed decision making and competence within a cognitive framework

According to the CGS report, testing is acceptable and therefore it is implied that autonomy is possible if, 'the person is either adult, or is able to appreciate not only the genetic facts of the matter but also the emotional and social consequences of the various possible test results.' Although recognizing that the assessment of ability to give informed consent is problematic, an argument has been made for the value of establishing methods of assessing children's competence to understand the implications of genetic testing (Clarke and Flinter, 1996). Concepts of an acceptable appreciation of consequences and a cognitive capacity to weigh such consequences have been used in research examining children's capacity to give informed consent. Michie and Marteau (1996a) review research by Kamps et al. (1987) in which children were considered able to consent if they were capable of appreciating, 'the nature, extent and probable consequences of the proposed treatment or procedure'.

Rational models have been rejected as inappropriate for the study of adults' decision-making (Michie and Marteau, 1996b; Shiloh, 1996). Michie and Marteau (1996b) discuss the basic assumption that, 'rational decision making involves the systematic weighing of alternatives and exploration of the consequences of each alternative to produce judgements and choices' and state that people rarely make decisions in this way. That there is no unambiguous or rational way to define burdens and risks (Lippman-Hand and Fraser, 1979, cited in Michie and Marteau, 1996b) and that no version of events or story can 'ever be neutral or value free' (Lippman and Wilfond, 1992, cited in Shiloh, 1996) have been raised as problems with a rational model. But these discussions of decision-making still occur within a theoretical framework which formulates a rational process involving the balancing or processing of discrete entities.

Representing competence in terms of an ability to appreciate the consequences of testing is problematic when, rather than being a neutral object, a 'consequence' is a concept that has to be constructed. An assessment of an adequate appreciation of a possible future is dependent on a value-based judgement regarding what an adequate appreciation entails. What is a more important consideration at one point in a person's life may be less so at another point, and some disadvantages may apply to some social groups more than to others. For example, the impact that higher insurance premiums could have on a person's life is potentially variable in relation to personal factors such as income level and assessments of the implications for oneself and family members of having greater difficulties in obtaining insurance. Social factors, such as whether reasonably priced private rented accommodation or suitable 'social housing' is available, are likely to alter over time and location and are therefore unpredictable. Within a cognitive representation of decision making, these value-laden stories are reduced to the variable of knowledge operating within a cognitive apparatus.

Even though a disparity between these models and 'the way that people actually make decisions' has been acknowledged, and the concept of 'a rational decision' has been criticized on the grounds that assessment of its components is not possible in any objective way, these concepts of normality and rationality still structure Michie and Marteau's reassessment of decision making in genetic counselling situations. A failure to follow the normative model is described as 'not necessarily irrational'. Rationality is then defined in terms of being adaptive, by suggesting that a non-normative decision-making process may allow the individual to decide upon a course of action in an uncertain situation. Even though it is impossible to define a decision as rational and to determine whether a normative model is being used when it is impossible to give a value to possible outcomes, or accurately assess their likelihood, it is still suggested that there is a normative process which exists independently of the values fed into it. 'In some cases, it may be that people are using normative decision-making processes, but in the context of questions and problems that are different to those defined by the researcher' (Michie and Marteau, 1996b).

The representation of the decision-making process that is employed requires the separation of process from entity and is dependent upon the metaphor of weighing. The process of weighing different values is represented as distinguishable from the values themselves and can still be defined as normative. Questions of how possible outcomes would have to be compared, and what an appreciation of a test result would entail, are not addressed. These representations have implications for the questions asked about children's 'capabilities' and the cognitive categories constructed to facilitate the empirical process of learning more about decision making. Those theories of informed decision making that involve notions of rationality ascribe greater legitimacy to certain versions of reality or ways of reaching decisions than to others, and thereby create the potential for the values of individuals or groups assessing competence to take precedence. In discussing the many potential claims for a suitable knowledge base for an objective or appropriate assessment of competence to consent, Alderson (1993) has noted that most publications on consent are making claims for 'the author's profession and its unique expertise'.

9. Talking about childhood testing – a discursive alternative to attitudes, competence and informed consent

Competence is assessed within an interaction in which the relevant issues and understandings are constructed. It is possible to argue that, rather than being a description of a mental state, understanding is demonstrated through social practices. The understanding that one person feels they have may be disputed by others (Potter and Wetherell, 1987). An assessment of competence to make decisions or a discussion of genetic testing constitutes a social activity in which children and adults may both be involved. A rejection of a model of the human subject as consisting of 'a context/content independent central processing device' allows a rejection of the idea of consistent and stable internal mechanisms and representations of the external world as determining a person's responses. A theoretical framework which does not assume the existence of cognitive representations underlying talk would emphasize the situated nature of any talk through which decision making, the giving of informed consent, outcomes of counselling or competence to make decisions was assessed.

Edwards has questioned the relationship between talk and internal representations of the world that is assumed in much research examining children's understandings or conceptual development:

'The particular idea with which I take issue is that adults and children possess underlying cognitive representations that are expressed in talk. I have no objection to there being mechanisms underlying the production of talk. The problem is with talk's conceptual content – that is, models of reality, versions of the world, explanations, that talk expresses: talk as a window (a dirty window perhaps) on the mind. This is the sort of conception of the relation between talk and mind that is in Piaget's (1934/1928, 1926/1929, 1932) early work, especially in the studies based on interviews and children's talk, and also in many later and current studies of children's thought.' (Edwards, 1993).

Edwards explores children's concepts as 'formulations that occur within, and are to some extent constituted as, situated discursive practices'. Rather than an investigation of competence which locates the relevant phenomena as 'firmly within the child' (Burman, 1994), the subject of study could be a definition of the accomplishment of competence within social interactions. Parents, children and clinicians could be understood as drawing upon culturally available resources to create situationally coherent arguments as they are positioned in various discursive practices.

10. Discussion

Rational theories of the human subject and the notion of objectivity have been employed in arguments that policy should be based on empirical psychological research. By arguing for objectivity rather than paternalism or prejudice, a position in which empiricism allows a shift of power away from health professionals is created. It is supposed that objectivity and therefore neutrality allow a more equal grounding of debate. On the basis of empirical research it has been argued that parents appear more favourable to testing than professionals (Michie, 1996). In a debate largely restricted to health professionals, giving a voice to those living with a disorder or the threat of themselves or others developing a disorder must be considered to be paramount. The concept of rationality implicit to the cognitive subject, however, and the associated statistical discourse in much psychological research (Spicer, 1995) have limited the ways in which psychology can represent the positions of parents and children in relation to the issue of childhood testing. The recognition of inconsistencies between the findings of research studies carried out at different times or involving different populations, or apparent contradictions between what people do and say, or what people say on differing occasions then require explanation in terms of factors influencing decisions (Shiloh, 1996), conflicting needs and wishes of differing levels of cognitive–affective and interpersonal functioning (Kessler, 1980) and distinctions between attitudes, intentions, values and motives (Shiloh, 1996). These explanations do little to allow a way forward in addressing or debating the rights of parents and children.

By treating attitudes, beliefs and understandings as the possessions of individuals, it becomes possible to avoid a discussion of either the values inherent in constructing future outcomes of genetic testing or those constituting informed consent. Representations of autonomous adults and competent children requesting testing also fail to address the implications of the subject positions available to parents and children and the various moral discourses of responsibility that surround parenting.

Psychological research will not provide objective solutions to questions of how we should conceptualize and deal with issues of power in relationships between parent, child, health professional and state. As neutrality in investigative practice is an impossibility, the debate and research of childhood testing are unavoidably value-laden activities. The construction of tests of competence and 'objective' knowledge bases concerning the consequences of testing allows the genetic testing of adults and children to proceed without the more difficult issue of power being addressed.

References

Alderson, P. (1993) Children's Consent to Surgery. *Open University Press*, Milton Keynes.
Burman, E. (1994) *Deconstructing Developmental Psychology*. Routledge, London.
Burman, E. and Parker I. (eds) (1993) Introduction. Discourse analysis: the turn to text. In: *Discourse analytic research: repertoires and readings of texts in action*. Routledge, London, pp. 1–13.
Clarke, A. and Flinter F. (1996) The genetic testing of children: a clinical perspective. In: *The troubled helix: social and psychological implications of the new genetics*. (eds T. Marteau and M. Richards) Cambridge University Press, Cambridge, pp. 165–176.
Clinical Genetics Society (1994) Report on the Genetic Testing of Children. *J. Med. Genet.* 31: 785–797.
Costigan, J. M. and Fidler E. (1995) *Down Syndrome families: Interpreting Their Social Worlds*. Paper presented at conference *Understanding the Social World: Towards an Integrated Approach*, University of Huddersfield.
Danziger, K. (1990) *Constructing the subject: historical origins of psychological research*, Cambridge University Press, Cambridge.
Edwards, D., Ashmore, M. *et al.* (1995) *Death and Furniture: the rhetoric, politics and theology of bottom line arguments against relativism. Hist, Hum. Sci.* 8(2): 25–49.
Edwards, D. (1993) But what do children really think?: discourse analysis and conceptual content in children's talk. *Cognition Instruction* 11(3&4): 207–225.
Eiser, J.R. (1995) Of matters of fact and psychological discourse. *Psychol. Health* 10: 267–271.
Gillet, G. (1995) The philosophical foundations of qualitative psychology. *The Psychologist* 8(4): 111–114.
Harre, R. and Stearns P. (1995) Introduction: psychology as discourse analysis. In: *Discursive Psychology in Practice*. Sage, London.
Kamps, W.A., Akkerboom, J.C. *et al.* (1987) Altruism and informed consent in chemotherapy trials of childhood cancer. *Loss Grief and Care* 1: 93–110.
Kessler, S. (1980) The psychological paradigm shift in genetic counselling. *Social Biol.* 27: 167–185.
Lippman-Hand, A. and Fraser F.C. (1979) Genetic counseling – the postcounseling period. II. Making reproductive choices. *Am. J. Med. Genet.* 4: 73–87.
Lippman, A. and Wilfond B.S. (1992) Twice-told tales: stories about genetic disorders. *Am. J. Hum. Genet.* 51: 936–937.
Marteau, T.M. (1994) The genetic testing of children. *J. Med. Genet.* 31: 743.
Marteau, T. and Johnstone M. (1987) Health psychology: The danger of neglecting psychological models. *Bull. Br. Psychological Soc.* 40: 82–85.
Michie, S. (1996) Predictive genetic testing in children: paternalism or empiricism? In: *The Troubled Helix: Social and Psychological Implications of the New Genetics* (eds T. Marteau and M. Richards) Cambridge University Press, Cambridge, pp. 177–183.
Michie, S., Bobrow, M. *et al.* (1996) Genetic testing of children: psychological impact. British Human Genetic Conference, University of York. *J. Med. Genet.* 33: (Suppl. 1), abstract.
Michie, S. and Marteau T. (1996a) Predictive genetic testing in children: The need for psychological research. *Br. J. Health Psychol.* 1: 3–14.
Michie, S. and Marteau T. (1996b) Genetic counselling: some issues of theory and practice. In: *The Troubled Helix: Social and Psychological Implications of the New Genetics*. (eds T. Marteau and M. Richards). Cambridge University Press, Cambridge, pp. 104–122.

Potter, J. and Wetherell M. (1987) *Discourse Analysis And Social Psychology: Beyond Attitudes And Behaviour*. Sage, London.

Shiloh, S. (1996) Decision-making in the context of genetic risk. In: *The Troubled Helix: Social and Psychological Implications of the New Genetics*. (eds T. Marteau and M. Richards). Cambridge University Press, Cambridge, pp. 82–103.

Shweder, R. (1991) *Thinking Through Cultures*. Chicago University Press, Chicago, IL.

Spicer (1995) Individual discourses in health psychology. *Psychol. Health* 10: 291–294.

Stainton Rogers, R. and Stainton Rogers W. (1992) *Stories of Childhood: Shifting Agendas of Child Concern*. Harvester Wheatsheaf, Hemel Hempstead.

Woolgar, S. (1988) *Science the Very Idea*. Tavistock Publications, London.

Yardley, L. (1996) Reconciling discursive and materialist perspectives on health and illness: A reconstruction of the biopsychosocial approach. *Theory Psychol.* 6(3): 485–508.

17

Predictive and carrier testing of children: professional dilemmas for clinical geneticists*

Alison Chapple, Carl May and Peter Campion

Abstract

Recent advances in molecular genetics mean that doctors are now in a position to test for certain genetic disorders that may develop later in life; similarly, carrier testing has become more frequent. This has prompted debate about the ethical justification of presymptomatic testing of adults. The presymptomatic testing of children is an ethical issue that is even more complex. However, those providing genetic counselling on such topics must consider not only the wishes of the parents but also the rights of the child. This paper reports on a qualitative study of thirty families attending an out-patient Regional Genetics Clinic, and discusses the problems that the genetic counsellor faces in negotiating the everyday exigencies of such work. In particular, we focus on the question of how guidelines or protocols might help counsellors deal with these routine, practical dilemmas. The paper examines the way in which these dilemmas are managed by geneticists and families. We explore the potential conflict between guidelines for genetic counselling practice and the interests and anxieties experienced by families.

1. Introduction

Living in a 'post-Hippocratic' era, doctors are no longer accepted as the 'benign, authoritarian, paternalistic decision maker, taking full responsibility for the welfare of his or her patients' (Pellegrino and Thomasma, 1988). Today people are

---

* Reproduced from *Eur. J. Genet. Soc.* (2 (1996): 28–38) by kind permission of the authors, editor and publishers. Minor changes have been made to punctuation and to phrasing.

---

better educated and are less deferential to doctors, who are 'criticized for abusing their medical power by controlling or oppressing their patients' (Lupton, 1994). Doctors now have to take into account the wishes of the patient, and if the patient is a child, consider the wishes of the parents, or in some cases the wishes of guardians, courts and ethics committees. In the case of children, doctors have the moral responsibility of involving parents in decisions, and Pellegrino and Thomasma (1988) say that, 'They should abide by their decisions within the limits set by good moral practice'. However, as technology and molecular biology move forward rapidly, and as recent legal changes have increasingly defined the 'rights' of the child and the obligations of parents, it is not necessarily clear how 'good moral practice' should be defined. Montgomery (1994) has argued that although health professionals are morally obliged to avoid harming those for whom they are responsible, are obliged to do patients good and are supposed to follow principles of justice and autonomy, the correct course of action may not always be obvious. This is well illustrated in the field of clinical genetics, where geneticists and other genetic counsellors face numerous moral dilemmas, and where the 'correct' course of action may be contested.

Recent advances in molecular genetics mean that doctors are now in a position to test people for a number of genetic diseases that may develop later in life. There has been much debate about the ethical justification of presymptomatic predictive testing of adults for disorders such as Huntington's disease (HD), when no cure is possible (Ball et al., 1994; Davison et al., 1994). It is recognized that a positive test result may have devastating psychological consequences and may even lead to suicide (Terrenoire, 1992).

However, about 10–15% of adults at risk do ask for testing, for a number of reasons, and tests are now available. Similarly, there has been much debate about the advantages and disadvantages of carrier testing in adults. In this case people carry a faulty gene or chromosome, but it causes them no health problem. However, such faulty genes or chromosomes may result in problems for their descendants. People may want to be tested to see if they are a carrier of such a genetic disease in order to make decisions about family planning. However, it is well established that there may be adverse socio-psychological effects of testing. For example, Skirton (1995) points out that the information that a person is a carrier may 'alter that person's entire self-image and change forever the way he or she thinks of him or herself'.

The presymptomatic testing of children is an issue that is even more complex. The International Huntington Association has recommended that children should not be tested for HD (Wertz et al., 1994). However, there are situations when other diseases run in families, when it may be reasonable to agree to the presymptomatic testing of children or to agree to the testing of older children for unaffected carrier status.

There are both advantages and disadvantages associated with presymptomatic predictive testing in childhood (Clarke et al., 1994; Wertz et al., 1994). The most important reason for testing in some cases is the chance of immediate medical benefit which may prevent the disease developing and provide early treatment. Testing may also be of benefit to minors when they are making reproductive or career decisions. It may also be of help to other members of the family in DNA linkage analysis. Several members of a family may need to be tested to secure reliable pre-natal diagnosis. Childhood testing may also be a benefit in that it reduces family uncertainty about the future, and may reduce anxiety in the short term. However there may be major disadvantages in testing children. For example, a positive test result may damage a child's self esteem, may lead to stigmatization, may affect relationships within the family, may lead to difficulties in obtaining life and health insurances, and may affect employment opportunities. Moreover, testing in childhood affects children's potential autonomy, by removing the opportunity for a considered choice at a later date, where adolescents or young adults can consider the implications of testing for their own lives (Clarke, 1994).

The Working Party of the Clinical Genetics Society (Clarke *et al.*, 1994), carried out a questionnaire survey of about 3000 health professionals in Britain, and they found that there was no consensus concerning the wisdom of testing children for late-onset diseases for which no useful medical treatment would be available. The majority of genetics co-workers asserted that parents should not necessarily have the right to request the genetic testing of their child for such diseases. In contrast to this view, the majority of paediatricians and haematologists thought that parents should have this 'right'. Clinical geneticists were divided on the issue. Clarke and colleagues also found a wide range of views on the question of testing children for gene carrier status. The Working Party elicited the views of groups concerned with genetic diseases – while some favoured testing, others indicated that genetic tests should not be performed in childhood unless there was clear medical benefit.

On the basis of this evidence, the Working Party concluded that predictive testing for an adult-onset disorder should not generally be undertaken if the child is healthy and if there are no medical interventions established as useful that can be offered in the event of a positive test result. Also the Working Party recommended that children should not be tested to determine carrier status, where this would be of purely reproductive significance to the child in the future. Wertz *et al.*, 1994, have suggested similar guidelines, asserting that, 'If no clear benefits exist, parents should restrain their desire to know'. They go on to say that 'There is ordinarily no ethical justification for testing before the age of 11 or 12 years in the absence of proven medical benefits'.

Until now there has been uncertainty and disagreement about whether testing should be carried out in certain situations. In the United Kingdom, recent legislation – The Children Act 1989 – has assigned children the right to be consulted about their treatment (Booth, 1994; Department of Health, 1991; Pithers, 1994). The Act attempts to strike a new balance between the role of the state, the responsibilities of parents and the rights of children (Lyon and Partin, 1995). It also means that parents have a legal obligation to act in their child's best interest, to make choices that promote their child's welfare, and can no longer exercise their parental rights in whatever ways they please (Montgomery, 1994). Thus, Clarke *et al.* (1994) observe that: 'Parental decisions about testing for genetic disorders should therefore be made according to whether the child will benefit, not in order to relieve the anxieties of the parents'. However, the Working Party recommendations and the implementation of The Children Act have not relieved geneticists and others concerned with genetic counselling of the difficult and uncertain situations that they encounter in their work.

The results of the study reported here indicate that some cases are extraordinarily complex and suggest some of the conflicting pressures exerted on the geneticist during consultations. The geneticist has to weigh up the ability of the child to understand the situation, and to become involved in decision making. As Montgomery (1994) points out:

'A young person who lacks the requisite understanding will be unable to consent even if they are 17. In addition, a younger child who can understand will be able to give a valid consent even though they have not reached the statutory age.'

As we have suggested elsewhere, lay conceptions of genetic risk are drawn from a variety of sources. Ideas about the probability of inheriting genetic conditions are highly complex and may not easily be grasped even when the counsellor devotes much time and energy to explaining them. Moreover, ideas about risk are intimately linked to culpability, and both demand action of the parents (Chapple et al., 1995a,b).

In such circumstances, definitions of what is 'right' for the child are contingent not only on the geneticist's moral view but also on the extent to which parents and others contest their definition of the choices available. Some parents feel that their children should be protected from the burdens of decision making, or that it is wrong to seek consent from children for medical procedures (Alderson, 1992). However, as Freeman and McDonnell (1987) point out, 'not all parents are models of care and concern for their children'. Some parents may want to do the best thing for themselves or other members of the family. Moreover, they may not understand the implications of testing, even where the geneticist attempts to explain these in detail. At every stage in the process of genetic counselling, notions of best interests may be contested. In this context, it cannot be assumed that the parents' 'best interests' and those of their children are the same.

The three case studies described below illustrate some of the dilemmas that are faced by genetic counsellors when parents seek testing for their children. Such case studies are useful because, as Label (1995) points out, there is a need for explication of the circumstances and concerns that prompt requests for the genetic testing of minors and we need to know more about what actually happens during the process of genetic counselling. In addition, as Clarke et al. (1994) note, 'For some disorders, there is insufficient evidence to know whether a diagnosis in childhood is helpful in the medical management of the possibly (not yet) affected child'.

2. Study group and method

This study was conducted in the North of England in 1994 and 1995. Thirty families attending a Regional Genetics out-patient clinic were invited to take part in a research project exploring the process of genetic counselling. With prior consent, the patients and their families were video recorded as they talked to the geneticist. After each consultation the researcher (AC) interviewed the geneticist to ascertain what he thought of the consultation. Within the next few days the researcher also interviewed the family members who attended the clinic appointment, to obtain their views about what had taken place. Adults were interviewed individually, without the presence of children. Teenage children were also interviewed, with their own and their parents' consent. These interviews were repeated 6 months later. In all, 140 interviews were tape-recorded, transcribed and analysed, using procedures broadly in accordance with those set out by Silverman (1993). One of the advantages of this research design was that the researcher could compare the video recordings with respondents' accounts of what had actually taken place in the consultation.

Patients came to the clinic with a variety of problems, and most parents were seeking a diagnosis for an existing, visible, condition. However, three sets of parents were explicitly seeking testing for their children, and it is these families that are the focus of this paper. Two sets of parents sought predictive testing for their children, while the third couple wanted tests carried out to determine the carrier status of their children. It should be noted that in the discussions which follow, all individuals and institutions have been assigned pseudonyms to protect the anonymity of patients and professionals.

2.1 A mother with myotonic dystrophy

Mrs A had a classic case of adult myotonic dystrophy. She slept most of the day, and according to her husband had difficulty in opening her eyes. She had cataracts, difficulty with her grip and poor balance. She also suffered some pain in her shoulders. She did not develop the disease until she left school, which is typical. Mr and Mrs A came to the clinic with their two children, aged 6 and 9. At an earlier visit they had requested predictive testing for the children. Children who have parents with myotonic dystrophy are not routinely tested because there is no medical treatment that can be given to prevent or ameliorate the condition. The Working Party of the Clinical Genetics Society (Clarke et al., 1994) found that over half of the geneticists questioned in their survey said that they would not test a 5-year-old child whose parents had myotonic dystrophy and who wanted to know the genetic status of their child.

This family attended the clinic as the result of a critical incident concerning dental treatment for their son. Myotonic dystrophy presents a possibly lethal risk to the patient if anaesthetics containing paralysing agents are administered, because of the acute respiratory sensitivity that is one of the effects of this disorder. Because their son had a 50/50 (i.e. 50% or 1-in-2) probability of inherited myotonic dystrophy, the dentist had refused to use an anaesthetic to remove a damaged tooth, and had insisted that the child attended hospital for the required dental treatment under the supervision of an anaesthetist. By the time the tooth was finally removed in hospital, the child had suffered 2 weeks of great pain, and as a consequence of his distress the parents had lost a number of nights sleep.

Shortly before this episode, the family had experienced a tragedy which they attributed to myotonic dystrophy. Mrs A's father had suffered from this disease, and had recently died after suffering a heart attack. They believed that this had been caused by the eight injections of local anaesthetic that he had been given during a dental extraction. Because of these events, Mr and Mrs A were determined that their children should be tested, so that if it were found that one of them had myotonic dystrophy, and if that child was in an accident or needed dental extractions, they could warn health professionals that an anaesthetic might be dangerous.

This couple felt that they had good reasons for their request for testing. They had seen their son suffer for 2 weeks and they had witnessed the early death of Mrs A's father. In addition both parents were convinced that at least one child would have the disease. Even though the geneticist had clearly told the parents that there was a 50/50 chance that each of the children would be affected, the father said to the researcher:

Mr A: 'Myotonic dystrophy, it has never skipped a generation. It has never skipped a generation. The grandfather had it. Frances's father had it. Frances has got it. Her brother has got it. It's never skipped. So, to me, it looks as though it could be positive. Frances is convinced it could be positive. She said, 'There is no way, all of a sudden it is going to skip one generation'.

Thus these parents were very anxious about their children. They had experienced anxiety and distress due to the death of Mrs A's father, and they were convinced that their children had inherited myotonic dystrophy.

However, the geneticist was reluctant to test such young children without careful consideration of the problems associated with such early testing. Mr A said that at their first visit to the clinic, the geneticist had told them that there was a disadvantage with such early testing. He reported that the doctor had said that 'It's a big thing to carry around for years and years'. The researcher (AC) asked the father if any other points had been discussed.

AC: When you went to see Dr P. did he give any other reasons why they shouldn't have the test?

Mr A: No, not really. It was just that they were so young, and I think if we didn't have a good cause for having it done, he wouldn't have done it. He is a good doctor. You can sit there and talk to that man. He was smashing. He knew we had a good reason for doing it, and that's why he did it. If we'd said, 'We just want to have it done to make sure', I think he'd have said, 'No, we'll leave it until they are old enough to decide'.

This family was video recorded when they returned to the clinic with the children for testing. Their son showed no sign of understanding the consultation. Afterwards, the parents explained that he thought that he was having a blood test to see if he had a broken arm. Their 9-year-old daughter had some idea of why she was having a blood test, but its implications were not discussed with her during the consultation. After their first visit to the clinic, the parents had been asked to reconsider their request for testing. They were now sure that they were doing the right thing. Thus, early in the second consultation blood was taken from the children and was sent to the laboratory for testing.

After the clinic appointment the researcher interviewed the geneticist and discussed the issue of testing these young children for myotonic dystrophy. The geneticist said:

PP: Now the situation when I first met them was that they were referred by the muscle clinic because they wanted testing for the two children, which is an ethical issue, and exercising the minds of clinical geneticists in general. It has been difficult to reach any consensus on what one should do. Should one be offering to test children born with diseases that are predominantly adult-onset, when they are minors, and do not fully understand what you are testing them for?

AC: They may prefer not to know.

PP: They may prefer not to know, or make the decision for themselves when they are older.

AC: How did you come to this decision whether or not to test the children? Did you come to it together?

PP: Yes. They came asking for the test, really. Obviously what we did was go over the pros and cons. This family is in an awfully difficult situation because Frances' father died suddenly after dental extraction . . . In view of that I phoned my colleagues in Borchester, who have a lot of experience of myotonic dystrophy, and asked them what they were doing. I spoke to a Research Fellow there, and she was saying, you know, they probably are in some circumstances, anyway, testing the children if requested.

AC: So there are no national guidelines?

PP: There are no national guidelines on this. It's a bit of a subjective decision . . . it's an extremely difficult situation . . . Well, there is a report. The [Clinical Genetics] Society did institute a Working Party, but again they didn't come up with very strong (guidelines); you know, I think there was general agreement with things like Huntington's . . . then there are those in the grey area, like myotonic dystrophy, where there might be implications, it's an extremely difficult situation.

AC: Apart from anaesthetic, are there any other problems that come to mind for which it might be useful to know (if they had the gene)?

PP: Probably not in this situation. There is no specific treatment for myotonic dystrophy, for children. They are probably not going to develop any symptoms for some years.

AC: What about insurance applications? Did you have to discuss those sort of things, or not?

PP: Well, I can't remember whether they were discussed directly. I don't know how an insurance company would weight having a gene for myotonic dystrophy. You *can* go through life with absolutely no problems.

AC: Would they be obliged to disclose to the insurance company that they had had this test?

PP: Errr . . . no. I think it is an extremely difficult area, and nobody is clear. However, on the form they will be asked their family history, and they may well be asked, 'Have you had any tests?' What kinds of tests are we talking about? Normally when you think of a test, it means you've got symptoms, and therefore you have the test to see the cause of those symptoms. So, you know, these terms are extremely loosely worded, on insurance forms, and I don't think they would necessarily have to declare that they had a gene test for this particular condition. So, it's a difficult issue, but I would say I didn't feel strongly one way or the other in this family. If it had been Huntington's and they had wanted the test I would have dug my heels in and said, 'No'. Ummmm ... in this situation, given the experience they had, I felt it was probably reasonable to accede to their request. Having given them a few months to think about it, and having asked them how they viewed matters, it was fairly clear that the kids had come along expecting a blood test.

Later in the interview the doctor concluded:

PP: So, on the whole, although I feel this is such a difficult situation, it is ethically a difficult one, as to whether we should be testing children under these circumstances, and although I felt uneasy about it, I felt easier about it after this session, because I felt they had considered it, and under the circumstances it was not an unreasonable request. I don't think I have acted unethically in taking blood samples. I hope the tests are negative, but we'll see.

It transpired that the tests were negative, and when the researcher called at the house for the second interview the parents expressed their relief and joy.

This case study illustrates some of the dilemmas faced by geneticists. The doctor felt that he was in an uncertain situation. He listened to the experiences of the family who had lived with myotonic dystrophy and was aware of the impact that the lack of knowledge was having on family life. However, there was no attempt to explain events to the 6-year-old child. Alderson (1992), in her study of childrens' consent to surgery, found that some children as young as 3 years old could understand procedures if they were explained simply and carefully. Franklin (1994) also found that children as young as 8 were capable of making important decisions and understanding their consequences. In this context, she suggests that:

'An ideal situation would involve the professional, the parents and the child, with the child making a major contribution. But this rarely happens'.

Franklin points out that many professionals have little training in communication, particularly with children. While this may be so, the parents clearly did not want to involve their children. Mr A said that he wanted the children to finish their childhood without worrying. Thus the geneticist had to balance these demands when coming to his decision to test the children. There can be no simple moral protocol that can tell the doctor how to proceed in such a situation. Guidelines may help, particularly if they are drawn up by geneticists as well as those concerned with biomedical ethics, because as Callahan (1980) points out.

'Those in ethics have learned, mainly by hard experience, that they will make little sense to practising physicians if they are not fully aware of the experience and details of clinical practice. It is simply not enough for those in ethics to roll up to the bedside the ghost of Immanuel Kant, John Stuart Mill, or G.E. Moore and provide instant moral diagnoses.'

The Working Party of the Clinical Genetics Society recognized that circumstances might arise when it could be appropriate to test a child for an adult-onset disease, even when there appeared to be no immediate medical benefit (Clarke *et al.*, 1994). The report asserts that:

'Despite our judgement that it is wise to avoid predictive tests in childhood, we recognise that experience and information may accumulate over time, and the apparent disadvantages of predictive testing for HD may not apply in practice to all other late-onset disorders or even to Huntington's disease in every situation and for all time'.

Testing the children in this family did not result in immediate medical benefit. However, in view of the possibly lethal effects of anaesthetic in such cases, not testing may have had fatal consequences had other circumstances led to the need for surgery. The next two cases are of interest because in both, guidelines (Clarke *et al.*, 1994; Wertz *et al.*, 1994) suggest that it might be inappropriate to test the young children involved. However, while in the case of family M the geneticist resisted pressure from the mother to test the young children, in the case of family T. the geneticist agreed to the parents' request for the testing of their young children.

2.2 A family with balanced 13 and 14 chromosome translocation

Mr and Mrs M came to the clinic with six of their 13 children. Both parents left school at 15 years of age, and both were unemployed blue collar workers. Some time ago, Mrs M had given birth to a child with Down's syndrome. When this child's chromosomes were examined, the laboratory staff found Trisomy 21 as expected, but in addition they found a balanced 13,14 translocation. Mrs M was checked and was found to be a carrier of the 13,14 translocation. In other words, she has a fused chromosome 13 and 14, so that all her genetic material is organized into 45 chromosomes instead of 46. This means that although she is a carrier, she is healthy herself. Her offspring may be normal but they may also be carriers, or they may be born with severe abnormalities due to Trisomy 14 or Trisomy 13 (Patau's syndrome).

The genetics of this problem are complex, and when the family arrived at the consultation it was clear that they did not fully understand the situation. The mother was uncertain which chromosomes were involved. She did know that the problem made it more likely that she might miscarry or have an affected child. Mrs M had already had six of her older children tested for the chromosome abnormality; two were found to be carriers. Now she wanted three of the other children to be tested. She said that they would feel more relaxed if they knew whether or not they were carriers. The geneticist explained the genetics of the condition, with the use of detailed diagrams of chromosomes. He asked Jane, the 15-year-old girl, if she understood what he had been saying. The mother responded, and said that Jane would feel more relaxed during pregnancy if she knew whether or not she was a carrier. The doctor again asked Jane if she understood everything. Jane nodded, but the mother repeated that the children would feel more relaxed if they knew. The geneticist replied:

PP: Yes, but it depends on whether they want to know now, or whether . . .

The mother broke in with:

Mrs M: Get it over with.

When the mother again stressed that she thought all her children needed the test, the geneticist said

PP: The thing is, to have this test, it is quite useful to have it at a time when it means something to you. If you have the test and you don't quite understand what it is all about, and then you forget about it, and you know . . . it may be more helpful and stick in your mind (if you have it later), especially if it is an abnormal result.

No other disadvantages of testing were discussed. In the end, the geneticist agreed to test the eldest girl, Jane, but he said that the other children, aged 11 and 13 should return at a later date.

A few days later, the researcher visited the family. It became clear that the parents did not see any major disadvantage in testing all the children. The researcher said

AC: Dr P. didn't want to test the younger ones. Why do you think he was anxious about it?

Mrs M: Well, their veins are very small. That was one of the reasons. They might not like the needle. Little children don't like it. I don't know, it would have been a lot easier (to have got them all tested).

Later in the interview the researcher asked Mrs M how she felt about her visit to the clinic. She replied:

Mrs M: I wish they had done more of the kids' blood tests, because it would have been over. I can understand how with the little ones it might have hurt them, but it would have been over quick. They are used to having needles for inoculations. There's not a lot of difference. It would have been over with.

AC: Was there any other reason why he didn't want to do it?

Mrs M: They wouldn't understand, why they were having it done, mainly. But they have so many things done they don't understand. They have inoculations, polio . . . Alice had one a few weeks back. I doubt if she really understands what polio or anything like that is . . . so really . . .

AC: So no one has suggested any other problem of having the test done?

Mrs M: Just that your veins are small. I mean the first prick hurts, and then it's all over, isn't it?

Clearly Mrs M did not understand how a positive test result might affect her children in an adverse way, although she told the researcher that when her eldest daughter found out that she was a carrier it affected her relationships with boyfriends. Mrs M wanted the children tested for her own peace of mind, and she actively contested the geneticist's version of what was 'right'. It was clear that the consultant found this difficult:

AC: They might choose not to be tested just because it might affect their choice of partner?

PP: Well . . .

AC: Is that something you might discuss with them?

PP: That's a difficult one. Yes . . . I don't know. I don't know whether in that situation. I think that's an interesting question. I don't think it's a question that people have looked at. It's not a question that is usually raised with us. It's not a question I usually raise with them. I'm not sure when people who are carriers, of whatever, actually tell their prospective partners about the issue.

It should be noted that conditions in the clinic were not ideal for thoughtful discussion. During the clinic appointment, Jane was busy looking after the younger children. In particular she was trying to amuse her younger brother who suffered from Down syndrome.

Throughout this consultation, little was said about the possible disadvantages of testing. In the USA in the 1970s some of those who were found to be carriers of sickle cell trait, and who were perfectly healthy, experienced discrimination at work and from insurance companies that raised their premiums (Nuffield Council on Bioethics, 1993). As we have noted, possible damage to social relationships was also not discussed. Given time constraints at the clinic, perhaps this is understandable, but it is a misunderstanding to think that adults or children necessarily receive sufficient information to make informed decisions. Moreover, their 'lay' understandings of expert knowledge may subsequently be framed in such a way as to damage relationships within families (Chapple and May, 1996).

2.3 The T family: a mother with polyposis coli

Mr and Mrs T also attended the Genetics Clinic because they wanted tests to be carried out on their three young children, aged 18 months, 3 years and 6 years. Mrs T had left school at 18, and had worked as a nursing auxiliary. Mr T had left school at 16 and was an architect's assistant. There was a family history of polyposis coli and this condition almost invariably leads to a carpet of small tumours appearing over the interior surface of the large intestine, ultimately leading to cancers of the colon. Mrs T had previously had part of her bowel surgically removed to prevent cancer. The three children had a 50% risk of inheriting the gene for polyposis coli, and the parents were determined to have them tested to see whether or not they were at risk. Children do not develop this disease before the age of about 12, and thus screening with sigmoidoscopy does not start until about that age. Therfore, testing before the age of 10 would not benefit the children in any obvious manner.

During the clinic appointment, the geneticist explained that he would prefer to test the children when they were older. (Even though some geneticists say that they would test young children for polyposis coli if the parents requested it, current guidelines suggest that it is better to wait until the children are at least 10 years old.) The geneticist explained that the test is not completely accurate, because the fault cannot be seen in the gene itself and because it involves looking for markers that are situated near the gene. The geneticist said that it might be wiser to wait until a more accurate test was available. At this point in the consultation the attending nurse intervened, suggesting that parents might find it hard to accept the situation if the current test showed that the children only had a slight chance of developing polyposis coli and a subsequent, more accurate test revealed that, after all, one or more of the children were almost certainly going to develop the disease. The nurse also pointed out that, if one child was shown to be at risk of developing the disease, the parents might treat that child differently. Mrs T responded by saying that she would probably be 'harder on the one that had it'.

Mr and Mrs T explained why they wanted the test carried out immediately. Mrs T said that it would be easier to test all three children now, because there would be no need to explain why the blood test was being done. Mr T said that if his children were going to develop the disease it would be better to tell them earlier rather than later, when perhaps they were in the middle of exams. He also said that if he knew that the children were at risk he would be able to 'make people listen'. By this he meant that he would be able to remind medical professionals that his children needed regular examinations for polyposis coli.

The nurse then asked the parents if they were asking for the test for the benefit of the children or for their own benefit. The father replied that it was for his benefit really. He said, 'I'd like to know'. At the end of the consultation the geneticist asked Mr and Mrs T to reconsider their request. The nurse said that she would not make another appointment, but if they decided to return with the children for testing they should ring the clinic to make the appointment. The nurse explained to the researcher that if the parents have to make the appointment themselves it requires more determination to have the tests carried out, than if an appointment is sent by the clinic.

The video recording of the consultation and interview transcripts suggest that the main reason that the parents wanted to know about the genetic status of their children was that they could not tolerate the uncertain situation. After the clinic appointment the researcher (AC) talked to Mrs T at home.

AC: You want all the children tested?

Mrs T: All at the same time. Now it's just a waiting game.

AC: Was the uncertainty bothering you an awful lot?

Mrs T: Yes, we want them tested. We want to know for *us*.

Mr T expressed the same feelings. The researcher said:

AC: Are you still determined to have the tests?

Mr T: Yes. It's just I need to know really. We are looking forward to going back in a couple of months time.

AC: You had a quick chat to Dr P about the pros and cons of testing, didn't you?

Mr T: Yes.

AC: As far as the cons are concerned, do you think there any major problems about testing?

Mr T: Not at all, no. The children have injections anyway. It's the same really. You know, a couple of months and it will be all over. What *I'll* gain from it, [will] outweigh what Dr P was saying.

AC: You said it would put your mind at rest. So you can keep an eye on them?

Mr T: Yes. I think my wife feels the same.

AC: You don't think you would treat the children differently?

Mr T: Oh no. It's a gradual approach, up to when they could have an operation, or whatever. Rather than just getting to about 9 or 10 years old, and then a bombshell come down and hospital.

Clearly here was a dilemma. The geneticist did not want to test these young children. One of the children was only 18 months old. The baby might be labelled as having the genetic predisposition to a serious disease, and later more accurate tests might reveal that a mistake had been made. The geneticist told the researcher that he had tested one other child for polyposis coli in the past 2 years. In that case the parents had asked for the test when the child was aged 7, but the geneticist had 'put them off a bit', and eventually the test was done when the child was 9. However, in the case discussed in this paper, Mr and Mrs T were articulate, and persuasive. The consultant said:

PP: I mean they admitted it (the test) was mainly being done for their benefit, rather than necessarily for the children's benefit at the moment. On the other hand they made, I think, a valid argument that if they knew which ones were at high risk or low risk, then they could start introducing their fears to them, which I think is true, because after all this is not something that is not going to affect them until adult life, it's not like Huntington's. This is something that is going to affect them in childhood, and so before they reach their teens they are going to be offered sigmoidoscopies, which are not pleasant, a bit of an invasion, so to introduce that to them, I think is a good idea. To introduce the idea that Mum has had a tummy operation sort of thing, is not a bad thing . . . to explain things as they reach certain developmental stages, when they can take certain things on board. So I think that argument is valid . . . It was helpful that Jennifer (the nurse) came in as well, because it was heavy going. It's useful for someone to come in and put a different angle, so that I can relax for a bit, and . . .

AC: Think?

PP: Yes.

Mr and Mrs T said that they understood all that the geneticist had told them. Mrs T said:

Mrs T: He tended to repeat the same thing over and over again. We'd already understood what he had said, and then he kept repeating it.

AC: So had he told you all that the first time you went? This was the second time, wasn't it?

Mrs T: Yes, he has sort of explained it. He explained all the markers. He seemed to explain that quite a lot. I think he went over it too many times.

However, even though Mr and Mrs T clearly understood the information about genes and markers, the implications of testing the children at such a young age may not have been so clear to them. The father had said, 'The children have injections anyway. It's the same really'. The consultant appeared to be reasonably happy with his decision to do the tests if requested. At the end of the interview the consultant described the problem as he saw it.

AC: And how did you feel at the end of that session?

PP: Ummmm . . . I found it quite heavy going, because I find these sorts of ethical issues quite difficult. I suppose from my own perspective, what would I do? I don't know what I would do really. I suppose I would be inclined to wait until the children are a little older. On the other hand, I can see their point of view, and I think their arguments are reasonably valid. I think therefore we should do the test if that's what they want. I think most of the issues were rehearsed with them.

Eventually the children were all tested, and results showed that they had not inherited the gene for polyposis coli.

3. Discussion

The three cases discussed in this paper illustrate the dilemmas faced by geneticists and other genetic counsellors when parents want their young children tested for conditions that they may or may not develop in the future, or for carrier status. In his effort to promote what he considered to be the best interests of these children, the geneticist was able to consult the report of the Working Party of the Clinical Genetics Society (Clarke *et al.*, 1994), and to consider their recommendations. At the same time, he had to consider the wishes and arguments of the parents of the children and assess whether or not the children themselves could understand the issues involved. It is possible that this doctor felt that he was under greater pressure to test the young children when he was negotiating with an articulate middle-class couple (Mr and Mrs T), than when he was dealing with the less-educated family (Mr and Mrs M). However, both Mr and Mrs T, and the parents of the children with myotonic dystrophy (Mr and Mrs A) expressed compelling personal reasons for requesting testing, and as Gaylin (1982) suggests, 'We must continue to presume that parents are acting in the best interests of their children, until proved otherwise'.

Alderson (1990) points out that reasoning in bioethics can in some ways work against the interests of patients. She says:

'There is the risk, especially when discussing individual patient's dilemmas, that ordinary people's abilities will be underestimated, and that decisions powerfully affecting their lives will be moved outside their control . . . Patient's own meanings will be lost, and issues of vital importance to them excluded.'

She goes on to say that,

'Moral dilemmas are too important to be left to academic experts; they also have to be understood through direct experience'.

Fox (1988), has noted that traditional Chinese medicine stresses the importance of looking at the actual situations in which ethical problems occur. She observes that thinking in an abstract way about moral or social questions runs the risk of what Chinese scholars historically have called, 'playing with emptiness'. Since genetics is a relatively young field, and since new genes and genetic markers are being identified all the time, counsellors inevitably find themselves in situations where there are no guidelines or recommendations to follow. Looking at these case studies it appears that this British geneticist, rather like those concerned with medical morality in China, meets and analyses each situation as 'lived-in experiences'. For Fox, this is preferable to the 'ordered, cerebral, armchair inquiry', that is given precedence in American bioethics. The members of the Working Party of the Clinical Genetics Society themselves clearly recognize the importance of looking at the actual situation in which ethical problems occur, when they say that although decisions must be made on the basis of the best interests of the child, 'no blanket policy would be defensible' (Clarke et al., 1994).

Clements (1985) calls for the 'rejection of ethical essentialism and the acceptance of the uniqueness and diversity of cases'. However, while recognizing that in some ways each case is unique, many cases have aspects in common; in situations where uncertainty exists about how to proceed, guidelines can be most helpful. We suggest that further research is needed to explore the concerns that prompt requests for the genetic testing of minors, to understand the anxieties felt by both parents and children when there is a genetic disease in the family, and to examine the doubts and concerns of geneticists and other genetic counsellors. As we have shown in this paper, the situational ethics that these encounters present are structured through the participants' perceptions of what is 'good' or 'right' for the child, but these perceptions are not uniform because parents actively contested the geneticist's definitions of the child's best interests. Their dilemma was one where, especially in circumstances characterized by great uncertainty, the parents' own 'best interests' also demand attention.

Acknowledgements

The study reported in this paper is supported by the Medical Research Council (Grant G9320295). We gratefully acknowledge their support. We would like to thank the staff and patients at the clinic concerned for their time and candour. We are also grateful to Lucy Frith for her useful comments on this paper.

References

Alderson, P. (1990) *Choosing for Children: Parents' Consent to Surgery.* Oxford University Press, Oxford.
Alderson, P. (1992) In the genes or in the stars? Children's competence to consent. *J. Med. Ethics* 18: 119–124.
Ball, D., Tyler, A. and Harper, P. (1994) Predictive testing of adults and children. Lessons from Huntington's disease. In: *Genetic Counselling. Practice and Principles.* (ed A. Clarke). Routledge, London.
Booth, B. (1994) A Guiding Act? *Nursing Times* 90 (8): 30.
Callahan, D. (1980) Shattuck Lecture – Contemporary biomedical ethics. *N. Engl. J. Med.* 302 (22):1228–1233.

Chapple, A. and May, C. (1996) Genetic knowledge and family relationships: Two case studies. *Health and Social Care in the Community* **4** (3): 166–171.

Chapple, A., May, C. and Campion, P. (1995a) Parental guilt: The part played by the clinical geneticist. *J. Genet. Counseling* **4** (3): 179–191.

Chapple, A., May, C. and Campion, P. (1995b) Lay understanding of genetic disease: A British study of families attending a genetic counseling service. *J. Genet. Counseling* **4** (4): 281–300.

Clarke, A. (1994) Introduction. In: *Genetic Counselling. Practice and Principles* (ed A. Clarke). Routledge, London.

Clarke, A., Fielding, D., Kerzin-Storrar, L., Middleton-Price, H., Montgomery, J., Payne, H., Simonoff, E., and Tyler, A. (1994) *The Genetic Testing of Children*. Report of a Working Party of the Clinical Genetics Society. Clinical Genetics Society, Birmingham.

Clements, C. (1985) Bioethical essentialism and scientific population thinking. *Perspect. Biol. Med.* **28** (2):188–207.

Davison, C., Macintyre, S., and Davey Smith, G. (1994) The potential social impact of predictive genetic testing for susceptibility to common chronic diseases: a review and proposed research agenda. *Sociol. Health Illness* **16** (3): 340–371.

Department of Health (1991) *The Children Act 1989*. HMSO, London.

Fox, R. (1988) *Essays in Medical Sociology: Journeys into the Field,* 2nd edn. Transaction Books, New Brunswick.

Franklin, P. (1994) Straight talking. *Nursing Times* **90** (8): 33.

Freeman, J. and McDonnell, K. (1987) *Tough Decisions: A Casebook in Medical Ethics*. Oxford University Press, Oxford.

Gaylin, W. (1982) Who speaks for the child? In: *Who Speaks for the Child. The Problems of Proxy Consent* (eds W. Gaylin and R. Macklin). Plenum Press, London.

Label, R. (1995) Genetic testing for children and adolescents. *J. Am. Med. Assoc.* **273** (14): 1089.

Lupton, D. (1994) *Medicine as Culture*. Sage, London.

Lyon, C. and Partin, N. (1995) Children's rights and the Children Act (1989). In: *The Handbook of Childrens Rights: Comparative Policy and Practice* (ed B. Franklin) Routledge, London.

Montgomery, J. (1994) The rights and interests of children and those with mental handicap. In: *Genetic Counselling. Practice and Principles* (ed A. Clarke). Routledge, London.

Nuffield Council on Bioethics (1993) *Genetic Screening. Ethical Issues*. Nuffield Council on Bioethics, London.

Pellegrino, E. and Thomasma, D. (1988) *For the Patient's Good*. Oxford University Press, New York.

Pithers, D. (1994) Acting fair. *Nursing Times* **90** (8): 32.

Skirton, H. (1995) Psychological implications of advances in genetics – 1. Carrier testing. *Professional Nurse* **10** (8): 496–8.

Silverman, D. (1993) *Interpreting Qualitative Data*. Sage, London.

Terrenoire, G. (1992) Huntington's disease and the ethics of genetic prediction. *J. Med. Ethics* **18**: 79–85.

Wertz, D., Fanos, J. and Reilly, P. (1994) Genetic testing for children and adolescents. Who decides? *J. Am. Med. Assoc* **272** (11): 878–881.

18

Complementary methodologies in the evaluation of newborn screening for Duchenne muscular dystrophy

Evelyn Parsons and Don Bradley

1. Introduction

The presymptomatic screening of healthy newborn infant boys for Duchenne muscular dystrophy (DMD), a currently untreatable genetic disease, raises a number of very significant social, psychological and ethical issues. Clearly there can be no debate concerning the rights of the child to be involved in the decision making process, as arises in the discussion about genetic testing of children for some other disorders, but the issue of informed consent and the impact of a presymptomatic diagnosis on the parent/child relationship and family stability are just as significant.

DMD is a lethal genetic disease that is transmitted to boys by perfectly healthy women, who carry the defective gene on one of their two X chromosomes. Affected boys are healthy at birth and only begin to show the first signs of muscle wasting and weakness when they are 3 or 4 years old. They experience increasing difficulty walking and are usually wheelchair bound by 11–12 years old. The muscle deterioration is continuous and boys usually die in their late teens or early twenties as a result of either a chest infection or heart failure. The gene for DMD was identified in 1987 (Koenig *et al.*, 1987) but unfortunately no effective treatment is yet available.

Screening the general population in the newborn period has been possible since the mid-1970s, but until recent advances in molecular genetics the only medical intervention available to women found to be carriers, or at risk of being carriers, was to terminate all subsequent male pregnancies, of which more than 50% would

---

The Genetic Testing of Children, A.J. Clarke (ed.).
© 1998 BIOS Scientific Publishers Ltd, Oxford.

be unaffected. However, the 1990s have seen advances in genetic diagnostic techniques so that accurate pre-natal diagnosis, using chorion villus sampling is now available in most families. This ability to offer choice in future pregnancies to the majority of index women, and their female relatives, was one of the two main arguments for introducing newborn screening for DMD into Wales (Bradley *et al.*, 1993). In addition, many identified families feel that they have the right to know if their son has the disease, they have taken advantage of early physiotherapy and the opportunity of planning at a practical level for the future of their affected son.

We were very aware that screening presymptomatically in the newborn period did not fit the traditional criteria for such programmes (Haggard, 1990; Wilson and Jungner, 1968) and fears were expressed that both the early mother/baby relationship and family stability might be threatened. These fears have never been substantiated by research evidence and therefore could be attributed to what Oliver (1990) refers to as the 'psychological imagination': the projection by so-called 'normal' members of the population of their own imagined response should such a situation occur to them. Screening, therefore, was introduced into Wales in July 1990 with three safeguards:

(1) The programme incorporated a psychosocial evaluation of the impact of screening on both families and health professionals. Although other screening programmes operate world-wide (van Ommen and Scheuerbrandt, 1993), none have contained this element.
(2) It was made an opt-in test on the basis of informed parental consent.
(3) A protocol was developed detailing the procedure for handling a positive screening elevation and on-going primary health care support for identified families (Bowman, 1993; Parsons *et al.*, 1996).

The relationship between the social sciences and medicine has, at times, been strained, with evidence that the former have offered a highly critical appraisal, highlighting the negative aspects of medicalization (Freidson, 1970; Illich, 1977; Kennedy, 1983; Zola, 1972) and the marginalization of the sociologist (Freidson, 1983). The last decade, however, has seen an increasing acceptance of the contribution multi-disciplinary research can make in evaluating health care services and providing data for cost-benefit analyses. The incorporation of a medical sociologist into the newborn screening programme to undertake the evaluation of the psychosocial impact of the programme of newborn screening for DMD in Wales is an example of such acceptance. This chapter describes how the research was designed, and demonstrates the value of adopting complementary methodologies.

2. The psychosocial evaluation

No stipulation had been made by the research funding body (The Muscular Dystrophy Group of Great Britain) about the nature and structure of the original psychosocial evaluation. There were, however, certain practical constraints which had to be taken into account. First, although other programmes are operating world-wide there is only minimal published research in this area and none related to the psychosocial impact of newborn screening for DMD. Looking wider, there have been studies of parental response to newborn screening for cystic fibrosis (Al-Jader *et al.*, 1990; Boland and Thompson, 1990) but these have been retrospective and involved larger cohorts than the anticipated 4/5 affected boys per year in Wales. Second, the funding only provided for a half-time researcher responsible for both the day-to-day administration of the programme and the psychosocial evaluation. Third, although it was anticipated that only 4/5 confirmed positive results would be identified each year, and potentially a similar number of transient elevations, the population of Wales is very scattered, with over 200 miles between the research centre and the furthermost districts. It was therefore necessary to design the research to take account of the small cohort, the problems of geography and the limited research resources.

There are three parameters that need to be defined at the outset of any research: the research question/s to be answered, the sample population/s to be studied and the research methodology.

2.1 The research questions

There are certain standard questions that need to be addressed in any screening programme:

(1) Can the cases be found?
(2) Is there any benefit in screening?
(3) What potential harm will be done?

It has been shown that screening for DMD in the newborn period is technically possible (Zellweger and Antonik, 1975) but no analysis of the benefits and harms that result has been performed. The key questions for this study therefore centre around the psychosocial impact of screening.

2.2 The sample populations

It was decided that the main focus of the research would be a study of families with a confirmed positive result and smaller studies on three other sample populations: families in which the newborn boy had a transient elevation of blood creatine kinase (CK), families who refuse the screening test (approximately 5% of families refuse screening each year) and health professionals who either offer the test (midwives) or are involved with identified families (consultant paediatricians, health visitors and General Practitioners – GPs). This chapter specifically discusses the research designed to evaluate the response of families with a confirmed positive result, details of the other studies will be published elsewhere.

2.3 The research methodology

In very broad terms, there are two different methodological approaches to collecting research data: quantitative and qualitative. The former aims to measure and enumerate the impact of an intervention on specific indicators; the latter describes, or classifies, the experience of respondents using less structured research tools – thereby encouraging the collection of rich, in-depth data that is more respondent-generated.

Baum (1995) discusses the innate preference of medical science to align itself with the positivist paradigm and the belief that 'phenomena can be reduced to their constituent parts, measured and then causal relationships deduced'. A quantitative research protocol for newborn screening might take the following format:

Indicators: Measure/score of mother/baby relationship
                   Measure/score of family stability
                   Measure of levels of anxiety using standardized test
                   Attitudes to screening
                   Reproductive intentions
Control Groups: A matched group of mothers in the general population; a group of mothers who have a son diagnosed later
Research Tools: Structured questionnaire consisting of standardized psychological measures and closed questions, administered at fixed time intervals
Principle: Every family would be asked the same questions at the same time interval and compared with controls to measure the impact of the intervention

In contrast, a qualitative study might have the following design:

Indicators: A number of themes would be identified at the outset to focus data collection, for example the nature of the mother/baby relationship, family stability, responses to the diagnosis, reproductive intentions, attitudes to screening etc., but it would be anticipated that other themes would emerge as a result of reflective evaluation of the data
Research Tools: Unstructured/Semi-structured interviews
                   Case Studies and field notes generated by the researcher
                   Data generated by other health professionals would also be used, to supplement the researchers primary field notes and offer useful 'triangulation'
Principles: The social construction of reality by each family is unique and therefore their experience of a common life event cannot be measured using pre-designed categories, at fixed time intervals.

There were strong arguments for adopting a purely qualitative design. The lack of previous research meant that there was a need 'to provide insights into aspects of reality otherwise not plumbed' (Freidson, 1983). The sample size would be small (approximately 10–15 over the three years of the initial funding), so it would be difficult to establish findings that would have useful levels of statistical significance. The topic area was potentially very sensitive, and it has been argued that 'private' accounts are far better collected using qualitative methods (Cornwell, 1984). The 1990s, however, have seen an increasing recognition that there is a need for integrated methodologies that are eclectic (Baum, 1995). Pope and Mays (1995) similarly recognize that the dichotomy between the qualitative and quantitative approaches has been overstated. The use of complementary methodologies, incorporating elements from both approaches appropriate to our specific setting, was therefore attractive. There were responses, such as the family's attitude to newborn screening, which could be enumerated, whilst issues such as their awareness and response to the onset of symptoms could be better explored in a less structured way.

The initial phase of research consisted entirely of data generated by the researcher, that is, primary data:

(1) An informal home visit 2/3 weeks after the confirmed positive result and then annually, generating comprehensive field notes.
(2) Two semi-structured interviews (one 6 months after confirmation of the elevation of CK and one when the boy was 4 years old and becoming symptomatic). The interview schedule incorporated structured questions (e.g. attitudes to screening, scored evaluation of how the diagnosis was handled, a closed question on relationship with partner, adjective check lists to assess current emotions and two standardized measures of anxiety and general well-being) together with topic areas to be explored in greater depth (e.g. the reason why they chose to have the test, how they responded to the news, whether they have ever regretted screening, their account of events since that time, whether they feel that their relationship with their son has changed and their future expectations etc.).

Although this protocol appeared to be quite robust, we soon realized that many of the families' spontaneous emotions and vacillations in attitude would be missed because of fixed time interval interventions. It is a truism that respondents' most interesting moments never occur when an interview is scheduled, they happen either before and have been forgotten, or after and are never recorded. This evident gap in our research led us to explore the use of secondary data sources, in particular, the potential input from health visitors. In the newborn period they are the main primary health care support for the family, they have been involved with them before the screening test results are known and, as part of their professional role, are concerned with the mother/baby relationship, the baby's development and the family unit. Just as the primary data used complementary methods, so did our design for secondary data collection. A semi-structured questionnaire was introduced for health visitors which contained:

- A numerical assessment of the mother/baby relationship before and at two points after the screening result, using seven indicators (Salaryia and Cater, 1984).
- Questions about each family's response to the diagnosis.
- A closed question about their own attitude to screening.
- Closed questions about their assessment of the information and the support they had received.

In addition, we have developed a new, qualitative method of collecting secondary data which we have called the 'Technique of Professional Accounting' (TPA). This technique involves including accounts of events in the case studies from health professionals who are working with the family (for example health visitors, physiotherapists, family care officers, clinicians, specialist genetic fieldworkers, social workers). The data are collected by regular telephone contacts with the health professional concerned and are focussed on five main areas: (i) the family's on-going response to the diagnosis and the onset of clinical symptoms; (ii) the nature of the mother/baby relationship; (iii) the impact on family stability; (iv) attitudes to screening and (v) reproductive intentions and patterning. In the first few months after the confirmed result, the researcher and the health visitor are usually in weekly contact, later there are fewer contacts but other professionals become involved. Overall the number of professional accounts in 1 year amount on average to 25 per family.

The addition of secondary data to each family case study has added a new dimension to the research. We have not only achieved an axis 'Structured/Semi-Structured' but also 'Primary/Secondary' (*Fig. 1*). The interrelationship between primary and secondary sources in terms of each research topic is shown in *Table 1*.

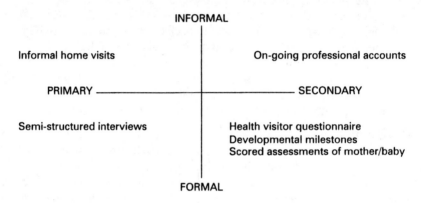

Figure 1. Complementary data collection methods

Table 1. The interrelationship between primary and secondary data sources by research topic

Research topic	Primary data	Secondary data
Response to diagnosis	Interviews – open and closed questions Adjective check lists	Professional accounts (TPA)
Mother/baby relationship	Interviews – open and closed questions Adjective check lists Health visitor questionnaire	Professional accounts (TPA) (esp. of over-protectiveness and under-stimulation). Numerical assessment
Family stability	Interviews – closed and open questions Health visitor questionnaire	Professional accounts (TPA)
Attitudes to screening	Interviews – closed questions Health visitor questionnaire	Professional accounts (TPA)
Reproductive patterning	Interviews – open questions Reproductive behaviour	Professional accounts (TPA)

The use of secondary data sources means that the on-going responses of a family can be captured without excessive researcher intrusion and in addition rich, qualitative data can be generated relatively economically. It should be noted that this technique is only effective, and ethically acceptable, if both parties are aware that research data are being collected in this way and have given their consent.

3. Complementary methodologies discussed

In a research setting where it is possible to adopt complementary methodologies, such as the newborn screening programme for DMD, the great advantage is that one gets the best of both worlds. We have successfully collected substantive quantitative data which is enhanced by rich, in-depth qualitative accounts, both primary and secondary. We have feedback from the families, quantitative and qualitative, at fixed time intervals which enables important between-family comparisons, but we also have reports on the on-going response of the family through TPA. In addition, mixed methodological research allows a degree of flexibility which can reduce non-response rates. The social interactionist acceptance that individual patterns of social action are uniquely constructed fails to recognize that respondents are equally unique in terms of the way they prefer to generate research data about themselves. The semi-structured, informal interview may be methodologically sound, but is it necessarily the research tool all respondents prefer? We have demonstrated in our research programme, in which the majority of families have been involved in both elements of the research, that there are those who find the closed questions in our questionnaires too restrictive and there are also those who do not feel comfortable talking in a less structured way in interviews about very emotional issues. There is one family who have opted out of the primary data collection process, but who are still willing to be involved with the research via secondary sources.

In the next section we will focus on two particular areas and, drawing on the research, demonstrate how the use of complementary methods has broadened the scope of our research data.

(All names have been changed and certain personal details have been constructed to retain anonymity.)

4. Complementary methodologies in practice

4.1 The qualitative/quantitative data axis

Example 1: The value of quantification: an assessment of the Welsh protocol.
While the diagnosis of DMD is devastating to a family at any time, we recognized that although the message could not be changed, the way families received the news could be strategically planned. During the first 2 years of the programme we developed a specific protocol for handling any elevated screening test result (full details are reported in Parsons *et al.*, 1996).

We have been able to interview two groups of families, 25 who had the news broken about the diagnosis of Duchenne using the protocol and 16 where it was not used. The respondents were asked to rank their level of satisfaction with how the situation had been handled from 'Poor' (1) to 'Excellent' (5). A statistically significant difference was found with 88% of families where the protocol had been used saying it had been 'well' or 'excellently' handled (score of 4 or 5) whilst only 19% did so where the protocol had not been in place.

The detailed and often graphic accounts from families about those early medical encounters had been translated into meaningful statistics by the addition of three structured questions in the interview.

Example 2: Attitudes to screening: the reality of vacillation. In each family questionnaire there is a structured question about their attitude to screening, offering the options: 'In Favour', 'Against', 'Unsure'. There have been three families who have consistently ticked 'In Favour', and during the two semi-structured interviews they have remained consistent, even when asked if there have been any short periods of regret. It has been the qualitative accounts that have provided a valuable insight into periods of uncertainty. This is illustrated in the following case study.

Case Study 1: Seven weeks after the elevated CK result had been confirmed the health visitor said she had met Jane and her mother in town, and although they were anxious waiting for the genetic test results, they had been very positive about screening saying 'that it was far better to know these things than find them out later'. Two weeks later the health visitor received a telephone call to say that the family were 'at the end of their tether' and were regretting ever having agreed to the test. The addition of the professional account enabled us to put this response into context: that week Jane's sister had announced she was pregnant and Jane's father was admitted to hospital with cancer.

Once highlighted, these emotional vacillations have been identified as a new theme to be extrapolated from other family accounts and to link with a discussion (below) about the 'diagnostic crisis'.

Example 3: The interview as a social process: does it preclude certain expressions of emotion? Conventional methodological wisdom has argued that the semi-structured interview enables respondents' 'private' accounts (Cornwell, 1984) to be explored. There is within the qualitative tradition a concern not to define the reality of one set of accounts above another, but underlying this is the implicit assumption that 'more' will be revealed. The following case studies offer a challenge to this position.

Case Study 2: Judy's son was 2 years old when the researcher and health visitor made the annual home visit. This consisted of a short structured questionnaire (sent ahead in the post) and an informal discussion. We left after an hour feeling that Judy was 'a bit down', but the health visitor did not feel that anything had changed. Later, when the questionnaire was analysed, it revealed significant indications of depression on both the General Health Questionnaire and the Spielberger (STAIT) anxiety measures and major worries that had not been verbalized were also identified on the adjective check list.

Case Study 3: Penny and Ian both wanted to be involved with the research and Ian had taken time off work for the first interview. They organized a cup of tea in the garden whilst we talked, both of them having already completed the structured questionnaire. They talked about their experience of the diagnosis and their relationship with their affected son, Gareth. Penny said she was finding things 'a bit of a strain' and Ian said 'It is just the upset of it all and thinking, what is Gareth is going to be like? – that's the most worrying part'. No stronger emotions were expressed verbally. The profile from the questionnaire gave a far more graphic picture of their emotional state. In response to the adjective check list for current emotions Penny had ticked: 'angry, no-one understands, afraid, always wanting to cry' and had added her own word 'heart-broken'. Ian had ticked: 'why me ?, resentful, and afraid'.

In both case studies far more emotion had been expressed in the structured questionnaire than during the qualitative interview. There are two possible explanations. First, because the researcher has become a 'professional friend' they respond positively to the visit as they would to any other guest and during tea in the garden deep emotional hurts are not expressed in the same way as when they are alone. Second, the quantitative element of the study offers the opportunity, with its wide range of adjectives (all previously generated by other respondents during previous research in DMD) to identify some of the underlying emotions which the parents are still not able to verbalize.

Clearly, allowing respondents to generate both qualitative and quantitative data enables the researcher to develop a far more comprehensive picture of the impact of newborn screening.

4.2 The primary/secondary data axis

Example 1: Health visitors' professional accounts: the mother/baby relationship. One area of major concern surrounding newborn screening for DMD is its potential impact on the mother/baby relationship. This is assessed in several ways: a standardized scoring key completed by health visitors (Salaryia and Cater, 1984), structured questions in both the family and the health visitor's questionnaire, and as a topic area explored by the researcher during the interviews. This however is all supplemented by process data generated through professional accounts.

Case Study 4: Two weeks after the disclosure of the newborn screening test result, the health visitor mentioned that she thought Gwen's attitude to her baby had changed:

'. . . . it is much more ambivalent . . . she is grasping him to her and looking him in the face, I am sure that her look is different'.

Two weeks later she was convinced:

'She has definitely changed in the way she is handling Rory. Before the result it was 'pass the parcel', he was always passed around, but now he is hidden and she is the only one touching him.'

The following week she reported:

'I would say that there has been some change back in terms of handling and touch, he is not so far away from her, but since the query her touch has been more general than before, that nuzzling of the head has gone, there is more nuzzling again but it is still not back to the normal position.'

It took a further week before she was sure that, 'things have settled back to normal'.

This case study illustrates the value of the qualitative, secondary accounts. It also demonstrates a significant element of 'control' – in terms of reliability and validity – that can be offered by health visitors who have known the family before the initial query was raised. They provide important base-line data on the prior mother/baby relationship, family stability etc. They can also make significant comparisons in terms of levels of anxiety, expectations etc., with babies from other families in the same geographical area, of the same age, who have not had a positive result.

Example 2: Professional accounts: identifying the 'diagnostic crisis'. The family's response to the diagnosis is recorded by the researcher at the two fixed time interval interviews. It has already been argued that the use of professional accounts enables data that are generated at other times to be recorded and have been responsive to vacillations in emotions and critical incidents in the family.

Case Study 5: Penny and her husband were living with her mother when the diagnosis was made. Although there had been a certain amount of distress expressed, the health visitor felt that kinship support had been effective in 'dissipating' it, that was until 36 weeks later when Penny showed signs of depression and became very aggressive in her behaviour. The health visitor felt it was a delayed reaction to the diagnosis, combined with the frustration of living at home and Penny's arm being in plaster. She received medication and the events of that week were never mentioned again, certainly not when the researcher called some 15 weeks later.

Case Study 6: It was the health visitor's account that gave a significant insight into the immediate response of Paula and John:

> 'They are just devastated, John has not gone back to work and they have just been sitting by the crib bawling their eyes out. Paula was a bit depressed before the news and is even worse now.'

One week later Paula's response was: '1 in 4000 – why did it have to be mine?'

The following week Paula was beginning to explore her emotions and realized that one of her biggest problems was the fact that he was going to die:

> 'How am I going to tell him that he will not be here when he is 20 and it's all my fault?'

She had decided not to tell their neighbours and friends because she did not want him to be treated any differently, but she was finding it hard to cope when people looked in the pram and said, 'What a lovely baby'. She said, 'I just want to scream at them: 'yes he is but he is going to die!'.

It was the addition of the TPA that enabled these vivid responses to be recorded and the theme of the 'diagnostic crisis' to be identified. This theme, with its chronological variability and range of expression, is reported elsewhere (Parsons and Bradley, 1994).

5. Summary

In summary, we have identified 6 advantages of the TPA:

(1) It provides process data about events that are often lost in respondents' retrospective accounts, including vacillation in emotions and attitudes.

(2) It enables respondents' accounts to be dated and set in their social context.

(3) It enables the generation of important research data without unnecessary researcher intrusion and in a relatively economical way.

(4) It has eliminated non-participation in the research: all families with a confirmed positive test result are involved, at one level or another, with the research programme.

(5) The use of the health visitor provides an important control in terms of baseline and comparative data.

(6) The use of health professionals from different specialities, describing a family's response to the same event, offers a degree of 'triangulation'.

In addition, a number of indirect benefits have been identified:

(1) Health professionals have recognized that regular contact with the researcher has been a source of specialist information and support for the family.

(2) It provides a valuable indicator that the family is receiving on-going support from the primary health care team.

(3) The focused nature of the research has offered health professionals a framework for their on-going relationship with the family.

(4) The research has raised health professionals' awareness about the issues raised by screening, and consequently families have gained in terms of more integrated support. Preliminary analysis indicates that families with a disclosure in the newborn period receive much stronger support from the primary health care team than families who have experienced a later, traditional diagnosis. The full implications of this difference have yet to be explored.

6. Conclusion

The final structure of the psychosocial evaluation was reflexively developed during its early years, with several quantitative measures being introduced as a result of feedback from the qualitative elements. This research programme has demonstrated that, where the use of complementary methodologies is feasible, it enables a better understanding of the richness of the world and confirms Baum's contention that 'methods should explore rather than deny diversity' (Baum, 1995).

References

Al-Jader, L.N., Goodchild, M.C., Ryley, H.C. *et al.* (1990) Attitudes of parents of cystic fibrosis children towards neonatal screening and prenatal diagnosis. *J. Med. Genet.* 27: 658–659.

Baum, F. (1995) Research public health: behind the qualitative–quantitative methodological debate. *Soc. Sci. Med.* 40: 459–468.

Boland, C. and Thompson, N.L. (1988) Effects of newborn screening for cystic fibrosis on reported maternal behaviour. *Arch. Dis. Child.* 65: 1240–1244.

Bowman, J.E. (1993) Screening newborn infants for Duchenne muscular dystrophy. *BMJ* 306: 349.

Bradley, D.M., Parsons, E.P. and Clarke, A.J. (1993) Experience with screening newborns for Duchenne muscular dystrophy in Wales. *BMJ* 306: 357–360.

Cornwell, J. (1984) *Hard Earned Lives: Accounts of Health and Illness from East London.* Tavistock, London.

Freidson, E. (1970) *The Profession of Medicine: A Study of the Sociology of Applied Knowledge.* Harper Row, New York, NY.

Freidson, E. (1983) Viewpoint: Sociology and medicine: a polemic. *Sociol. Health* 5: 208–219.

Haggard, M.P. (1990) Hearing screening in children. *Arch. Dis. Child.* 65: 1193–1195.

Illich, I. (1977) *Limits to Medicine.* Penguin, Harmondsworth.

Kennedy, I. (1983) *The Unmasking of Medicine.* Granada, London.

Koenig, M., Hoffmann, E.P., Bertelson, C.K. *et al.* (1987) Complete cloning of the Duchenne muscular dystrophy (DMD) cDNA and preliminary genomic organisation of the DMD gene in mouse and affected individuals. *Cell* 50: 509–517.

Oliver, M. (1990) *The Politics of Disablement.* Macmillan, Basingstoke.

van Ommen, G.J.B. and Scheuerbrandt, G. (1993) Neonatal screening for muscular dystrophy. Consensus recommendation of the 14th Workshop sponsored by the European Neuromuscular Centre (EMNC). *Neuromusc. Disord.* 3: 231–239.

Parsons, E.P. and Bradley, D.M. (1994) 'They must have got it wrong': the experience of families identified by a newborn screening programme for Duchenne muscular dystrophy. In: *Qualitative studies in health and medicine* (eds M. Bloor, M and P. Taraborrelli). Avebury, Aldershot, pp. 43–59.

Parsons, E., Bradley D. and Clarke A. (1996) Disclosure of Duchenne muscular dystrophy after newborn screening. *Arch. Dis. Child.* 74: 550–553.

Pope, C. and Mays, N. (1995) Researching the parts other methods cannot reach: an introduction to qualitative methods in heath and health services research. *BMJ* 311: 42–45.

Salaryia, E.M. and Cater, J.I. (1984) Mother child relationship – first score. *J. Adv. Nurs.* 9: 589–595.

Wilson, J.M.C. and Jungner, G. (1968) *Principles and practice of screening for disease.* Publ. Hlth. Paper, WHO No. 34.

Zellweger, H. and Antonik, A. (1975) Newborn screening for Duchenne muscular dystrophy. *Pediatr.* 55: 30–34.

Zola, I.K. (1972) Medicine as an institution of social control. *Soc. Rev.* 20: 487–504.

19

Childhood, genetics, ethics and the social context

Priscilla Alderson

1. Introduction

How much do ethics and genetics have in common? This chapter suggests that current understandings of ethics and of scientific, clinical and counselling work in genetics have common characteristics. More critical awareness of these common threads, and more attention to the social context, would deepen understanding in all these disciplines and increase their efficacy. Underlying patterns of thought in social attitudes towards childhood and eugenics will be considered, to show how ethics and genetics reflect and reinforce such attitudes.

The example of Robert illustrates how attention to the social context can easily be lost. Robert had no physical or mental disability. However, his family lived in two small damp rooms and he often had ear infections. When he began to attend school, he was soon in trouble for not obeying his teacher, and for not learning or relating positively to other children because his hearing had become so poor. Soon he was 'kept in' during break times and he grew more lonely and restless. His parents began to 'ground' him after school, and he became sad and angry. The teachers started to talk of hyperactivity, fragile X and Asperger's syndrome, and referral to a special school. The case of Robert provides one of many examples of how social conditions are reframed into supposedly genetic ones.

2. Policy from the past

Today's health services for those affected by genetic conditions can have great benefits as amply illustrated through this book. However, a darker side of genetics is linked to trends from the past which can still be influential, although they are now followed less consciously and less publicly.

One typical example of its time is a Medical Research Council (MRC) Report from

The Genetic Testing of Children, A.J. Clarke (ed.).
© 1998 BIOS Scientific Publishers Ltd, Oxford.

1932. The premise of the report is that crime, 'indicates a congenital fault of mental organization... a failure of the brain to function properly'. Menstruation is posited as a cause of insanity, and crime is '...attributable to irritation of the ovaries or uterus, a disease by which the chaste and modest woman is transformed into a raging fury of lust'. Facial and other anomalies of the 'criminal type' were thought to resemble 'the Mongoloid or sometimes the Negroid', being morally and physically 'peculiar'.

Psychologists quoted in the report had examined numerous women prisoners, noting their abnormal genitalia, menstrual changes and 'orgies' which included attempts to escape from prison. Many case studies of working class adolescents noted ugly and 'fierce' features, the 'marks of a hopeless moral reprobate'. These 'very dangerous' girls, the MRC Report recommended, must be sterilized and permanently confined in colonies 'under the guardianship' of psychologists who should be allocated 'hunting grounds' in working class areas from which to select their cases (Pailthorpe, 1932).

Are attitudes now very different? The 1990s are only a single academic generation from the report, because the experts quoted, such as Cyril Burt, were teaching in the 1950s when many of today's professors were trained. Similar views pervade present society, on 'problem adolescents', on 'congenital faults' that are thought to affect morality and behaviour, on the importance of researching and controlling 'abnormality', and on biology as destiny without reference to social influences. Overt misogyny and racism are far less respectable today, but prejudice against disability and difference is still pronounced (Crawshaw, 1994; Nuffield Council on Bioethics, 1993; Oliver, 1990), and working class and black young people are greatly over-represented in psychiatric and penal systems (Lansdown and Newell, 1994). The following sections consider how current attitudes in genetics, ethics and society reflect or challenge those earlier views on inherited conditions.

3. Today's attitudes towards childhood and adolescence

Genetics is inevitably influenced by its social context, including attitudes towards babies and children. Our society is somewhat hostile to children. Of course, there are many loving relationships between adults and children, as there are loving relationships between individual black and white people. But just as there are racist structures in society, there are also ageist ones set against young people. Ageism is almost always discussed in relation to old people. We do not even have a word for 'childism', possibly because it is so endemic and accepted that it is rarely noticed.

Here are a few examples of numerous negative responses to young people. Governments that have ratified the United Nations (UN) Convention on the Rights of the Child (United Nations, 1989) report regularly on their progress to the UN. The UN was very critical of the British government for its many failures to respect and protect children (United Nations, 1995). In over 300 pages, ways are documented in which law, policy and practice in the UK have to be changed if standards are to meet those in the Convention (Lansdown and Newell, 1994). Problems for young people are increasing (Wilkinson, 1994) and one common theme is exclusion from mainstream society: increases in school exclusions, homelessness, and unemployment among young people; rising suicide rates among young men; one in three children living in poverty; cuts in benefits for young people; the fear of strangers and dangers keeping children away from public places.

Party political manifestoes constitute a revealing measure of public opinion, advocating policies which they believe will attract the most votes. The two main British parties both emphasize measures to control and punish young people, such as a proposal (June 1996) for an evening curfew for everyone under a certain age, although the vast majority are not criminals. Young people are constantly criticized and derided in the mass media, and tend to be stereotyped as either victims or villains, instead of being seen as the complex and often competent individuals they actually are. Adolescence is usually framed as a 'problem' age, as if this is a biological inevitability. It is not certain that teenagers really experience more problems than other age groups, and the problems they have can more clearly be traced to social structures than to their personal qualities or physiology (Musgrove, 1984; Phoenix, 1991).

Another exclusion is in social research. Children are hidden inside 'the school' or 'the family'; adults' views about them are surveyed but we know almost nothing about children's own views and experiences (Mayall, 1994). In Western societies concerned with cost rather than value, children tend to be seen as expensive dependents, as a threat or a nuisance (Holt, 1975). Anthropologists find that abortion and active child abuse, which includes unrealistic demands for 'the perfect child' and excessive pressures on children to succeed, are endemic in wealthy societies. In contrast, passive inadvertent child neglect and infanticide through poverty is endemic in poorer societies where children are loved in less conditional ways (Scheper-Hughes, 1987). Different treatment of children and attitudes towards them arise from deep assumptions about differences between adults and children. Human beings are divided into two unrealistic extremes (see *Table 1*), with the left-hand-side column (of *Table 1*) assigned to children and, until recently, to women, and the right-hand-side column to adults and, especially, to professionals.

Trends in attitudes towards children and teenagers, with the 'them-and-us' opposition between adults and younger people, have been summarized because these can

Table 1. Assumed child–adult dichotomies

Feeling	Thinking
Emotion	Reason
Body	Mind
Ignorant	Knowing
Inexperienced	Experienced
Foolish	Wise
Volatile	Stable
Weak	Strong
Dependent	Protective
Unreliable	Trustworthy
Immature	Mature
Irresponsible	Responsible
Child	Adult
(Female	Male)

influence values and policies in genetic research and in clinical practice and counselling. If, however unconsciously, children are seen as a threat or a drain on market economies and on adults' freedoms, then children with genetic impairments are likely to be viewed still less favourably. Such attitudes can then generate and license programmes to prevent the birth of impaired children, and to control teenagers and young adults who carry certain, 'unacceptable' traits and disorders (Foucault, 1970). Arguably, to divert billions of dollars, and the accompanying public concern and hope, into genetic research and away from social research and reform detracts from addressing the real problems of young people (Watt, 1996).

Genetic research is presented as a means of furthering children's interests in the future through the development of gene therapy. The newsletters of self-help associations and family support groups concerned with genetic disorders celebrate research as liberating children from extremely distressing conditions and their siblings from passing on disorders to their own children. Yet how much are these reports intended to further researchers' interests and to increase their funding, support and prestige? Researchers assume that they serve the families' interests, but when genetic knowledge is used mainly to detect and terminate affected pregnancies, the benefits to people who have muscular dystrophy, for example, are less clear. When asked, people usually say that they would rather have some life than none at all. There is a tragic contradiction in the newsletters between reports which celebrate the lives of affected children (when they win medals and awards or go to university) and reports which support research primarily used to prevent their lives. The research may detect a relevant gene long before any 'treatment' other than termination of pregnancy is available. Disabled people are concerned that by the time a treatment is developed, so few of them will have survived prenatal screening programmes, and the few that do so are likely to be so stigmatized, that investment in developing 'cures' will no longer be considered worthwhile. It is probably more economical and efficient to select unaffected embryos for *in vitro* fertilization, rather than to try to tamper with genes after conception. In any case, whilst gene manipulation for disorders such as cystic fibrosis or muscular dystrophy will not necessarily alter the person concerned, with other disorders this may not be so. Is Down's syndrome or fragile X an essential part of the 'actual' person? Who would they be without these conditions? Is the ethical and most rewarding approach to these conditions a social rather than a scientific one: to accept affected people as they are and to value them as full members of society?

Questions arising from a realistic view of social attitudes towards children in late twentieth century Western society include the following: How much do current attitudes towards children promote the human genome programme? And how much does the programme, along with many other factors, affect attitudes towards children?

4. Research about childhood

Current social research about childhood has three main areas.

(i) Analysis of popular and academic theories of childhood (James and Prout, 1990) shows how philosophy and developmental psychology are misled, for example, by pre-Darwinian notions of evolution (Morss, 1990), by scientists' personal fantasies (Bradley, 1989), and by inaccurate assumptions, such as those in *Table 1*. Scholars are gradually unravelling and refuting such theories.

(ii) Large-scale studies of children's social and economic status demonstrate much disadvantage and exclusion (Qvortrup et al., 1994).

(iii) Small-scale surveys of individual children's views and experiences demonstrate that they have greater capacities than has generally been assumed (Alderson, 1993; Hutching and Moran Ellis, 1998).

The extensive research literature on childhood presents both an intellectual and an emotional challenge. Intellectually, we are invited to rethink numerous influential fallacies and to re-examine social realities. Emotionally, our initial response may be to feel anger and anxiety when our deepest assumptions about childhood are questioned, but we are challenged to develop more creative responses.

Perhaps inadvertently, ethicists and geneticists tend to assume, and thereby to reinforce, discredited theories about children's incapacities and limitations (Hughes, 1989). This is shown in debates about children's consent to genetic tests, when historical beliefs are quoted to justify future policies (Wertz, et al., 1994), instead of being critically examined. Ethicists accept refuted theories about children's limitations (Buchanan and Brock, 1990; Scarre, 1989), and question their status as human beings and even the value of their lives (Harris, 1985; Tooley, 1983). In both disciplines, when the human intellect is highly valued, professional and academic thinking implicitly, inevitably and perhaps inadvertently, undervalues the lives of young and disabled people of low intelligence.

Troubled by the animal aspects of human nature, ethicists have pursued two simplistic and unsatisfactory alternatives (Midgley, 1994). One is to perceive 'real' humanity as mainly consisting in our spiritual and intellectual aspects, the body being regarded as a problem. In this view, young children risk being classed as animals rather than humans, and forfeit their human rights. Childhood is seen as a long apprenticeship in becoming human, and even young adults' morality is still sometimes reduced to biological terms. For example, in 1996, a psychiatrist wrote a leader (which is supposed to summarize the most up-to-date conclusions) on 'teenage sex' in the *British Medical Journal*. She alleged that unreliable use of contraceptives may be attributed to 'cognitive immaturity'. 'We know that the frontal lobe of the brain, which deals with control of sexual drives as well as abstract reasoning and planning, is not completely myelinated until 14 or 15 years of age. Major changes in cognitive ability occur between early and late adolescence, most notably the capacity to reason abstractly, predict future consequences, and see things from different perspectives' (Stuart-Smith, 1996).

Genetics identifies with such reductionism in the attempts to improve and to 'perfect' the body, to trace the origins of behaviours such as violence to bodily material rather than to the mind or relationships or the social context, and to assume that people generally would prefer to select for high intelligence. Another response is to emphasize animal aspects of human nature in a negatively 'brutish' image of irrational humanity, and to stress the need to control this through such means as contract theory (Rawls, 1972), risk management (Thornton, 1993), and market forces instead of altruism (Hutton, 1996).

Contracts which are appropriate in business deals between strangers now invade personal relationships. Marriage is no longer acceptance 'for better, for worse, in sickness and in health'; it is becoming more like a business contract with a growing number of conditions, including checking out potential partners' genes. The parent–child relationship is perhaps the only remaining human relationship of unconditional acceptance, though this is increasingly provisional and intolerant, from pre-childhood in parents' antenatal choices to the end of childhood as illustrated by the growing numbers of homeless young people.

Although it is only one of many social movements, genetics arguably plays a key part in this move to redefine human relationships as contracts (Stacey, 1992). In promoting 'parental choice', genetics not only makes possible the selection and rejection of potential children, it also presents such choice as morally acceptable, even desirable and responsible, although abortion was illegal in Britain and considered reprehensible until 30 years ago.

Frequent reference to parents' guilt about passing on genetic defects represents a change away from former religious and fatalistic views about inheritance and human nature. It further implies that we 'ought' to pass on 'high' standards of body and mind to our children, and to expect high standards from them. Standards of what it counts to be an acceptable human being are rising, as more genetic conditions are identified for which pre-natal choices can be offered.

The 'contract society' which emphasizes formally negotiated and conditional relationships brings welcome new benefits and liberties, but also brings the loss of trust, tolerance and security. Contracts, being essentially bargains between equally informed and voluntary agents for their mutual reward, will especially disadvantage children. Knowledge and power (the elements of informed voluntary consent to contracts) are often unequally distributed between children and adults. To talk, as some ethicists do, of a contract between the fetus and the parent is to misunderstand the necessity for equality between the parties to a contract. The infinite variety and subtleties of parent–child relationships elude definition and slip between the hard grid of the formal contract. The salient qualities in parent–child relationships largely concern being, rather than doing, but to set a contract about the kind of person a son or daughter must be turns the child into a kind of 'super-pet' (Holt, 1975).

Popular notions of the child as a net drain on resources, and a burden rather than a blessing, are extended in antenatal genetic services when it is stated or implied: 'Children are so difficult to bring up [a telling phrase] that any extra defect or disability will make life hard, perhaps impossible, for the parents.' The assumption that life with a disability is too great a burden for a family to bear can implicitly over-estimate the effects of mental or physical impairments and under-estimate the social measures which help severely impaired people to lead a fulfilling life. It can express overt hostility to disabled people, or covert hostility in presupposing that society will reject them.

Some laws in Europe are moving away from seeing children as their parents' property, towards respecting children, and allowing parents to have rights only in so far as this enables them to fulfil their responsibilities (Montgomery, 1988, 1997). However, genetics is moving the other way in its emphasis on parents' choice, their right and ability to select and reject potential children (Brazier,

1996), and their right to know and publicize personal and sensitive information about their child (Bobrow *et al.*, 1991; Dalby, 1995).

Many geneticists and counsellors believe that they simply respond to society's demands for testing and screening, and that morally they are reactive rather than proactive. Yet by virtue of their relatively high status, derived from their alliance with scientific and innovative medicine, wittingly or not, they strongly influence public opinion, collectively through the mass media (e.g. statements made by academics such as Professors Steve Jones, Richard Dawkins and Lewis Wolpert; books such as Dawkins 1976), and individually when counselling new clients (Rapp, 1988). The antenatal period is an impressionable time for future parents, especially during a first pregnancy and for those with little experience of children or of disability. Along with medical advice, clients absorb social and moral advice from professionals they see as experts in a prestigious specialty. A pattern of parental choice and notions of potential children as perfect, acceptable or substandard products is promoted through antenatal screening. As consumerism overtakes the older, more fatalistic and religious world views, and as it colours the way society frames the questions raised by genetics, their public policy, morality and science mirror and fortify one another in ways that are especially apparent in the context of genetics.

5. Genetics and eugenics

Eugenics is defined in the Shorter Oxford English Dictionary as meaning 'pertaining or adapted to the production of fine offspring, the science which treats of this'. The aims and net effects of genetics and eugenics are similar. For example, currently in Britain, spina bifida is diagnosed in about 1000 fetuses each year, and about 900 of these pregnancies are terminated (O.P.C.S., 1994). Some adults with the condition argue that their lives are worthwhile and they criticize the current policy.

Because of its history, eugenics is usually dissociated from genetics by geneticists and ethicists. They argue that eugenics involved 'state' coercion, whereas genetics takes the opposite approach in respecting and increasing personal choice. Yet when the social realities of the two approaches is considered in some detail, the abstract contrasts become less convincing. Diane Paul illustrates how eugenics has a public 'front door' policy (Paul, 1992). This involves an overt public policy, government programmes, and open appeal to people's patriotic obligations. If anyone resists the eugenic programme then there is obvious coercion which enforces compliance. The public and private benefits are much publicized.

In contrast, genetics has a 'back door' approach when governments support eugenic practices in law and publicly funded hospitals, such as abortion of impaired fetuses and discouragement of 'risky' conceptions. Yet in Britain there has always been some political ambivalence towards abortion and infertility services, which have never been adequately funded or routinely provided in all districts (Pfeffer, 1993), and this is linked to a wariness about eugenics. Counselling services tend to be individual and semi-private (when compared with national nursing specialities). Clients are offered choice, and indirect persuasion through 'non-directive counselling' and values propagated through the mass media. The private benefits to the adults concerned are emphasized, and the discussion of the benefits to the state are confined mainly to medical journals.

In an abstract analysis, genetics and eugenics can look very different. However, in an analysis which takes account of the social context, it is immediately clear that the net results are the same: a highly organized national programme to ensure 'the production of fine offspring'. The second 'personal choice' approach can be seen as more effective than enforced eugenics. Power is less effective when people are aware of it, and is far more potent when people believe that they are choosing and not complying (Lukes, 1974; Rose, 1990). This is not to say, that individual and state interests are necessarily at odds, but that abstract analyses which ignore historical links can be misleading.

Antenatal screening and selection are claimed to be humane alternatives to infanticide, a view which has been accepted in all societies. This argument begs questions about choices and choosers. Families on subsistence level have far fewer options about sustaining impaired babies than have parents in wealthy societies with effective health services. The uncertain history of infanticide tells us nothing about the mothers' views, and whether they were consenting or distraught. Parents' grief at stillbirth and infant death, which 40 years ago went unmentioned in medical textbooks, is now widely understood as an intense experience. Suggestions that this is a modern phenomenon are refuted in classical Chinese poetry and, in England for example, in Ben Johnson's sixteenth century verses on his dead children (Johnson, 1954).

Antenatal screening is not experienced by women as wholly beneficial (Marteau, 1993). Concepts of percentile charts and measuring the fetus and child for 'normal' growth are very recent. A 'norm' meant a carpenter's set square until the late nineteenth century, when eugenicists began to apply the term to people, as recorded in the Oxford Etymological Dictionary of the English Language. Although physical differences have always been noted, differences in mental ability appear to have been largely unnoticed before the eighteenth century (Goodey, 1995), and these alone would not show in the early months so would not be a reason for infanticide. Measuring human physical and mental differences and attempts to correct and prevent difference are recent concerns (Foucault, 1976).

The next section considers whether an ethics dealing with abstractions or one dealing with everyday experience contributes most to the moral understanding of genetics.

6. Abstract or evidence-based ethics?

The ethicist Daniel Callahan claims that the 'blooming [sic] buzzing confusion' of actual experience 'cannot be ordered or morally interpreted without the help of abstractions ... The aim of ethical theory is to provide us with some general, high-level abstractions that will help us to make sense of experience' (Callahan, 1996). Ironically, he quotes William James' notion of the 'buzzing confusion' in which babies were supposed to exist. Psychologists have since demonstrated babies' powers to make sense of experience and have concluded: 'the buzzing confusion which William James once attributed to the infant's perceptual world lay not in [the infant] but in our own minds and recording techniques' (Schaffer, 1977). So if babies can manage to make sense without benefit of abstract philosophy, cannot adults also? How essential is abstract philosophy to our understanding?

Callahan asserts: 'We have all been better off standing firm, hanging on to our abstractions' (Callahan, 1996). He illustrates his point with the example of 'hard and exceptionless, highly abstract rules against ... racism', which have succeeded in 'banning and forbidding' racism. He criticizes sophisticated racists who plead for compromise 'in the name of context, ambiguity, multiculturalism, sensitivity or whatever other version of moral complexity can be thrown into the breach against the devil of abstractions'. Yet who can believe that racism has been really banned, or that this can happen only through 'hard rules' and not also through attention to personal experiences and beliefs?

All theories are imbued with values, and when ethicists or counsellors imagine that they can be 'value-free', existing in some curious time- and space-warp, and suppose that thereby they can think more clearly and technically, they are more vulnerable to influence from values held subconsciously (Grimshaw, 1986). Criticizing abstract ethics, the philosopher Bernard Williams considers that it 'makes people think that, without its very special obligation, there is only inclination; without its utter voluntariness, there is only force; without its ultimate pure justice, there is no justice. Its philosophical errors are only the most abstract expressions of a deeply rooted misconception of life' (Williams, 1985). Understanding depends on making the two approaches, which Callahan sets in opposition, work together. Yet Callahan seems to have set up a straw polarity, since he concludes by advocating 'a working back and forth between our abstract principles and the actualities of experience.. The problem is not with abstractions as such. It is in knowing which to keep, which to modify and which to abandon'.

Yet this begs the questions: Who are 'we'? Who has authority to 'change principles'? And how they will do so? The philosopher Mary Midgley gives a helpful analogy (Midgley, 1994). Schizophrenia (or any other condition which might be genetic) is like a mountain with tents set all around it. Each tent represents a discipline or way of understanding schizophrenia: psychiatry, history, anthropology, nursing, personal experience, poetry, ethics, religion, genetics, biochemistry, psychotherapy, and so on. Many of the tent dwellers acknowledge that their expertise offers only a partial view and they value learning from others. A few experts, Midgley names biological reductionists, illogically insist that only their tent can achieve the sole authentic view of the entire mountain and encompass or surpass all the other disciplines.

Purely technical or abstract reasoning in ethics or genetics risks excluding numerous other relevant considerations. The next part of this paper considers ways of understanding genetics which take greater account of the complex social context.

7. Reducing social into genetic concepts

The neurobiologist, Steven Rose, criticizes the way some geneticists try to reduce social concepts into genetic ones (Rose, 1995). In this kind of reductionism, dynamic processes become things, people's fluid transient reactions are seen as fixed entities, their many complex characteristics are separately identified as immutable parts of their character. Individuals' reactions to other people and events, which make sense when seen in context, are viewed in isolation as 'phenotypes' assumed to derive from their genotype, reducing the person to a mindless set of biological impulses. Inherited impairments are inaccurately called 'handicaps' as if they solely determine the person's level of ability and opportunity throughout life. This fatalism ignores the powerful effects of social circumstances, and the ways in which personal and political action can transform impaired people's lives. Someone using a wheelchair may live a very limited handicapped life, or a very fulfilled one (McNamara and Moreton, 1995; Oliver, 1990).

Steven Rose further considers how personal characteristics are reduced into impersonal concepts which are then seen as fixed (potentially or actually genetic) traits which drive the affected individual. This can encourage us to think in mechanistic ways about human beings. The cliché of a genetic 'blue print' can imply that anatomy is destiny, that each person's entire life is mapped out from birth by the genes, like the design which determines exactly how an engine will be built, and that any deviation is incorrect. In such biological determinism, free will, human agency, imagination and impulse are sidelined. Difference can too readily become 'abnormality' to be avoided, rather than diversity to be celebrated. In contrast, 'growth' suggests the unique individual branching out unexpectedly.

The expectation that ethics and genetics can or should be value-free and nondirective, in itself expresses a set of values. Morality then tends to be seen aesthetically as 'choices' and 'personal values' which have no strong roots in social consensus or in their impact for good or ill. Reverence for individual choice, without specifying the nature or effects of choices, disguises how powerful people tend to be the choosers and vulnerable ones the losers. When there are discrepancies of power and knowledge, a neutral position is not possible. When scientific and ethical expertise is valued much more highly than personal experience, such as living with the effects of genetic disorders, morality tends to take its meaning from abstract law and principles, philosophical or scientific ones, instead of from living relationships and experience. People then tend to be seen less as creators of their own lives, and more as relatively passive consumers of services, and of life itself when they are offered prenatal choices. Risks which were once accepted, or else undertaken in the expectation that resources and support would be found to cope with their effects, are now presented as the responsibility of isolated individuals, to be reduced or avoided. Rather than learning partly through their own sensations and bodily experiences, people are encouraged to turn to scientific and moral experts to interpret to them their own subcellular activity (Midgley, 1994).

When genetic counsellors counsel, they talk in terms of genetic disease, statistical risk and personal decision making; one measure of their effectiveness is change in their clients' behaviours. Professional and economic pressures combine with technological imperatives to act – to control, reduce, prevent, correct – and lead to the belief that it is better to do something than nothing. This has been contrasted with a 'feminine' approach of 'holding' together a tension of opposites, a more 'wait and see' ethos (Marquis, 1989; Ruddick, 1983). When ethics is linked with business management and risk calculation it is drawn into a similar ethos of control. This tends to put greater onus on any one who rejects decisive health interventions and, for example, decides not to abort a fetus with Down's, to justify their 'inaction', although for the woman, to continue with the pregnancy entails much more action than to end it. Common terms, such as 'positive' results when impairment is found, increase the implication that doing 'nothing' is negative. A genetic counsellor lecturing to midwives in 1996 said 'occasionally we lose a Down's", meaning 'occasionally a fetus with Down's is not detected pre-natally and is born at term'. This reverses centuries of midwifery values when 'loss' meant death and not birth.

Some ethicists elevate science to a moral authority, basing their understanding on the science of genetics, instead of viewing it from a more independent critical position that sees science as one (very effective and important) type of knowledge among other types. One of many examples of this practice is the Clothier Report on Gene Therapy (Department of Health, 1990). Abstract ethics can reinforce the very weaknesses and problems in medicine that cause most worry in people: impersonal reductionism, harsh utilitarianism, social control by experts who use confusing jargon, threats to the very freedoms which ethics purports to respect, a predominance of one discipline over all others, as the medicalization or the 'ethicalization' of people's lives.

Genetics needs to be complemented and enriched by ethics that takes account of shared moral values, including compassion, respect and care at the personal and political levels. Outward-looking ethics considers the impact of decisions not only on the individuals directly concerned but on everyone who might be affected, such as everyone affected by a certain genetic condition and their place in society and the ways they are perceived and treated.

8. Conclusion

The specialist chapters in this book on law and ethics are very clear and elegant, with most useful and orderly summaries. They are like well-organized kitchens where everything is allocated to the correct place. We know far more about a teaspoon when it is classified as cutlery and not mistaken for a tea bag. In contrast, attention to the social context can look messy and trivial. What is the point of taking out a teaspoon to look at it again? When the subject is not a teaspoon but, for example, a child with spina bifida, it is vital to review our own assumptions about children's needs and abilities, adults' responsibilities, the nature of genetic conditions and their actual effects, about what we know and do not know, and what we can learn from this child. People cannot be tidied away like cutlery.

Greater attention to the social context, besides questioning current assumptions about childhood, can complement analysis in ethics and law and increase critical awareness of common threads between ethics and genetics. It can question covert assumptions that we are hostages to our biology rather than competent agents. By deepening understanding between all the related disciplines the social context can help to increase the richness and relevance of their various contributions to understanding of genetics. It can question trends towards seeing humans as machines, helplessly driven and looking for experts' solutions and technical answers. Social awareness sees how the parts inform and are informed by the whole, in each individual and in society.

References

Alderson, P. (1993) *Children's Consent to Surgery*. Open University Press, Milton Keynes.

Bobrow, M. *et al.*, (1991) Ethical issues in clinical genetics: a report of a joint working party of the college committee on ethical issues in medicine and the college committee on clinical genetics. *J. Roy. Coll. Physicians London*. 25 (4):284–288.

Bradley, B. (1989) *Visions of Infancy: a Critical Introduction to Child Psychology*. Polity, Cambridge.

Brazier, M. (1996) Embryos, property and people. *Contemp. Rev. Obstet. Gynaecol*. 8: 50–53.

Buchanan, A. and Brock, D. (1990) *Deciding for Others*. Cambridge University Press, New York.

Callahan, D. (1996) Ethics without abstraction: squaring the circle. *J. Med. Ethics* 22: 69–71.

Crawshaw, M. (1994) *CERES News*, 14: 2–5.

Dalby, S. (1995) Genetics Interest Group response to the UK Clinical Genetics Society report 'The Genetic Testing of Children'. *J. Med. Genet*. 32: 490–491.

Dawkins, R. (1976) *The Selfish Gene*. Oxford University Press, Oxford.

Department of Health (1990) *Report of the Committee on Gene Therapy*. HMSO, London.

Foucault, M. (1970) *The History of Sexuality: Volume one, an Introduction*. Penguin, Harmondsworth.

Foucault, M. (1976) *The Birth of the Clinic*. Tavistock, London.

Goodey, C. (1995) Mental retardation. In: *A History of Clinical Psychiatry* (eds. R. Porter and G. Barnos) Atheline Press, London.

Grimshaw, J. (1986) *Feminist Philosophers*. Wheatsheaf, Brighton.

Grodin, M. and Glantz, L. (1994) *Children as Research Subjects*. Oxford University Press, New York, NY.

Harris, J. (1985) *The Value of Life*. Routledge, London.

Holt, J. (1975) *Escape from Childhood*. Penguin, Harmondsworth.

Hughes, J. (1989) Thinking about children. In: *Children, Parents and Politics*, (ed G. Scarre).Cambridge University Press, Cambridge.

Hutching, I. and Moran Ellis, J. (1998). *Children and Social Competence*. Falmer, London.

Hutton, W. (1996) *The state we're in*. Vintage, London.

James, A., Prout, A. (1990) *Constructing and reconstructing childhood*. Falmer, London.

Johnson, B. (1954) *Poems* (ed. G. Johnston). Routledge, London, pp. 14, 23.

Lansdown, G. and Newell, P. (1994) *UK Agenda for Children*. Children's Rights Development Office, London.

Lukes, S. (1974) *Power: a Radical View*. Macmillan, London.

Marquis, D. (1989) An ethical problem concerning recent therapeutic research on breast cancer. *Hypatia*, 4(2): 140–155.

Marteau, T. (1993) Psychological consequences of screening for Down's syndrome. *BMJ*, 307: 146–147.

Mayall, B. (1994) *Children's Childhoods: Researched and Experienced*. Falmer, London.

McNamara, S. and Moreton, G. (1995) *Changing behaviour*. David Fullton, London.

Midgley M. (1994) *The Ethical Primate: Humans, Freedom and Morality*. Routledge, London.

Montgomery, J. (1988) Children as property? *Modern Law Rev.* 51: 323–42.

Montgomery, J. (1997) *Health Care Law*. Oxford University Press, Oxford. (Especially the chapters on children's rights, competence and confidentiality.)

Morss, J. (1990) *The Biologising of Childhood: Developmental Psychology and the Darwinian Myth*. Lawrence Erlbaum Associates, Hove.

Musgrove, F. (1984) *Youth and Social Order*. Routledge, London.

Nuffield Council on Bioethics (1993) *Genetic Screening: Ethical Issues*. Nuffield Council, London.

Office of Population Censuses and Surveys, 1994. Health and Personal Social Services Statistics for England. H.M.S.O., London.

Oliver, M. (1990) *The Politics of Disability*. Methuen, London.

Pailthorpe, G. (1932) *Studies in the Psychology of Delinquency.* Report from the Medical Research Council, HMSO, London.

Paul, D. (1992) Eugenic anxieties, social realities and political choices. *Social Res.* 59: 663–683.

Pfeffer, N. (1993) *The Stork and the Syringe: a Political History of Reproductive Medicine.* Polity, Cambridge.

Phoenix, A. (1991) *Young Mothers?* Polity, Cambridge.

Qvortrup, J., Bardy, M., Sgritta, G. and Winterberger, H. (1994) *Childhood Matters: Social Theory, Practice and Politics.* Avebury, Aldershot.

Rapp, R. (1988) Chromosomes and communication: the discourse of genetic counseling. *Med. Anthropol. Quarterly (New Series)* 2: 143–157.

Rawls, J. (1972) *A Theory of Justice.* Oxford Univesity Press, New York, NY.

Rose, N. (1990) *Governing the Soul: the Shaping of the Private Self.* Routledge, London.

Rose, S. (1995) The rise of neurogenetic determinsim. *Nature* 373: 380–382.

Ruddick, S. (1983) *Maternal thinking.* In: *Mothering: Essays in Feminist Theory.* (ed J. Trebilot) Rowman and Allanheld, Totowa, NJ.

Scarre, G. (ed.) (1989) *Children, Parents and Politics,* Cambridge University Press, Cambridge.

Schaffer, R. (1977) *Mothering.* Fontana, Edinburgh.

Scheper-Hughes, N. (1987) *Child Survival: Anthropological Perspectives on the Treatment and Maltreatment of Children.* Reidel, Dordrecht.

Stacey, M. (1992) *Changing Human Reproduction: Social Science Perspectives.* Sage, London.

Stuart-Smith, S. (1996) Teenage sex. *BMJ,* 312 (7043):390–391.

Thornton, J. (1993) Genetics, information and choice. In: *Consent and the Reproductive Technologies.* (ed P. Alderson) Institute of Education, London, pp. 35–51.

Tooley, M. (1983) *Abortion and infanticide.* Oxford University Press, New York, NY.

United Nations (1989) *Convention on the Rights of the Child.* UN, Geneva.

United Nations Committee on the Rights of the Child (1995). *Consideration of reports of state parties: United Kingdom of Great Britain and Northern Ireland.* CRC/C/SR 205, January, UN, Geneva.

University of Surrey (1995) *First conference on Social Competence in Childhood.* University of Surrey, Guildford.

Watt, G. (1996) All together now: why social deprivation matters to everyone. *BMJ,* 312 (7037): 1026–1029.

Wertz, D., Fanos, J. and Reilly, P. (1994) Genetic testing for children and adolescents: who decides? *J. Am. Med. Assoc.,* 272 (11): 875–881.

Wilkinson, R. (1994) *Unfair Shares: the Effects of Widening Income Differences on the Welfare of the Young.* Barnardo's, Barkingside.

Williams, B. (1985) *Ethics and the Limits of Philosophy.* Fontana, Edinburgh, pp. 195–196.

20

Appropriate paternalism and the best interests of the child

Zarrina Kurtz

Abstract

Paternalism is exercised by parents in making decisions with regard to the genetic testing of children and in caring for an affected child, and by the state in the resources allocated and the type of services provided. Many factors influence the exercise of paternalism by parents such as their understanding of the disease condition in question and their capacity to handle the consequences for family relationships of the information they have. Factors that influence the exercise of paternalism by the state include the cost effectiveness in terms of measurable health gain of providing genetic services compared with other services, and of particular services such as family counselling within the genetic services overall.

In this field, what is regarded as appropriate paternalism may not be linked primarily to the best interests of the child. There are, in any case, likely to be a number of views on what are the best interests of the child. To date, children's own views in this regard have not been satisfactorily sought.

These considerations are discusssed, as well as the importance of the rights of even young children to be informed and consulted when there is a question of testing for genetic conditions, and of the proper inclusion of their feelings and thoughts along with the views of others in arriving at family decisions. The state has a similar duty to find ways of bringing the experiences of whole families to bear upon the exercise of its powers in matters concerning testing for genetic conditions and the provision of related services.

1. Paternalism

It is largely assumed that paternalism is the basic standpoint of both the parents of children who may be the subjects of genetic testing, and of the state by means

---

of the Department of Health and the provision of medical services. That assumption stands in making decisions about who to test and about other issues such as for what diseases or conditions to test, when and how to test, and how best to offer support and treatment for those with genetic conditions.

That paternalism is a satisfactory basis upon which to plan and provide services has been called into serious question by the promotion of the rights of children (Freeman, 1987). Paternalism can only be justified if the recipients of services are truly totally dependent. The consumer movement and the greater attention now paid to the opinions of users of services and of the lay public in health and welfare policy have also dented the largely paternalistic framework that has directed the provision of health and welfare services in the UK until recently. Possession of rights inherently supposes the autonomy of each person, and thereby contradicts a relationship of dependency of one upon another.

Children, however – certainly younger children – are dependent upon others for protection. Here it is helpful to regard their rights as not dependent on actual autonomy but on the capacity for it. Recognition of this would confine paternalism without totally eliminating it. As Freeman (1987) has put it:

'Parties to a hypothetical social contract would know that some human beings are less capable than others; they would know about variations in intelligence and strength, and they would know of the very limited capacities of small children and the rather fuller capacities of adolescents. They would be conversant with the evidence from cognitive psychology that children aged 14 (and perhaps even 12) are as capable of making decisions about their lives as adults are for them. They would bear in mind in particular how the actions of those with limited capacities might thwart their autonomy in time to come when their capacities were no longer limited. These considerations would lead to an acceptance of intervention in peoples' lives to protect them against irrational actions.'

Freeman calls this liberal paternalism but, in order to consider whether it is exercised appropriately, one has to consider what are its objectives. This is where 'best interests' come in.

It cannot just be assumed that the best interests of the child are what drives, or even should drive, clinical and service decisions. As can be seen in Chapter 23 by Dorothy Wertz, and may be imagined further, decisions about genetic testing are made differently in different countries and by different professional groups, and by parents and young people in different families and situations. It is unlikely that all these decisions are made on the sole basis of the best interests of the child. And even if they were, systematic differences would be found in the types of decisions that were made.

A child is a member of a family and the issues that arise in genetic testing of that one child have implications for the whole family. A hierarchy of responsibility for a child's best interests has been described (Kent, 1993). Closest to the child is the family, then the community, local government, state government, national government, international non-governmental organizations, and international governmental organizations. Ideally each agency supports those that are closer to the child but does not substitute for these other agencies. The principle of best interests is complex and is inextricably linked to the cultural context in which it is invoked. At and between each level of the hierarchy listed above, there may be differing views and tensions with regard to children's best interests. Because of the role of government agencies in deciding on priorities for health care and in determining the context in which clinicians treat patients, civil and political rights cannot be separated from economic, social and cultural rights. The position of the United Nations is that the two sets of rights are of equal importance. The UN Convention on the Rights of the Child, ratified by the UK in 1991, establishes only what have been called 'soft rights', which can be transformed into 'hard rights' by suitable national and local laws along with effective agencies for their implementation.

What is regarded as appropriate paternalism will depend upon whether one is working within the framework of the UN Convention in which children's best interests are to be *a* primary consideration, or *the* primary consideration as under English adoption law, or *the paramount* consideration under the Children Act 1989. The text of article 3(1) of the UN Convention reads:

> 'In all actions concerning children, whether undertaken by public or private social welfare institutions, courts of law, administrative authorities or legislative bodies, the best interests of the child shall be a primary consideration.'

There are even weaker formulations before children's best interests become part of a purely utilitarian calculation at large, that is that the child's best interests shall be a consideration, as in the Australian Family Law Act, for example (Parker, 1994). The implications of the differences in formulation are that the role of the child seems to change with each. Where the child's best interests are *paramount*, the attention is on the individual child. In the version in Article 3(1), the context may be quite different. First, it is not clear whether children as a class or children individually are intended to be the beneficiaries. Article 3(1) begins by referring to all actions concerning children (plural) but ends with the requirement that the best interests of the child (singular) be the primary consideration. In practice it is hard to see how the article can have anything other than a collective focus. Although courts of law often make decisions about individual children, the other decision-makers embraced by the Article often make decisions about groups of children. If article 3(1) is given the same individualistic focus as the paramountcy formulation, it will be of no use in situations such as that illustrated by the following example.

Since the NHS reforms of the last (Conservative) government, the provision of genetic services and what they should include are linked to the overall concepts of population health gain and obtaining the best value for the tax payers' money. This inevitably places children's interests in a utilitarian context and weights choices made by individuals regarding testing in favour of termination of an affected fetus or the decision not to risk having an affected child by avoiding pregnancy altogether. Services for people with genetic conditions would be regarded as extravagant within this framework, and it is difficult to argue in their favour because the health gain derived as a result of maximal quality of life is notoriously difficult to measure.

At least two levels of paternalism should be considered. First, the state exercises paternalism in health policy and in deciding whether or not to provide resources for genetic testing. Because resources are limited, it is likely that the state will require the maximum yield from these services, which will probably be provided because of the argument that significant public health benefits will accrue in terms of reduction in the incidence of genetic conditions and thereby in expensive treatments and care. Thus, it can be questioned as to whether it is appropriate paternalism on the part of the state to provide life-long care and support for, for example, a Down's syndrome child born to a mother who chooses not to terminate the pregnancy even though she knows that her baby will be affected.

In addition, health service policy may decide that testing will only be made available on the basis of certain criteria such as maternal age (in testing for Down's syndrome), whereby younger mothers may not be offered the chance to have screening tests and, hence, the opportunity to terminate an affected pregnancy. The recently reported case where a mother was not told that her unborn baby had spina bifida following fetal ultrasound scanning illustrates another aspect of this in that the mother was not informed of the abnormality found on the scan because it was assumed that, as a Catholic, termination would not be an option for her (Toynbee, 1996). In this case there was also a question of whether the test gave a false reading or was not correctly read or interpreted by the tester.

If we conduct the arguments concerning genetic testing of children solely around ideas of paternalism and best interests, we can sidestep the contribution that can and must be made by a consideration of rights. Paternalism combined with best interests can lead, for example, to a decision that children in a given situation should be tested to learn their genetic status because this would result in the greatest good for the greatest number of children..... a utilitarian view. But Dworkin's (1977) thesis of rights is that they embody equal concern and respect for each person; that if persons have moral rights to something, they are to be accorded those rights even if a utilitarian calculation shows that utility would be maximized by denying it to them. Rights are about equality as well as about autonomy, and so they trump countervailing utilitarian considerations.

2. Rights

Rights are important in that they ensure that children themselves will have a say in what are their best interests, and should ensure that parents take their children's views of their best interests into account alongside those of other members of the family. They should ensure, in addition, that the state also takes children's best interests into account in planning services and this requirement could be enforced legally if the state failed to do so. There remain large questions about the extent to which the state should police or intervene in matters of genetic testing in the child's best interests when this is contrary to decisions made by parents and by doctors. This is pertinent to the issue of the regulation of testing by private companies with open access to all comers. We have much to learn from the experience in managing genetic disease, the consequences of testing, and the handling of knowledge about carrier status among particular communities. One example is the Greek Cypriot community in London where there is a high incidence of thalassaemia which has seen a dramatic reduction because of genetic counselling and pre-natal diagnosis. Another example is that of the community of deaf people, many of whom value their deaf status and are little interested in preventing deafness in their children. The members of these communities exercise powerful paternalism, with particular views of what is in the best interests of the child.

Essentially, we do not know what the best interests of the child are in the majority of genetic testing situations. In forming government policy, there has been a gradual process of trying to work with an agreed parental view to inform the state provision and to develop guidelines for local services and professional practice. Considerable achievements have been made in this respect with regard to sickle cell disorder and Huntington's disease. State paternalism, however, does not end with decisions about genetic testing; it extends to who needs to know the results. When I was medical advisor to the Inner London Education Authority (ILEA) in the late 1980s, there was a strong directive to ensure that information about sickle cell disease was widely disseminated and that the profile of the disorder was raised throughout all schools in the authority. Given the high prevalence of populations at risk for sickle cell attending ILEA schools, this was regarded somewhat unthinkingly as a good thing and as counteracting any accusations of neglect of the needs of minority ethnic pupils. But it became clear very early in the process of implementation that young people from the groups at risk did not necessarily want their disease status to be better known; any benefits in terms of early recognition of sickle cell crises and more sympathetic handling of episodes of illness were felt to be more than offset by the likely adverse effects of labelling – in limiting an individual's choice of marriage partners and even of boyfriends or girlfriends.

Consideration of the best interests of the child is fraught with difficulties with regard to genetic testing. This is because what is learnt as a result of testing has profound implications for those who will care for the child – usually the parents and maybe siblings – and indeed for the state. The decision to undertake testing is made in the individual case almost without exception by parents, and at the population level decisions are made by the state through the allocation of resources for service provision and through the quality of service that is made available. To the extent that the consequences for the child are taken into account, up to now, these have largely been assumed for them by adults (parents, professionals, special interest voluntary groups). The best interests of the child, however, depend a great deal upon the support and capabilities of these adults, so that children themselves may sometimes see their own best interests in potential conflict with those of other members of their family.

3. Decision making

In the current situation where consultation with children and their consent is required when decisions are made that affect them, children themselves should have a much greater influence, not only in individual decisions about testing, but also with regard to the availability and practice of testing as a public service. However, testing is not always carried out at an age when children can be expected to be competent to give consent. The timing of testing differs according to the condition to be tested for and the balance of advantage to be had from testing at an earlier or later age. This question of timing will depend upon the state of our knowledge about the particular genetic disease and about the possibilities for prevention and treatment. Once again, assumptions are frequently made about the desirability of treatment and the extent and type of treatment, even though much of our knowledge is uncertain. The cost implications (not just financial but also broader social costs) are such that major questions should continue to be raised about each of these.

Parker (1994) describes how, according to rational choice theory, a determinate answer to any specific question relating to a decision will in general require that the following knowledge conditions are satisfied:

(1) all the options must be known
(2) all the possible outcomes of each option must be known
(3) the probabilities of each possible outcome occurring must be known
(4) the value to be attached to each outcome must be known.

It follows that objectively identical problems will be resolved differently (ignoring coincidence) if the decision-makers have different knowledge at any of these stages. The usual reason given for different views of children's best interests is that they are based only on different values. In fact, decisions on apparently similar facts by the same decision-maker or by two decision-makers with similar values, may diverge because of differences in knowledge about outcomes or options. And currently it is practically impossible to satisfy all four of the above conditions because:

(i) the range of possible outcomes for the child are a matter of pure speculation, even if the world stood still throughout the time of a child's growing up
(ii) even if two decision-makers happened to come up with the same range of possible outcomes, they would then need to assign some probability to each outcome occurring
(iii) there is the problem of attaching a value to all of the possible outcomes (as weighted by the probability of them occurring) – and should this take a long-term or a short-term perspective? Should health in one sense – perhaps in terms of improved mobility for example, be placed above another such as freedom from pain? These questions could not possibly be answered in a definitive manner.

That the child's wishes be taken into account in all matters of serious concern to him or her is written into the Children Act 1989 and is one of the cardinal principles in the Department of Health guidance on *The Welfare of Children and Young People in Hospital* (1991) which states:

'Young people should be kept as fully informed as possible about their rights. Even where young children do not have the required understanding, they should be provided with as much information as possible and their wishes ascertained and taken into account.'

An excellent guide to ways in which to encourage the participation of children in decisions about their health care and to maximize their informed consent offers helpful and detailed discussion of the range of important issues (Alderson and Montgomery, 1996).

There is no substitute for children themselves having an increasing say as they grow up, with the balance of decision making increasing in favour of the child over appropriate paternalism with their increasing maturity, but also being influenced by the particular family situation. This has been called dynamic self-determinism (Eckelaar, 1994). It creates space for self-development and its purpose is to bring the child to the threshold of adulthood with the maximum opportunities to form and to pursue life goals which reflect as closely as possible an autonomous choice. The balance between protection of the child and autonomy shifts towards autonomy as the child grows up, taking account of his or her need for protection in order to have the best chances to develop autonomy. Dynamic self-determinism allows the child to make mistakes and also to make choices that may be against his or her self-interest.

Acceptance of the idea of self-determinism implies that children's decisions should determine the outcome in matters concerning them, subject to two constraints. The first is its compatibility with the general law and the interests of others. The second might be when a child takes a decision that is contrary to his or her self-interest when dynamic self-determinism should be disapplied. In the second situation, the child's competence can be called into question legally. The emerging capability, which has been described as acting as a guide to Gillick competence in giving consent to treatment, parallels the concept of dynamic self-determinism (Mitchels and Prince, 1992).

4. Conclusion

In conclusion, the exercise of appropriate paternalism in the matter of genetic testing of children raises difficult issues for the health service, but children's rights demand that certain matters are pursued. These include improving the ways in which service planning and provision is informed by parents and community groups in their paternalistic role; the capability of services to offer full, accurate and up-to-date information and wide-ranging discussion to children and families; and continuity in the relationship between service providers and children and families, giving ready access to appropriate advice and help over time as different situations present themselves.

References

Alderson, P. and Montgomery, J. (1996) *Health Care Choices. Making Decisions with Children.* Institute for Public Policy Research, London.

Dworkin R. (1977) quoted in Freeman.

Eckelaar, J. (1994) The interests of the child and the child's wishes: the role of dynamic self-determinism. In:*The Best Interests of the Child: Reconciling Culture and Human Rights.* (ed P. Alston). Clarendon Press for UNICEF, Oxford, pp. 42-66.

Freeman, M.D.A. (1987) Taking children's rights seriously. *Child. & Soc.* 1(4): 299–319.

Kent G. (1993) Children's rights to adequate nutrition. *Int. J. Child. Rights* 1: 133–54.

Mitchels, B, and Prince, A. (1992) *The Children Act and Medical Practice.* Jordan Family Law, Bristol.

Parker S. (1994) The best interests of the child – principles and problems. In: *The Best Interests of the Child: Reconciling Culture And Human Rights* (ed P. Alston). Clarendon Press for UNICEF, Oxford, pp. 26–41.

Toynbee P. (1996) Litigation is the wrong medicine. *The Independent* 12th June, p. 13.

21

Predictive genetic screening and the concept of risk

Rogeer Hoedemaekers

1. Introduction

Generally speaking, the aim of screening is to provide more certainty about a risk factor threatening health. The information may reveal that an individual is at increased risk of getting a specific disease. Screening may also bring more certainty about a risk that has been suspected previously – perhaps on the grounds of a family history of genetic disease. However, it may well be that an individual is left with new uncertainties, for example there may be a potential for harm caused by knowledge of an increased risk factor. This raises the question whether one of the purposes of screening (and possible further tests), namely bringing more certainty, is really achieved.

This question can be asked in connection with genetic screening and testing, as with other types of screening and testing. Uncertainties about the supposed advantages and disadvantages of information about a genetic risk can make it very difficult for an individual – and also public authorities – to assess the real value of the information. It also throws doubt upon the supposedly rational character of decision-making processes in genetic screening, testing and counselling.

The concept of 'risk' plays a key role in medical genetics. It is not always clear, however, what 'risk' actually means or what it refers to. For example, in an extensive report on genetic screening published by the (US) Institute of Medicine, 'Assessing Genetic Risks' (Andrews *et al.*, 1994), we can find ambiguity in the very title of this report. Does risk refer to a medical risk, posed by an alteration in the genetic make-up of an individual? Or does the title refer to the psychological, economical or sociological risks of the genetic screening or testing procedure? If the latter alternative is meant, a distinction can be made between risks posed by the product, the test itself, and the risks inherent in the interpretation and communication of the test result. It could also refer to risks, individual or societal, generated by the possession of genetic information. So there is some irony in the fact that this report, in a chapter discussing the Research and Policy

The Genetic Testing of Children, A.J. Clarke (ed.).
© 1998 BIOS Scientific Publishers Ltd, Oxford.

Agenda, recommends 'research on the meanings and implications of commonly used health terminology, and the changes in application of that terminology to the field of genetics' (p. 301). In the following we attempt to follow this recommendation with regard to the concept of risk.

Analysis of the concept of risk in a genetic context seems to be indicated if use of this concept creates ambiguities or confusion for the individual who is considering whether or not to undergo a predictive genetic test; such confusion may obstruct full understanding of the value of such a test, thus hindering proper assessment of the implications of the test. The objective of this article, then, is to analyse the concept of risk in the context of predictive genetic screening for asymptomatic disease and susceptibilities. It is assumed that this will lead to better understanding of the genetic screening and counselling procedure. Conceptual clarity and precision is, of course, also valuable in pretest education and counselling, and it should be an important prerequisite in written educational material. More conceptual precision and clarity may help the potential screenee to make better decisions.

Conceptual clarification will be carried out by combining recent literature on risk and risk assessment with current insights in medical genetics. Three major questions will be examined: (i) how is the notion of risk commonly used? (ii) how is it understood by potential participants in a genetic screening programme? and (iii) how is knowledge of genetic risk integrated in a person's life-project? Examination of these questions may help determine what conditions should be met to make genetic screening and testing an acceptable procedure.

Exploration of the above questions will be as follows: First we consider the need for further clarification of the concept of risk by presenting a survey of the most important 'risks' (Section 2). Then the concepts of uncertainty, possibility, probability and risk will be distinguished (Section 3). Also being at risk and taking a risk will be distinguished (Section 4).

After that the concept of genetic risk will be discussed (Section 5). This will be followed by a brief examination of differences between objective assessment of risk and more individual, subjective forms of risk assessment (Section 6). This will enable us to say something about the question in how far, and under what conditions, genetic information may be of value for a potential screening participant (Section 7). In conclusion we will sum up the main points (Section 8).

2. 'Risks' inherent in a genetic screening programme

(The concept 'risk' will be used here in the very broad sense as it is used in the Report of the US Institute of Medicine. Later in this chapter we will propose a more limited use of this concept. In order to distinguish the two concepts we have marked the usual, broad concept of 'risk' in this note by using inverted commas. In this section of the chapter, 'risk' is used in this broad sense, but we have only used inverted commas in this note, not elsewhere in the text.)

Before the introduction of a genetic test or screening programme attempts are usually made to assess its 'benefits and risks'. In this phrase the concept of risk is very broad, and seems to refer to all the potentially harmful consequences of genetic screening. A list of risks of a presymptomatic genetic testing procedure, even if not exhaustive, is impressive and comprises potential harm at an individual as well as at a societal level. (Major sources are: Andrews *et al.*, 1994; Clarke, 1994a; Clarke, 1994b; Hubbard and Wald 1993, Tibben, 1993.) In order to separate different types of risk we will distinguish seven different stages of such a testing procedure. For each stage in the procedure we will list some major risks currently under discussion.

(1) Raising awareness and education. Risks at this stage are that the written or oral information may be partial, one-sided or biased. Potential benefits may be overemphasized and risks may be downplayed, for example in the case of a commercial firm offering a screening test. This may lead to inappropriate or unhelpful decisions being made with harmful consequences. There is also a risk of direct or indirect pressure from family, environment or society which may also lead to 'wrong' choices.

(2) Pre-test counselling. Here also a major risk is that of biased or partial information. Potentially harmful psychological or psychosocial consequences may not be discussed, for example. The information may be biased by using a particular model of disease explanation or because of institutional or economic pressure. There is also a risk that wrong choices may be made because a potential screening participant lacks insight into his/her own feelings, values, norms and coping abilities, for example with regard to physical or behavioural effects (taking medication over long periods, change of lifestyle, undergoing periodic monitoring) or psychological effects (e.g. psychological adjustment to the genetic information provided).

(3) Decision to participate. This stage also brings risks. In the case of non-participation, an individual may be held responsible for not finding out their risk, or may be blamed for not being a responsible citizen or for being stupid. There may also be financial consequences. In the event of participation, there are the possibilities of inaccuracy and diagnostic errors, false positives or false negatives.

(4) Post-test counselling. This stage includes the interpretation, presentation, and communication of the test result. The result is often communicated as a statistical risk figure, representing the degree of probability that a given disease will manifest itself in the person tested. Communication poses a risk of inadequate or incorrect understanding of the test result, because the person tested may not fully understand numerical risks, for example, or because an individual receives the information at the wrong moment. There may also be inadequate support to help the tested individual to cope with the information received.

(5) Knowledge of the test result. This also presents important risks. In the case of a negative (i.e. favourable) test result, the test may not bring the expected emotional relief. Will one really be reassured? Will one feel guilt? Family members, relatives, peers may also respond in unexpected ways: for example, there may be hostility or resentment from other family members, whether affected or unaffected. When the test result is positive (unfavourable), will the individual be able to adjust to the genetic information? Will there be effects of denial or underestimation of the risk, or self-stigmatization, feelings of powerlessness or paralysis? Here too, there is always a risk that responses of the environment, relatives and family members are worse than expected, and this may bring risks of serious marital conflicts and separation.

(6) Follow-up procedures and support. It may not always be clear if future (long-term) needs for psychological or economic support will be met. This, again, may bring the risk of psychological maladaption and financial difficulties.

(7) Social response to predictive genetic testing. The problems that may arise at this stage have received a great deal of attention in the literature. Especially the short-term risks have been discussed at length. They include risks of violation of the private sphere, stigmatization, ascription of responsibility and blame, discrimination by insurers and employers, economic sanctions and the risk of abuse of genetic information by third parties. There are also more large-scale societal risks, such as the reinforcement of a reductive and individualized approach towards disease, the creation of an 'increased risk culture' leading to more anxiety and dependency on the medical system, and the generation of increased demand for medical advice and care.

The above list of risks may give the impression that 'risk' is interchangeable with 'danger'. Indeed, 'danger' and its synonyms are also in common use. In the above report of the Institute of Medicine the following words and concepts are used to denote possible disadvantages created by genetic screening procedures: harmful effects, potential for harm, possible harm, dangers, hazards, potential impact, psychological impact, implications of being at risk and, rather euphemistically, challenge (e.g. pp. 107, 164). Most of these terms have a negative connotation, but they are not used very systematically. The precise relationship between these terms and the concept of risk needs to be analysed.

3. 'Uncertainty' and 'possibility', 'probability' and 'risk'

We often find ourselves in situations where important decisions or choices have to be made without our being sure of what the consequences of our choices or decisions will be. In most of our choices there is a certain degree of uncertainty about future events, actions, developments or consequences. Uncertainty can be distinguished from risk in two ways. The first is that uncertainty may refer to either good or bad, desirable or undesirable future developments. The concept of risk, however, tends to be connected with negatively valued events: hazards, threats or dangers (Fischoff et al., 1981; Lupton, 1993). I may well be said to be 'uncertain' about a happy and healthy old age, but to say that one is 'at risk' of a healthy old age will probably raise eyebrows.

A more important difference between uncertainty and risk is that the concept of risk usually includes some form of probability assessment, whereas uncertainty includes a notion of possibility. We employ the concept of risk when we try to estimate or calculate the degree of probability with which uncertain events or developments will occur. Although there is as yet no consensus about a definition of risk (Slovic, 1987; Vlek, 1987), we believe that extensive discussion is not needed here and that for our purpose a definition quoted by Lupton (1993) is useful. Risk is defined as 'the probability that a potential harm or undesirable consequence will be realized' [This definition is quoted by Lupton (1993; p. 426 and 434) and is taken from The National Research Council Committee on Risk Perception and Communication, *Improving Risk Communication*, National Academy Press,

Washington, DC, 1989.] In practice, however, it is not easy to separate the notions of risk (probability) and uncertainty (possibility).

Two major forms of probability assessment are usually distinguished – an objective and scientific form of probability assessment, and a more personal, subjective and intuitive form (Heimann, 1990; Vlek, 1987). These forms of probability assessment are bound up with uncertainties, but there is a form of 'pure' uncertainty that can be distinguished from both. There are future events or occurrences, which may be foreseen or expected but the probability of which cannot be estimated. Such an event may or may not happen, but there is no way of assessing with any degree of certitude if, in what form, or when it will occur. In a situation where any probability estimate is pure guesswork, for example when there is no past experience one can rely on or when there is no relevant information, the concept of uncertainty is clearly appropriate (Lave, 1987). In this 'unpolluted' form, however, uncertainty may be rare. [The question how much evidence is needed to make (scientifically) valid judgements is an issue under discussion in the philosophy of science. This discussion is left out of consideration here. We merely wish to emphasize that there are basically two different methods of reaching a risk estimation. A subjective, intuitive approach based on everyday experience, simple logical rules and argumentation and a more objective method using scientific methods to calculate the probability that an event or occurrence will happen.]

Much more common is a form of uncertainty where awareness of a possibility that an event, occurrence or development will take place is combined with a subjective, intuitive and imprecise form of risk estimation (Heimann, 1990; Vlek, 1987). If, for example, someones rides on a bicycle in the City of London, some will say that it is possible that an accident will or may happen. Others will say that this person runs a risk that (s)he is run over by a careless car-driver, even though the exact probability of such an accident is unknown. They may base this on a general awareness that modern traffic has become dangerous, or on their own experience. The dividing-line between probability and possibility, between risk and uncertainty, seems unclear and difficult to draw here. Therefore we propose to introduce the concept "fuzzy risk" for this domain, where there seems to be a smooth transition or change from awareness of possibility to a sort of imprecise and intuitive form of risk assessment.

Although a clear distinction between pure uncertainty and a fuzzy risk concept may be difficult, we suggest that it is less problematic to distinguish pure uncertainty from a more objective scientific form of risk assessment determining the probability of a future event occurring [We do not suggest that the probability of a certain event can be determined completely objectively. Certain basic assumptions are always made (Katz, 1990)]. In the context of risk analysis and risk assessment a more scientific and objective form of probability estimation usually implies establishing a mathematical likelihood of an event or development occurring. The risk of an event or occurrence happening is calculated according to a particular model and is presented numerically, for example as a 1 in 100 000 chance. Yet, in spite of the apparent precision with which risk calculations are made, there are inherent uncertainties. Some of these uncertainties can be detected in the genetic counselling context, but before we turn to these we wish to connect the notion of danger and its synonyms with the concept of risk.

We propose that it is for the type of risk denoted as fuzzy risk that concepts like harm, danger or hazard are appropriate, as these notions convey both possibility and lack of precision. We take up a suggestion of Douglas here, also quoted by Lupton, who says that the use of risk is often preferred to danger in professional circles, because 'plain danger does not have the aura of science or afford the pretension of a possible precise calculation' (Lupton, 1993, p. 426). We therefore propose, for the sake of clarity for potential clients, that the concept of danger (or its equivalents) should be used where there can be no precise risk estimation – where the risk is fuzzy; the concept of (unqualified) risk, with its suggestion of more precision, is to be preferred when a more precisely calculated probability estimation has been, or can be, employed. Lack of precise knowledge about an expected outcome can be further emphasized by using combinations of words, like potential danger, potential hazard or potentially harmful consequences. Fuzzy risk may also be appropriately expressed by words like may and might. At first sight a combination such as 'potential risk' (see Andrews *et al.*, 1994, p. 235) does not seem to be very logical, until we realise that risk is used here as a synonym of danger or harm [See for example Lupton, who says that risk is in constant use as a synonym for danger in the context of public health (Lupton, 1993).] This combination also belongs to the domain of fuzzy risk.

Discrimination between calculated risk, fuzzy risk (or potential harm, danger or hazard) and uncertainty may be helpful in the context of genetic screening and testing. Clarity may be enhanced if pure uncertainties are termed uncertainties and if calculated risks are distinguished from fuzzy risks.

Both scientific risk-calculation and a more subjective and intuitive form of risk assessment can be found in counselling situations. Medical geneticists and genetic counsellors often employ the first type when a test result is communicated (numerical risk). However, a counsellor will not often be able to present other risks in numerical form. Precise and reliable risk estimates of undesired psychological and societal consequences of genetic screening or testing are not usually available, for example. This also applies to concerns such as a growing dependency on the medical system, an increased use of medical services, eugenic tendencies or a process like geneticization. Therefore these psychological and societal consequences can be classified as fuzzy risks. Referring to these anxieties as risks may be giving them too much weight. In our view terms conveying possibility should be used here instead. [We can refer here to the recommendations and background sections of the recent Report of a Working Party of the Clinical genetics Society (UK) on the genetic testing of children (Clarke, 1994b), where words like 'may', 'might', 'can', and 'potential' abound.] The use of concepts like potential harm, danger or hazard may help avoid giving them a spurious 'aura of precision'. This may help offering unbiased, more exact and precise information, especially in written educational or promotional material. It may help avoid the unintentional downplaying of risks or the overemphasis of potential dangers.

In genetic counselling at least four different forms of uncertainty can be linked with probability calculations. The first is, of course, the 50% chance that a specific parental copy (allele) of a particular gene is present in a child. Here, uncertainty means that there is no way of predicting (without testing) which of the two copies of the gene in either parent has been transmitted to the child: the chances are even. A second type of uncertainty is associated with the number of people who are being tested. Calculations may be based on experience with small groups, for example, and this means that there is a margin of uncertainty. More evidence (e.g., more people tested) may bring more precision. A margin of uncertainty is usually also present because risk calculation is often based on assumptions or presuppositions that may or may not be (quite) true (Lave, 1987).

A fourth form of uncertainty is connected with the interpretation of a statistical risk figure that is based on a large group or population; there may be uncertainty as to its precise significance for a particular individual person belonging to this risk group (Heimann, 1990; Lave, 1987).

A fifth uncertainty, connected with the fourth, is that the genetic counsellor cannot always know how the risk figure given to the individual is understood, interpreted and integrated in their life project. Numerical risk may convey the degree of probability to an expert with great precision, but this need not make much sense for the layperson, who often does not know how to handle numerical or calculated risk estimates, and tends to translate calculated risk figures in binary form, in terms of it may/will or it may not/will not happen (Lippman-Hand and Fraser, 1979; Tijmstra and Bajema, 1990). A counsellor will usually be fully aware of these uncertainties, but are they always clearly communicated to the prospective screening participant who considers undergoing a genetic test? Full information about the risks and the inherent uncertainties associated with screening may enhance the clarity of decision-making and may help the prospective screenee to make better decisions.

Use of the concept of risk can be misleading in situations where neither scientific nor intuitive probability estimates are possible. Here terms conveying uncertainty or possibility are appropriate. For example, a positive test result may give a precise probability estimate that a genetic alteration associated with a given late-onset disease is present. In spite of this precision there is often great uncertainty about the age of onset and/or the severity and the burden of the disease, because for these a scientific probability estimate is (as yet) often impossible. Although one could present a worst-case scenario, there is no way of determining the probability of this worst-case scenario in an individual case. This lack of precise knowledge should be expressed by using a term denoting uncertainty, rather than the concept of risk, as it brings out more clearly the lack of certitude or confidence with which the worst possible outcome is expected.

Perhaps superfluously, we wish to point out that general rules for the most appropriate use of concepts for genetic screening, testing and counselling are hardly possible. We suggest that the best and most appropriate terminology should be considered and selected for each genetic test, depending on the available scientific evidence with respect to foreseen or expected negative (unfavourable) consequences and uncertainties.

4. Being at risk and taking a risk

Modern technology often brings benefits, but can also generate new risks. If individuals are conscious of these risks, they may decide to take these risks or try to reduce or eliminate them. Consciously and voluntarily taking a risk means that responsibility for the potentially harmful consequences does not lie only with the originators or producers of the technology. Discrimination between being at risk and taking a risk may help define those who may be held responsible (and liable) for the harmful consequences of the application or use of technological devices. (We have argued that there is a difference in meaning between the concept of risk and the concept of uncertainty. From this it follows that there is also a difference between taking a risk and accepting an uncertainty, and between the creation of uncertainties and the creation of risks. Here, however, we will mainly focus on the difference between being at risk and taking a risk.)

A person may be said to be at risk of a particular disease: there is a given probability that this person will get this disease. A person may be completely unaware of this risk. But that does not mean that the risk is not there: the person is still at risk. If a person is told that (s)he is at risk, however, there are two possibilities. This person may not be able to do anything about this risk. Then we can only say that (s)he knows (s)he is at risk. But, if a person knows about this risk and is actually able to do something about it, the situation becomes different. This person can then decide either that (s)he will accept that risk or that (s)he will take measures to reduce or eliminate that risk. In the pursuit of a certain life-project the risk is taken into consideration and weighed against important values or benefits. Ideally, the taking of risks involves free choice, acceptance and the incorporation of a (new) risk in one's own life-project.

In genetic screening or testing the difference between being at risk and taking a risk is not so easy to define, partly because some of the new at risk situations (and uncertainties) are inherent in the testing procedure. Yet it is important to understand where an at risk situation (or uncertainty) already exists and when or where additional or extra at risk situations are generated. This will not only help ascribing responsibility and/or liability for potentially harmful consequences, but it can also help diminish or eliminate them. Discrimination between the generation of risks and the taking of risks is possible for each stage of the genetic testing and counselling procedure described above. By way of illustration we will give only a sketchy account of some of the risks.

An individual may be at risk of developing a disease with a genetic component. Such an individual may suspect that (s)he is at risk, because (s)he knows (s)he belongs to a group with an 'increased risk'. (S)he may also be unaware of this. If the individual already suspects an increased risk, (s)he may wish to know more about it and can go to a counsellor to get more information.

More information can make the individual really worried. The uncertainty here is that you do not know in advance if this is going to happen. A person who decides to get further information has to accept this uncertainty. However, there is some evidence that an offer of a genetic test for a specific disease can result in real worry for persons who are not already aware that they have an increased risk (e.g. Andrews et al., 1994). This is one of the dangers generated by the offer of testing, and it may result in individuals experiencing an additional burden of being at risk. It is the provider of the genetic test who has to take this risk, and, hence, responsibility for it.

Having received the information, an individual may decide to take a test. If (s)he is properly counselled, (s)he knows that in that case (s)he has to take some risks that are inherent in genetic testing. This may be, for example, the risk of diagnostic error – the risk of a false-positive or false-negative test result. If the individual tested is adequately informed and if all possible measures have been taken to prevent diagnostic error it could be said that the potential screenee simply has to take these risks. Thus it could be said that at least part of the responsibility comes to rest with the potential screenee. In case an individual refuses a genetic test, additional at risk situations (or uncertainties) are also created. Such an individual may be blamed for lack of responsibility or, in case the suspected disease develops, (s)he may be held responsible for the consequences.

Interpretation, communication and knowledge of test results also generates risks and uncertainties. Inadequate help in understanding the test result may lead to an inadequate understanding of the genetic information and hence to wrong decisions. For this a counsellor could be held accountable. But the situation is more complex here. If a screenee has been adequately informed about the uncertainties and risks of a genetic test and – on the basis of this information – has decided to take the test, it may be argued that this decision also involves taking responsibility for the negative consequences. However, we may question this argument, if pre-test counselling has not been adequate, if not enough is known about the negative consequences, or if not enough has been done to reduce or eliminate them. The question that can be raised is whether a test should be offered at all if not enough is known about the possible negative consequences or if not enough is done to avoid or reduce the risks and uncertainties.

A decision to undergo a test becomes even more problematic for an individual because there are uncertainties that neither the individual nor the counsellor can easily reduce or eliminate. It will, for example, often be impossible to foresee exactly how one will respond emotionally or how other family members may respond in the event of a positive (unfavourable) test result for a serous disease. It will often be impossible to foresee how far one can psychologically adjust to the information, or how family members will be able to cope with the genetic information, or how the social environment will react.

Therefore, simply defining and distinguishing between uncertainties and risks is not sufficient. At least three questions require an answer. The first is, 'which risks and uncertainties have to be accepted as inevitable because they are inherent in the genetic testing procedure?' The second question is, 'which of these unavoidable risks and uncertainties are acceptable and which are not?' and the third question that must be investigated is, 'which of these uncertainties and risks can be reduced or eliminated by further research?' We will return to these questions later. First the concept of genetic risk will be examined.

5. Genetic risk

As we have already suggested in the introduction, the concept of genetic risk is ambiguous and should be distinguished from a more general concept of risk referring to undesired consequences of a genetic screening or testing procedure. The concept of genetic risk must also be distinguished from the concept of genetic disease. For a more extensive discussion of the concepts of genetic risk and genetic disease we refer to Rogeer Hoedemaekers and Henk ten Have, (1998, in press).

In the history of medicine a distinction has grown between clinical manifestations of disease (complaints) and the causes of disease. Diseases came to be defined aetiologically. At first there was a simplistic emphasis on causes as single factors that directly caused single diseases (monocausal disease explanations); a functional or structural defect or abnormality tended to be seen as the cause of the disease. Since then, however, the awareness has grown that there are often many interacting factors, both internal and environmental, which all contribute to the development of disease. Diseases are seen as resulting from individual constitution, physiological laws and environmental influences. From this perspective the presence of a structural or functional abnormality cannot always be regarded as a specific diagnostic criterion for a disease as it is often but one contributory factor among others.

In the early phases of molecular genetics the focus was largely on the so-called monogenetic diseases, in which there is a strong association between a mutation in the particular gene and the development of clinical manifestations of the disease. In these contexts, there often does appear to be a monocausal relation between the gene mutation and the disease expression. These diseases were usually denoted as genetic diseases. More sophisticated and advanced DNA technology soon demonstrated that there are few diseases where there is a direct monocausal relationship between the presence of a genetic mutation and the expression of disease. There is growing scientific evidence that DNA sequences are only one single factor in a complex organic system in a complex environment in which there is mutual interaction between components and between components and the larger whole.

A concept like 'genetic disease' makes sense in a situation where a genetic mutation is a necessary and sufficient determinant of a disease. It makes less sense in cases where a genetic mutation is a necessary but insufficient determinant of a disease, but where other contributory (internal and external environmental) factors are needed for the development of disease. If one adheres to a holistic model of disease causation, it makes more sense to use the concept of genetic risk instead of genetic disease, not only in cases where other contributory (internal and/or external) environmental factors are needed, but also in cases where there is uncertainty about the role of other contributory factors of disease.

Genetic risks are often presented as 'increased risks'. A human being is a complex organism which is at risk of many different threats to health, both internal and external. A genetic risk may present an extra or additional risk, or, as it is usually termed, an increased risk, which is often presented in the form of a numerical risk. Geneticists may feel that this increased risk is in fact so high that the person with this genetic risk should know about this. Exclusive focus on the genetic risk factor may lead to two mistakes, however. There is a danger that a genetic risk is isolated and taken out of an intricate biological system where the overall functioning depends not only on genetic factors but also on many other components. Assessing the degree of increased risk for a given disease implies knowledge and estimation of the overall risk of getting this disease. Such calculations may be difficult, if not impossible, to perform. A comparison may illustrate this. A fully automated and computerized nuclear power plant is made up of hundreds of thousands of different components. Each of these components brings the possibility that something may go wrong in the system as a whole. The probability of each of these components breaking down and the extent of the damage that may be caused is estimated. Then scientists face the task of calculating the overall risk of all the components together. This may be difficult enough. If, for some reason or other, a less reliable component is installed, it constitutes an additional danger or, if the extra degree of danger can be calculated, an 'increased risk'. It might not make perfect sense to only focus on this additional risk factor, however. It also needs to be determined what this increased risk means for the whole system breaking down with disastrous consequences.

If disease is approached in a non-holistic manner, it is possible that the 'faulty' genetic component will be isolated and will not be seen as one component working together with many other components. But if it is considered as part of a complex whole, there is the extra difficulty of incorporating the uncertainty about the workings and faults of many other components of the whole. So the task of determining the actual overall risk for the whole 'system' is very complex, and perhaps impossible. Of course, an awareness of the complexity of disease processes exists among geneticists and medical professionals but it may not be assumed among the general public. It is important, however, that the lay person should also be aware of the complexity of disease processes. The lay person needs to be told about the uncertainties.

The second mistake is that a genetic risk is isolated and taken out of an individual's life-project or biography. Geneticists may be able to determine the presence or the degree of probability that an additional genetic risk is present. They take, so to speak, the additional at risk situation out of someone's life-project. Thus it may happen that, in pre-test counselling, a geneticist may think it sheer madness not to learn about an increased risk presented by a genetic alteration, but from the perspective of an individual's life-project a refusal to undergo a genetic test and to learn about this risk may make perfect sense. Individual preferences, values, beliefs and attitudes may play an important role in giving genetic risks a place in an individual's life-project. Isolating a genetic risk from this whole may lead to misunderstanding of the value of the genetic information on the part of the medical practitioner or counsellor. It may lead, for example, to the overemphasis of the genetic risk in comparison with all the other risks and potential dangers an individual runs. Ideally, however, a counsellor tries to understand an individual's life project and will put the isolated additional risk back into this life-project.

6. Risk perception, risk selection and risk taking

Before we turn to the question of the value of genetic risk information, we will briefly examine how an individual perceives and selects risks in everyday life and how decisions about risks are often made.

Risks continually threaten our own or other people's health and/or safety. We also continually put our own or other individuals' health and safety at risk. In our daily lives we accept and take many risks as a normal part of life. Which risks we take is partly dependent on the social or cultural context. A society's world view, experiences, perceived benefits and scale of values has an influence on which risks are judged as serious or not (Douglas, 1985). Also the value attached to important goals in a society and expert or authoritative assessment of risk may play a role. Thus the risk of a certain amount of environmental pollution is accepted, because we do not wish to give up our affluence.

Risk selection also occurs at an individual level. Risks crowd in from all sides. Attending to all of them would probably be paralyzing. This is one of the reasons why people tend to select risks (Douglas, 1985). One difficulty is that an individual in our modern industrial society often does not personally experience the external and collective risks posed by this society. We tend to rely on expert knowledge and we accept their risk assessments in the degree that they can provide reliable and convincing evidence (Heimann, 1990; Wilson and Crouch, 1987).

At the individual level apparent inconsistencies can be detected. We tend to neglect certain at risk situations (living in an earthquake area, sports-risks) and are extremely conscious of others (smoking). We take substantial risks, for example, because we are used to them (traffic), or because we are too familiar with them (home accidents). We also tend to take risks that we feel are largely under our control or that are taken voluntarily (e.g. not wearing safety-belts) (Slovic, 1987). Furthermore, people often see themselves as personally immune to hazards: nearly everyone believes him/herself to be a better driver than others, to live longer, less likely to be hurt by power tools. From the perspective of an individual's experience, risks look very small (Fischoff et al., 1981).

There may be a considerable difference between the risk assessment of experts and the risk perception and assessment of lay persons (Parsons and Atkinson, 1992). A distinction is usually made between a so-called objective and a subjective concept of risk, or, in other words, between a statistical risk concept, which presents a mathematical interpretation of probablity, and a personal, more intuitive estimation of potential danger. When a statistical probability estimation is given that a certain event will occur, an estimation which is usually determined for a certain group or population, a person is forced to ask: what does this risk mean to me in my situation? It is obvious that this involves subjective value-judgements (Heimann, 1990; Tverski and Kahneman, 1982; Vlek, 1987), and that this so-called 'perceived risk' may be different from a statistically calculated risk. Statistical figures require interpretation, but most people are not very good at interpreting numerical estimates of probability. This may lead to problems in communication between risk experts and lay persons. Lay persons, for instance, tend to think in binary terms (it may or may not happen), so for them an increased risk of 2.3% may be perceived in terms of a 50–50 chance of an event taking place. (Lippman-Hand and Fraser, 1979). If no help is given to understand numeric risk, misinterpretation with potentially harmful consequences may result.

A risk situation usually involves several options, and decisions to follow a particular course of action have to be made. Risk taking is also influenced by one's awareness and assessment of risks involved in the alternative courses of action one has in an at risk situation. Therefore it is important that the risks and uncertainties inherent in the options are clear. Ignorance of these risks may well lead to misperception and miscalculation of the potentially harmful consequences. This may lead to misjudgement of one's coping abilities. This implies that completeness of information is an important requirement for adequate risk taking (Pauker and Pauker, 1987; Pfeifer and Kenner, 1986; Pochin, 1982).

Decision making in situations of risk and uncertainty has received considerable attention (Kahneman and Tversky, 1982). Acceptable risk problems are often approached as decision problems. Sometimes decisions must be taken about one specific risk, while often there is a 'risk–risk' situation of having to choose between two different risks. In the latter case a choice must be made among alternative courses of action, which may all have potentially harmful consequences, directly or indirectly, short-term or long-term. Such a choice usually involves the following steps: specifying the goals by which the desirability of consequences must be measured (1), defining the possible options, which may include doing nothing (2), identifying the possible consequences of each option and their probability of occurring in case a certain option is adopted (3). Then the desirability of the various consequences must be established (4) and weighting of options and selecting the best one (5). The underlying assumption is that if this route is followed, one makes a choice that can be justified rationally. The attractiveness of possible options can be determined in each situation (Fischoff et al., 1981). However, the difficulties included in taking decisions about risks are considerable and include uncertainty about how to define or frame the problem that requires a decision, difficulties in determining and assessing the facts of the matter (often major uncertainties remain), difficulties in assessing the relevant values, what weight is to be given to each uncertainty and what weight must be given to individual beliefs, attitudes, intuitions and emotions in each decision-making process (Eddy, 1982; Fischoff et al., 1981).

It is obvious that the uncertainties that are usually involved in a decision-making process cannot always be resolved by rational decision making. Uncertainties will remain (as is usually the case in important life-decisions) (Kahneman and Tversky, 1982). Some simply have to be accepted. But a decision-making procedure may help reduce them, so that an important decision is not just a gamble, but becomes a 'calculated gamble' – which may help an individual accept the consequences of his/her choices more easily.

7. The value of genetic information

Numerous genes, in which mutations play an important role in both monogenic and polygenic, multifactorial or complex diseases have already been located, and the potential benefits of such knowledge have frequently been pointed out. The identification of genes helps us to understand their function, and this may lead to the development of therapies to prevent manifestation of disease. Such understanding may help to develop drugs that can compensate for the genetic mutation. It may also permit scientists to develop tests to detect individuals at risk of developing disease before they reveal clinical symptoms, either in themselves or in their offspring, so that preventive or therapeutic action can be undertaken. A knowledge of risk factors may also help individuals to prevent disease by changing lifestyles or habits. And, last but not least, the knowledge that a genetic risk factor is absent may, it is assumed, bring reassurance.

Besides benefits, genetic screening also presents uncertainties, potentially harmful consequences and risks, both in the case of participation and in the case of non-participation. This makes the determination of the value of genetic information complex. Problems may even increase if the value of genetic screening and testing is also judged in the light of other risks threatening our health and longevity. It should be remembered, though, that decisions in medical matters are often laden with uncertainties and risks. Also, a knowledge of genetic risk factors is not solely burdensome, however, in that it presents us with potential benefits as well as with potential harms, risks and uncertainties.

The question about the value of genetic information can be approached from two different perspectives. First, there is the pragmatic question: what sort of consequences does genetic risk information have for me and my family and relatives? The assumption here is that rational decisions can be taken by balancing benefits and risks or benefits and uncertainties, and that emotions or intuitions need not interfere with the decision-making process. Determination of the value of genetic information includes the question when these risks and/or uncertainties are acceptable for an individual in view of the supposed benefits of genetic knowledge (Hepburn, 1996).

From a different, more biographical or personal level there is the question, 'what does information about a particular genetic risk mean to a particular individual?' This question is more concerned with an individual's whole life-project and an answer is connected with an individual's attitude towards risk taking, the relative importance of genetic health risks among all the other hazards threatening life and safety, and an individual's preferences and values. It involves the consideration of one's beliefs, world-view and attitudes towards life and death. Here, individual feelings, beliefs and intuitions are not seen as interfering with rational decision-making but as forming part of a decision-making process – they are incorporated into it. Indeed, reason and emotion cannot be clearly separated in this context.

We assume that answers to the first question will be relevant for the second, more personal approach and that a general (rational) answer about the value of genetic information is not possible. Benefits, uncertainties, potential harm and risks can be different for different genetic risk factors, and their value is dependent on an individual's ability to incorporate this information within his/her life-project. So the proper question should be: what does information about this specific genetic risk mean for this particular person in this particular situation?

In order to answer this question adequately several requirements should (ideally) be met:

(1) An individual needs to know which uncertainties going with genetic screening/testing are inevitable and have to be accepted, which can be reduced and which can be eliminated for each step in the genetic screening/ testing process.
(2) Wherever possible, probability estimates should replace uncertainties.
(3) Understanding of the meaning of a statistical genetic risk figure.
(4) There should be insight into one's individual preferences, important values and norms and possible inconsistencies in order to decide if genetic risk information is really needed.

(5) Understanding of one's motives for seeking genetic information is needed (e.g., anticipatory regret, better safe than sorry).

(6) Understanding of the psychological and/or psychosocial consequences of participation or non-participation in a genetic testing procedure is also needed, both for the prospective screening participant and his/her partner and family.

(7) Insight or understanding of one's own and one's family's coping abilities is required in order to be able to judge if the genetic information is potentially beneficial or harmful.

(8) Understanding of social and economic consequences (e.g., insurance, employment, discrimination, stigmatization, blame, misuse of genetic information by third parties).

(9) Understanding of the value of genetic risk information for relatives or family.

(10) Understanding of the (subtle) forms of coercion created by societal or group values and norms which may determine preferences.

These requirements seem to indicate that, for individuals to come to terms with risks requires only a rational and detached balancing of risks and benefits. As we have already suggested, however, various other, not-so-rational factors may influence risk decisions. The first of these is the role of emotions and/or intuitions in risk decisions, which must not be underestimated. It is becoming apparent, for example, that for many genetic testing programmes, both population-based and family-based (e.g. for cystic fibrosis and Huntington's disease), the number of persons interested in testing is considerably smaller than geneticists and medical professionals had expected. Some sort of intuitive pre-judgement about the value of genetic risks is apparently made. The group of individuals who request a test is perhaps (partly) a self-selected group. They may already have decided that the value of the genetic test is great enough to accept possible risks, even if they have not yet been fully informed about all these risks. Others have decided not to take the test. This decision may also be based on an intuitive perception of genetic risk, their value system and/or their personal attitude towards risk taking. This already throws some doubt on the use of the (supposedly) rational decision-model with its balancing of advantages and disadvantages.

The second factor is the individual's attitude towards risk-taking. Differences between individuals may be considerable. One of the assumptions prevalent in modern western societies, and also detectable in genetic screening literature, is that we should try to eliminate or reduce risks wherever possible. This attitude, however, may not appeal to the 'risk-all', the person who is attracted to risk taking. Nor will it appeal to those people who simply do not care, or those who have a laissez-faire attitude or a fatalistic attitude towards the vicissitudes of life. But it may appeal to those who are so fearful of risk that they try to avoid exposure to every conceivable hazard. It will also appeal to the devotees of 'all risks' insurance, who wish to control any conceivable risk (including rain on holiday in Spain). Genetic testing may appeal to those who do not wish to take any unneccessary chances and it may appeal to those who prefer a more rational form of decision-making to intuitive choices.

Besides intuitive or emotional factors and different attitudes towards risk-taking there is the process of risk-selection: the perception and selection of genetic risks in comparison with other health or safety risks. Giving attention to some risks and ignoring others is a process that is not always carried out rationally. It may, for example, be influenced by societal values and priorities. Perhaps a disproportionate degree of attention is at present given to individual health risks. Health is an important value in modern Western societies and the prevention of suffering is high on our priority list (Taylor, 1989). Another important feature of Western societies is a desire to control and to manage risks. The rapidly expanding insurance industry may be evidence for that. These tendencies may help create a need for health surveillance and monitoring. A certain predilection for a technical approach to reducing life's problems and worries may also play a role. Reflection on the value of genetic information may include the question as to whether these tendencies have too strong a hold on individual lives.

Finally it should be noted that religious or societal beliefs may also be factors influencing an individual's attitude towards life's risks and, hence, may help determine the value of genetic knowledge. Belief in God, for example, may lead to greater acceptance of life's risks. Also, the possibilities for controlling fate and chance that have been created by modern science and technology may play a role. Full discussion of these factors is not possible here. Suffice it to say that for a full appreciation of the value of genetic knowledge for an individual, an analysis of their personal, religious and societal beliefs, values and priorities must be undertaken.

It may well be that the real value of genetic information lies in the very process of risk selection. Genetic information tells us about an additional risk run by an individual. This knowledge may be necessary to understand in what way his/her individual life-project can go wrong. This means that the identification of a genetic risk may be important so that a better selection between risks can be made. Thus the information may help to take certain risks and avoid others. It may help individuals to make better choices. This may be even more true if the uncertainties about possible outcomes – benefits, risks and potentially harmful consequences – can be reduced, eliminated or at least calculated.

There may be considerable differences in the perception of genetic risk between geneticists (and medical professionals) and potential test subjects. These differences may result from the lack of understanding of individual life-projects, but they will make it difficult to establish in advance if genetic information is of value for the projects of all the potential participants in testing. However, accepting that genetic information is valuable for some individuals, it is not immediately obvious either that it will also be of benefit for all individuals or that large-scale genetic testing programmes are needed.

If the numbers of subjects accepting the offer of new genetic tests appears 'inadequate' to those making the tests available, this lack of 'compliance' should not inevitably lead geneticists and genetic service laboratories to devise stategies to enhance test acceptance. The general public may have an intuition that the value of genetic information is limited. As we believe to have illustrated, information about a genetic risk factor may lead to a 'risk-shift', to other kinds of suffering, psychological and/or social, and this may be one reason to limit the provision of genetic testing to those who request it and who are already more fully aware of the risks they run.

For those who decide to have a genetic diagnostic test, the risks should be acceptable and reasonable. For providers of genetic diagnostic tests this has impli-cations. These cannot be fully elaborated on here, but are in our view aptly sum-marized in the following quotation provided by the (US) National Commission on Product Safety (Beauchamp and Bowie, 1993). In view of the increasing ten-dency to refer to screening participants as customers, and in view of the rapidly expanding domain of commercial genetic screening and testing, it does not seem inappropriate to regard a genetic diagnostic test as a product falling within the domain of the quotation.

'Risks of bodily harm to users are not unreasonable when consumers under-stand that risks exist, can appraise their probability and severity, know how to cope with them, and voluntarily accept them to get benefits they could not obtain in less risky ways. When there is a risk of this character, consumers have reason-able opportunity to protect themselves; and public authorities should hesitate to substitute their value judgements about the desirability of the risk for those of the consumers who choose to incur it. But preventable risk is not reasonable (a) when consumers do not know that it exists; or (b) when, though aware of it, consumers are unable to estimate its frequency and severity; or (c) when consumers do not know how to cope with it, and, hence are likely to incur harm unnecessarily; or (d) when risk is unnecessary in that it could be reduced or eliminated at a cost in money or in the performance of the product that consumers would willingly incur if they knew the facts and were given the choice'.

In view of what we have argued earlier, we propose that, for the screening par-ticipant, taking a reasonable or acceptable risk means that uncertainties, potential dangers and risks are taken knowingly and voluntarily. This does not only require the provision of information about all the uncertainties, risks and potential dan-gers inherent in a genetic screening, testing and counselling procedure. Acceptable risk also means an adequate assessment or calculation of the probabil-ity and severity of the adverse outcomes, a proper understanding of the differ-ences in meaning between uncertainty, potential harm and calculated risk, an exploration of one's coping abilities and the elimination of unnessary risks.

In the last few years a process of commercialization of genetic testing has started. Commercial biotechnology firms expect the most profit from predictive tests that can be conducted on large numbers of people. They assume there will be a great increase in susceptibility screening for multifactorial diseases and multi-plex testing for a number of diseases and disorders for which there is no clear treatment. Also, genetic testing and screening services will be increasingly carried out by commercial laboratories (Andrews et al., 1994; Hubbard and Wald, 1993; Joyce, 1987), and it is predicted that genetic screening, testing and counselling will become part of common practice and routine care. However, although the need for proper education, and pre- and post-test counselling is continually stressed as a way to overcome most of the threats that accompany genetic screen-ing, there seems to be a real possibility that less and less time will be spent on counselling procedures as these are usually the most expensive part of a testing or screening procedure (US Congress, 1992; Andrews et al., 1994). Moreover, it has been pointed out that primary health care providers may have neither the exper-tise, the willingness nor the time to do this adequately.

In view of what we have argued, this prospect for large-scale genetic screening and testing is worrying. If more and more genetic tests and services are marketed with little or only partial information in advance, through a further increase of genetic screening for asymptomatic diseases or susceptibilities, and if these tests become part of routine medical practice, then the uncertainties involved in decision-making based on genetic information may become unacceptably high.

8. Conclusion

By way of summary we wish to present the following conclusions:

(1) The concept of risk is commonly used in a broad sense, and comprises a variety of risks. This may lead to misunderstanding of potentially harmful or undesired consequences on the part of the prospective screenee.

(2) In the context of genetic screening and testing a distinction between uncertainty, potential harm (or fuzzy risk) and calculated risk may contribute to give a prospective screening participant a better understanding of the degree of certitude with which certain developments or consequences are expected.

(3) Analysis of risks, harms and uncertainties generated by genetic screening, testing and counselling can be facilitated with the help of the notions of being at risk and taking a risk, and may help define responsibilities of those offering tests as well as those being tested.

(4) In a holistic model of disease causation – emphasizing the presence of many different, interacting, internal and environmental factors – the concept of genetic risk is to be preferred to the concept of genetic disease. From a holistic perspective, a genetic variant associated with a disease is one among many other risk factors. There is a danger that such genetic risk factors may be isolated from the biological and biographical system within which it operates. This may lead to a misunderstanding by an individual of the overall risks and challenges to their health that they are having to confront; this may lead to a one-sided approach to reducing or eliminating the medical (specifically genetic) risks.

(5) It is often assumed that decision-making in at risk situations is a process involving a rational balancing of advantages and disadvantages. This rational model of decision-making tends to underestimate the biographical element: the role of personal beliefs, emotions, intuitions and attitudes towards risk. Also, many uncertainties inherent in a genetic screening or testing procedure throw doubt on this rational decision-making model.

(6) A general answer about the value of genetic information is not possible. The question that should be asked is, 'what information about a specific genetic risk means for *this* particular person in *this* particular situation?' Any answer should take the biographical element into account.

(7) An important step towards better decision-making is more research in order to convert the uncertainties and the intuitive understanding of the potential harms arising from genetic screening and testing into probabilities.

(8) Human geneticists herald the era of predictive medicine (e.g. Dausset, 1996). Apparently we have to live not only with people who pose risks and people who take risks, but also with people who wish to tell us about our risks. But is it not Shakespeare who already gave a note of warning? As Macbeth said to

his three 'counsellors', who were memorable for reasons other than the gift of certainty and the creation of conceptual clarity:

'Stay, you imperfect speakers, tell me more.' (Macbeth, I, iii, 70).

Not only the present lack of precise information, but also Macbeth's fate may help us realize that we should perhaps not be too eager to learn what genetic risks and dangers are in store for us.

Acknowledgements

We are grateful to the Commission of the European Community for funding the EUROSCREEN project, of which this paper forms a part. We also express gratitude for useful remarks received from members of the Department of Ethics, Philosophy and History of Medicine of the Medical School of the Catholic University of Nijmegen, Henk ten Have, Wim Dekkers, Godelieve van Heteren and Marcel Verwey. We are also indebted to members of the Euroscreen subgroup focusing on philosophical aspects of genetic screening, Ingmar Porn, Veiko Launis, Kris Dierickx and, especially, Jurgen Husted. We also wish to thank Mairi Levitt and Angus Clarke for valuable comment.

References

Andrews, L.B., Fullarton, J.E., Holtzman, N.A. and Motulsky, A.G. (1994) *Assessing genetic risks: implications for health and social policy*. Institute of Medicine, National Academy Press, Washington, DC.

Beauchamp, T.L. and Bowie N.E. (1993) (eds). *Ethical Theory and Business*. Prentice Hall Inc. Englewood Cliffs, NJ.

van den Berghe, H. (1987) Impact of genetics on society. In: Genetic risk, risk perception, and decision making (eds G. Evers-Kiebooms, J-J. Cassiman, H. van den Berghe and G. d' Ydewalle). Alan R. Liss, New York, NY, pp. 1–5.

Clarke, A. (1994a) *Genetic counselling: practice and principles*. Routledge, London.

Clarke, A. (1994b) The genetic testing of children. Working party of the clinical genetics society (UK), *J.Med.Genet.* 31(10): 785–797.

Dausset, J. (1996) Predictive medicine. Eur. J. Obstetrics gynecol. Reproduct. Biol. 65: 29–32.

Douglas, M. (1985) *Risk acceptibility according to the social sciences*. Russel Sage Foundation, New York, NY.

Eddy, D.M. (1982) Probabilistic reasoning in clinical medicine: problems and opportunities. In: *Judgement under uncertainty: heuristics and biases* (eds D. Kahneman, P. Slovic and A. Tverski). Cambridge University Press, Cambridge, pp. 249–267.

Fischoff, B., Lichtenstein, S. and Slovic, S. (1981) *Acceptable risk*. Cambridge University Press, Cambridge.

Hubbard, R. and Wald, E. (1993) *Exploding the gene myth*. Beacon Press, Boston, MA.

Heimann, W. (1990) *Das Risiko in menschlichem Leben*. Evangelischer Presseverband für Bayern e.V., München.

Hepburn, E.R. (1996) Genetic testing and early diagnosis and intervention: boon or burden? *J. Med. Ethics* 22: 105–110.

Hoedemaekers, R. and ten Have, H. (1998) Genetic health and genetic disease. In: *Genes and Morality* (eds V. Launis *et al.*). Rodspi, Amsterdam/Atlanta (in press).

Joyce, C. (1987) Genes reach the medical market. *New Scientist* 115 (16 July): 45–49.

Kahneman, D. and Tversky, A. (1982) The psychology of preferences, *Scientific American* 246 (1): 136–142.

Katz, L. (1990) The assumption of risk argument, *Social Phil. Policy*, 7 (2): 138–169.

Lave, L.B. (1987) Health and safety risk analyses: information for better decisions. *Science* 236 (17 April): 291–295.

Lippman-Hand, A. and Fraser, F.C. (1979) Genetic counselling – The postcounselling period: 1. Parents' perceptions of uncertainty. *Am. J. Med. Genet.* 4: 51-71.

Lupton, D. (1993) Risk as moral danger: The social and political functions of risk discourse in public health. *Int. J health services* 23(3): 425–435.

Parsons, E. and Atkinson, P. (1992) Lay constructions of genetic risk. *Sociology of Health and Illness* 14(4): 437–455.

Pauker, S.G. and Pauker, S.P. (1987) Prescriptive models to support decision making in genetics. In: *Genetic risk, risk perception and decision making* (eds G. Evers-Kiebooms, J-J. Cassiman, H. Van den Berghe and G. d' Ydewalle) Alan R. Liss, New York, NY, pp. 279–296.

Pfeiffer, K.P. and Kenner, T. (1986) The risk concept in medicine – statistical and epidemiological aspects: a case report for applied mathematics in cardiology. *Theoret. Med.* 7:259–268.

Pochin, E. (1982) Risk and medical ethics. *J. Med.Ethics* 8: 180–184.

Slovic, P. (1987) Perception of risk. *Science* 236(17 April): 280–285.

Taylor, C. (1989) *Sources of the self: the making of the modern identity*. Cambridge University Press, Cambridge.

Tibben, A. (1993) What is knowledge but grieving? On psychological effects of presymptomatic DNA-testing for Huntington's disease. *Universiteitsdrukkerij Erasmus, Rotterdam*.

Tijmstra, T. and Bajema, C. (1990) 'Je zult die ene maar zijn'; risicobeleving en keuze-gedrag rond medische technologie. *Ned Tijdschr Geneeskd* 134: 1884–1885.

Tversky, A. and Kahneman, D. (1982) Judgement under uncertainty: heuristics and biases. In: *Judgement under uncertainty: heuristics and biases*, (eds D. Kahneman, P. Slovic and A. Tverski), Cambridge University Press, Cambridge, pp. 3–20.

US Congress, Office of Technology Assessment (1992) *Cystic fibrosis and DNA tests: Implications of carrier screening* OTA Printing office, Washington, DC.

Vlek, C. (1987) Risk assessment, risk perception and decision-making about courses of action involving genetic risk: an overview of concepts and methods. In: *Genetic Risk, Risk Perception and Decision-making*, (eds G. Evers-Kiebooms, J-J. Cassiman, H. Van den Berghe and G. d' Ydewalle), Alan R. Liss, New York, NY, pp. 171–207.

Wilson, R. and Crouch, E.A.C. (1987) Risk assessment and comparisons: an introduction. *Science* 236: 267–270.

22

Commercial testing

Shirley Dalby

1. Introduction

Considerable concern has been expressed at the emergence, and probable growth, of commercial companies offering genetic testing. Discussion of this issue requires examination of three interrelated questions. Firstly, what are the major differences between the public and commercial sectors? Secondly, what are the issues giving rise to the concern? Thirdly, who, and under what circumstances, may use this service? It is only after considering these questions that recommendations and decisions can be made as to whether interventions and controls are neccessary.

2. Differences between the public and commercial sectors

In practice, all genetic testing has a commercial element – somebody has to pay for it. Within a state system such as the National Health Service (NHS) in the UK, the general population tends to believe that it is entitled to any treatment that is technologically possible. This is not necessarily so, and health economists require evidence of benefit before recommending that a costly service be provided. This is primarily a matter of balancing and then prioritizing both the costs and benefits of a service, which poses ethical problems in relation to genetic testing. In a large number of cases people may make decisions, particularly reproductive decisions about whether to have a child or whether to make use of pre-natal diagnostic testing services, which may save the health and social services considerable amounts of money. Other outcomes of genetic services may be more difficult to quantify, as with factors influencing quality of life. However, the NHS is guided by the principle of caring for patients – financial factors are constraints but not the raison d'etre of health services. Clinicians and other medical personnel are concerned with improving the health and well being of their patients – they are particularly opposed to doing them any avoidable harm. Thus, within a health service, the provision of genetic testing is governed by medical ethics, the assessment of an individual clinician, and finances (for instance, testing may be technically possible for a particular rare disorder but not available because of the expense). Within certain parameters, patients can be offered counselling and support to help them make the decision which they believe to be best for them – not for the financial advantage of another individual or an organization.

The Genetic Testing of Children, A.J. Clarke (ed.).
© 1998 BIOS Scientific Publishers Ltd, Oxford.

The position of clinicians in private practice is unenviable. Medical ethics remain the same, but are circumscribed by more immediate financial constraints. The patient requesting a test, whether paying personally or through an insurance company, is likely to have a limited budget. The opportunities for counselling and for ensuring that any decision is made on the basis of fully informed consent may be curtailed. It would be interesting to know whether insurance companies, or individuals, would be happy to pay a substantial sum for consultations which resulted in the person deciding not to take a test.

Commercial companies, by their very nature, are primarily concerned with making a profit. Because of concern about this, they have so far acted responsibly, offering information leaflets and opportunities for discussion to their clients. However, profits can only be made if there is sufficient uptake of a service and an adequate level of charges. The implications for promotion and provision are obvious – the general public will have to be made aware of the service and its potential benefits to them, and the demand for testing for the more common disorders is likely to make them both reasonably priced and profitable.

3. Causes for concern

The alarm raised at the prospect of commercial testing is an extension of the anxieties already in existence relating to the social repercussions and the ethical difficulties raised by the provision of genetic testing. Genetic testing permits glimpses into the future, and such predictions – as a science rather than an art – are still relatively new and often produce hostile, knee-jerk reactions. The pro-life organizations regard such predictions as an encouragement to abortion; some disabled people see them as a form of eugenics; the media is provided with both human-interest and sensational material; insurance companies visualize more accurate actuarial tables and therefore more profit; the general public is in turn fascinated and frightened because, for the most part, it is ill-informed. Those who are well informed are often ambivalent. Testing for conditions for which there is effective treatment raises relatively few difficulties, but the majority of genetic disorders are not in this category and pose real dilemmas. These issues are dealt with in detail elsewhere in this volume, so it is sufficient to say here that the long term repercussions of genetic testing for an individual are not yet known, that individuals are infinitely variable in their emotional and coping capacities, and the response of society through discrimination in such areas as employment, insurance, housing, and reproduction is in a state of flux. Those who have responsibility in this field – doctors, scientists, ethicists, politicians and others – are therefore disposed towards proceeding with caution and employing as many safeguards as possible until experience and research provide a firmer foundation on which to base future policies.

The four areas of particular concern:

3.1 The appropriateness of particular tests

(i) People with no family history of a condition may request testing for it when they read about it in a newspaper or come across it elsewhere in the media.

There is a long list of conditions for which testing could be provided, and a test for one particular disorder in these circumstances is likely to do little other than generate inappropriate anxieties. Counselling and the provision of accurate risk information to the general population is far more appropriate than testing. Will commercial enterprises perhaps be happy simply to test without ensuring that the testing is appropriate?

(ii) Many common disorders are caused by a complex interaction between an individual's genetic constitution and environmental factors; these conditions are influenced by genetic factors without being determined by them. For such disorders, information obtained from a genetic test may do little more than to moderately increase or decrease the risk faced by an individual in comparison to that faced by the general population. Such modifications of risk information are difficult for many people to comprehend, can generate much inappropriate stress, and will often not lead to significant benefits because the capacity to improve outcomes through changing a person's lifestyle may be very limited. Yet testing for many of the most common, and therefore potentially most profitable, conditions is likely to fall into this category – some cancers, heart disease and Alzheimer's disease. Will the commercial sector test for them despite the drawbacks?

3.2 Testing of children

The particular issues of testing children, as described in other chapters, become even more important if there is to be no independent advocate for the child. Will biotechnology corporations carry out tests on children regardless of the likely age of onset of the condition? Will they test for carrier status just on the request of the parent without attempting to involve the child? Will it be obligatory for the result to be given to the GP so that there can be follow-up?

3.3 Information and understanding

Genetic information is complicated. People need to know about the likely age of onset of the condition, about its progression and likely effects, about the mode of inheritance, and about the significance of the various possible test results. Experience has shown that it is often necessary to go over this information more than once and to provide a written explanation. Even in these circumstances people can misinterpret test results, too optimistically or too pessimistically, with unfortunate consequences. Accurate information is crucial both to the decision to take a test and to subsequent adaptation to the test results. Will commercial enterprises be able to provide this information in a user-friendly fashion and in a variety of formats?

3.4 Counselling

Information on its own is almost always insufficient. Regardless of the level of support from family and friends, most people benefit from the opportunity to discuss the issues with an outsider who has experience of counselling others in similar situations. A considered choice – and hence, hopefully, the optimum adaptation to the test result, if the person chooses to proceed with testing – is most likely to result from a careful exploration of all the options, the possible outcomes and their likely consequences for the individual and for those around him or her. This discussion will focus on the wishes of the individual in the light of these considerations; it will take into account the wishes of others but will not be driven by them. It is doubtful whether the costs of such counselling will be commercially viable in the long term.

4. Who are the customers?

As discussed earlier, there may be some requests for testing from people with no family history of a condition. However, it is much more likely that demand will come from people who have reason to be concerned about a particular condition and who, for some reason, are unable or unwilling to obtain testing through the normal channels. The Genetic Interest Group undertook a simple survey of its members to discover the circumstances under which people would consider requesting genetic tests through commercial companies, and what services they thought it would be necessary and/or desirable for those companies to offer. The survey was necessarily biased as, on the whole, GIG members are well informed of the issues. The respondents, however, were asked to reply in general terms rather than from a personal point of view in order to gain some insight into attitudes rather than their personal intentions. The results are based on 34 replies.

(1) What are the minimum requirements for pretest information?

Detailed leaflet with a specialist phone line service	43%
Face-to-face discussion and written information at a registered centre	34%
Detailed leaflet alone	22%
Detailed video	2%

On the other hand:

(2) What method of information provision do you favour?
(1 is most favoured, 4 the least; number given is mean score)

Face-to-face and in writing at a registered centre	1.32
Leaflet and specialized phone line	2.09
Video	3.19
Leaflet alone	3.3

(3) How important is pre-test counselling for exploring the potential impact of a test?

Essential	76%
Important	15%
Desirable	6%
Not important	3%

(4) How important is the availability of post-test counselling and support?

Essential	79%
Important	15%
Desirable	6%
Not important	0%

Questions 5–9 asked whether it would be reasonable to consider the use of commercial testing if it had been refused under the Health Service.....

(5) .... if the individual requesting testing had a psychiatric problem ?

Yes	12%
Maybe	50%
No	35%

(6) .... if the problem was ethical, such as testing an asymptomatic child?

Yes	26%
Maybe	44%
No	30%

(7) ....if the test had a low accuracy rate ?

Yes	29%
Maybe	42%
No	29%

(8) .... if there were delays in obtaining testing ?

Yes	56%
Maybe	35%
No	9%

(9) .... to avoid counselling procedures ?

Yes	18%
Maybe	29%
No	53%

Finally:

(10) Should there be a compulsory link between testing and the provision of relevant pre- and post-test information, counselling and support procedures of a certain minimum quality?

Yes	97%
No	3%

When interpreting these results, it has to be remembered that most respondents were trying very hard to give an overview rather than just an individual opinion. There were, therefore, a number of caveats, such as 'it depends on the condition' or 'it depends on individual circumstances'. Nevertheless, the survey shows clearly that the majority of respondents considered face-to-face counselling to be very important, perhaps essential, and that other methods such as the provision of written information or telephone counselling were regarded as a bare minimum rather than a desirable level of service. Having said that, if the service was not available through the NHS – or if there were too many limitations or restrictions on NHS provision – then many respondents would be prepared to consider looking elsewhere. Most respondents supported the principle that individuals and families should be responsible for their own decisions. In particular, it should be noted that only 30% would definitely not consider it reasonable to use commercial channels for testing asymptomatic children – although this could lead to problems: if adults are prone to misinterpret test results, then even more mistakes may be made when they are relayed to a child. Furthermore, only 29% would definitely not consider commercial testing for tests with a low accuracy rate. It would appear that, for some people, any information is better than none.

The responses to the final question show that, regardless of what the respondents would be prepared to do if they had to, they would overwhelmingly want genetic testing services to be provided in the best possible way.

5. Conclusions

It would seem that those with the responsibility for the provision of genetic testing and for the ethical review of genetics services and those who might wish to make use of these services are in almost total agreement as to the way in which these services should be provided. The proposal that regulation should be used to impose certain minimum standards for the provision of genetic services appears to be supported very broadly by service providers and service users.

This accord between service providers and users, however, is based on the assumption that services will indeed be available. Financial constraints causing long delays or difficulties in obtaining tests for very rare conditions will cause clients to look elsewhere. The same would be true if professional concerns about ethical issues were to prevent or delay testing; if the desire for a test is strong enough then people will ignore the concerns and the advice of professionals.

Another important point, although it was not included in the survey, is that commercial testing may allow people to obtain a result without this being recorded in their medical records. This could be a potential bonus when individuals consider such matters as house purchase, life insurance and, for the future, long-term care insurance.

If the gap in the market is large enough, and if there is sufficient profit to be made, then commercial enterprise will cater for it.

23

International perspectives

Dorothy C. Wertz

Abstract

This paper reports the results of conjoint surveys of stakeholders' views, including a 37-nation survey of 2903 geneticists, and surveys of 499 United States primary care physicians, 409 parents visiting genetics clinics in the US and Canada, and 988 members of the US public. In the US, parents, public, and primary care physicians favoured parents' rights to have their children tested for adult-onset conditions such as Alzheimer disease and Huntington disease. The majority of US geneticists would refuse testing, as would geneticists in northern/western Europe and English-speaking countries. In Asia, Latin America, and southern/eastern Europe, the majority of geneticists would accede to parental requests. This paper discusses the future of testing in the US in view of cultural beliefs in autonomy and the movement of genetics into general medicine.

1. Introduction

1.1 Autonomy as a pre-eminent cultural value

The United States has always been the land of individual autonomy. Writing in 1835, Alexis de Toqueville noted Americans' refusal to be bound by traditions, their innovativeness in politics, seemingly boundless optimism, and belief in science. 'Under their hand, political principles, laws, and human institutions seem malleable, capable of being shaped and combined at will. As they go forward, the barriers which imprisoned society and behind which they were born are lowered; old opinions, which for centuries had been controlling the world, vanish; a course almost without limits, a field without horizon, is revealed; the human spirit rushes forward and traverses them in every direction' (de Toqueville, 1951).

Americans also believed that individuals should control their own interests and that society cannot, and should not, try to interpret these interests. 'Every individual, private person, society, community, or nation is the only lawful judge of its own interest, and, provided it does not harm the interests of others, nobody has the right to interfere' (Schleifer, 1980).

One reason for the individualistic approach to health care in the US is that medicine developed into a profession much later in the United States than in Europe. Licensing laws were largely ignored until the end of the nineteenth century, because those who called themselves doctors demanded the right to freedom of practice, just as their patients demanded freedom of choice. Medicine developed in an entrepreneurial, commercial climate. Full professionalization, with speciality boards and standards for medical education, was not completed until the 1930s (Shryock, 1962).

This overriding belief in autonomy affects approaches to testing children in two ways. First, parents think that they should have the right to have their children tested, without interference from doctors or the state, because they, not the doctors, are the most knowledgeable judges of the child's 'best interests'. Many believe that knowledge in itself is good, even if there is no immediate treatment. Second, in a somewhat contrary vein, there is increasing recognition that children and adolescents themselves are entitled to autonomy in some aspects of health care. Courts in some states have ruled that adolescents below the age of legal majority (18 years) may obtain contraceptive prescriptions, abortions, tests and treatments for sexually transmitted diseases, including HIV, and treatment for addiction, without parental knowledge or consent (Ambuel and Rappoport, 1992). Adolescents living independently from their parents, college students, pregnant teenagers, or teenage parents (even if living at home) are considered to be 'emancipated minors' in many states and are regarded as adults in terms of health care decisions (Bennett, 1976; Ewald, 1982; Holder, 1989; Wadlington, 1973). Although courts have varied in setting the 'age of discretion' from 14 to 18 years, they have generally regarded 15 years as the age at which minors can consent independently to medical treatment (Holder, 1977, 1988). Federal regulations governing research accord a measure of autonomy to children at considerably younger ages. A child's 'assent' is required if the child is old enough to understand the general outlines of a research procedure (Code of Federal Regulations 46.408). Although the regulations set no fixed age, most psychologists, ethicists and courts agree that children of 7 years and older are usually capable of assent (Brock, 1989; Buchanan and Brock, 1989; Leiken, 1982). Although the child's assent is not legally binding, the National Commission for the Protection of Human Subjects recommended that 'a child's objection to participation in research should be binding unless the intervention holds out a prospect of direct benefits' (National Commission for Protection of Human Subjects, 1977).

1.2 Professional statements urging caution

The increasing recognition of children's autonomy has led to a conflict with traditional beliefs about parental autonomy. One of the strongest arguments put forward by bioethicists against testing children for adult-onset disorders is that premature testing denies the future autonomy of the child (Brock, 1989; Brock and Buchanan, 1989). Persons who have already been tested as children cannot make their own decisions about testing once they become adults.

This is an important point, because most adults in the US appear to have decided against genetic testing, at least for the present. About 16% of adults at risk have chosen to be tested for the Huntington gene (Bloch *et al.*, 1989; Tyler *et al.*, 1992). Except for pregnant women (concerned about the baby's health), the US public has evinced little interest in cystic fibrosis (CF) carrier tests. Between 1% and 77% (mostly pregnant) have accepted free offers of carrier testing for CF in the US pilot studies (Science, 1996). The uptake for breast cancer genetic testing has also been well below expectations (Science, 1997).

Although there are many potential benefits to testing asymptomatic children, the potential harms are such that many professional groups, including the American Medical Association (1995), American Society of Human Genetics (1995), National Society of Genetic Counselors (1995), and Institute of Medicine (1994), have urged caution. Although everyone agrees that children should be tested if there is a direct medical benefit in the near future, the various professional statements suggest that, in the absence of a direct medical benefit, testing for later-onset disorders or for carrier status is usually best postponed until adulthood.

We have summarized potential psychosocial harms elsewhere (Wertz, *et al.*, 1994), as have our colleagues in the UK (Clarke and Flinter, 1996; Harper and Clarke, 1990). A child tested prematurely may become a family scapegoat for having a 'bad' gene. Parents may favour the other children and withdraw family resources from the tested child. This is not entirely hypothetical. Several geneticists responding to the survey, reported below, said that parents had requested testing of children for the Huntington gene, in order to decide whether to save money for the child's college education. If the child had the gene, the parents would not waste family resources. Children may also be stigmatized outside the family, by schools, by peers, by prospective partners, and ultimately by employers or insurers.

Even children who do not have the gene in question may suffer from testing. They may feel 'survivor guilt' if a sibling has or will develop a genetic disorder, while they remain healthy (Fanos, 1997; Fanos and Nickerson, 1991). Their parents may not trust the test results and may continue to believe that the child really has the gene after all, a phenomenon known as the 'vulnerable child syndrome' (Green and Solnit, 1964). Although 'vulnerable children' are usually the survivors of serious illness, whose worried parents are forever looking for signs of a new bout of disease, a similar permanent worry could attach to children who have had genetic testing.

The various professional statements urge the health care provider to weigh the benefits and harms of testing in each case. Unfortunately, this may be impossible in the absence of research on the psychosocial effects of testing. To date, the only North American research on 'children' (actually adolescents age 15 and over) who have been tested reports a follow-up of Tay-Sachs carrier testing in high schools in Montreal, Canada. These studies suggest that learning that one is a carrier does not ordinarily damage self-image or lead to social stigmatization (Clow and Scriver, 1977a, b). One worrisome finding, however, suggests that some (7%) non-carriers would not mate with a carrier. A more recent high school study of CF carrier screening suggests a greater likelihood of peer stigmatization than did the Tay-Sachs studies (Page *et al.*, 1992).

2. Survey methods

In order to find out how the various stakeholders viewed testing of children, we surveyed 4609 genetics services providers in 37 nations, including 1538 in the US, 846 primary care physicians (paediatricians, obstetricians, family practitioners) in the US, 718 first-time visitors to 12 genetics clinics in the US and Canada, and 1000 members of the US public. The 37 nations included all those known to have at least ten practising geneticists. A geneticist colleague in each country identified geneticists and distributed and collected the anonymous questionnaires. In the US, Canada, and UK, Master's-level genetic counsellors and genetic nurses were included. The primary care physicians, all certified by their speciality board (Paediatrics, Obstetrics, or Family Practice), were randomly selected from the American Medical Association's Physician Masterfile, which lists all US physicians regardless of whether they are members of the AMA.

The questionnaires included questions about both professional and personal socio-demographic data and 50 questions about ethical situations related to clinical genetics, mostly presented in the form of case vignettes. Respondents were asked what they would do (or, for patients, what they thought the doctor should do), from a checklist of possible actions, and also to explain, in their own words, why they had chosen this course of action.

Geneticist, primary care physician, and patient and public questionnaires covered the same issues, but primary care physician questionnaires used common medical parlance instead of genetic terminology (e.g., Turner's syndrome instead of 45,X). Patient and public questionnaires were shorter and used simpler statements. The geneticist questionnaire was translated into 12 languages, but colleagues in Scandinavia, Belgium, the Netherlands, Italy, Greece, and Portugal preferred English.

Geneticists and primary care physicians received three waves of mailings with telephone follow-up. Parents received the questionnaires by mail before their first clinic visit, in order to ensure that their views were not influenced by genetic counselling. Members of the public were surveyed by printed questionnaires delivered in person, using a quota sample developed by Roper-Starch Worldwide, a professional survey company.

In all, 2903 (63%) geneticists around the world returned questionnaires, including 1084 (70%) of those who received questionnaires in the US and 102 (47%) in the UK. 499 (59%) primary care physicians, 473 (66%) patients, and 988 (99%) members of the public responded. Most patients lived in the US (93%), were women (91%) and white (89%). They had a median of 13 years of education; 13% had not finished high school, and 28% were college graduates (including 11% with some postgraduate training). Most were in working class occupations (clerical or sales (29%), service (19%), or production factory work (8%), with a minority in professional or managerial work (30%)). The majority of their spouses were also working class, with 31% in factory production work, 23% in clerical or sales, and 15% in service occupations. Median income was US $25 000-45 000. This group appeared to be closer to 'middle America' than the college-educated or consumer support groups sometimes surveyed about 'genetic discrimination' (Lapham *et al.*, 1996).

In the following tables we report responses of the 409 who were parents of minor children. The median age of child patients was 5 years, and the most frequently mentioned conditions were neurofibromatosis, Marfan syndrome, Down syndrome, cleft/lip palate, and cancer syndromes. Clinics contributing the most patients to the study were in Providence, RI (30%), Buffalo, NY (20%), Detroit, MI (12%), and Rochester, MN (12%).

3. Survey results

3.1 Testing children: US views

Large majorities of geneticists, primary care physicians, and parents favoured testing for familial hypercholesterolaemia and cancer syndromes (*Table 1*). About half of geneticists and three-quarters of primary care physicians would test for genetic susceptibility to alcoholism. (Many US physicians believe that alcoholism may be prevented if susceptible people are identified early, but there is little evidence for this). About one-quarter of geneticists would test for Huntington's disease or Alzheimer's disease. Fewer women than men would test for these conditions (Wertz, 1997).

Respondents were asked to explain, in their own words, their responses regarding testing. We recorded two reasons behind each write-in comment, using categories developed in an earlier survey. In giving reasons for their responses, 66% said they wished to avoid harm, through early treatment or prevention for familial hypercholesterolaemia, cancer, or possible alcoholism, or to avoid harm from premature

Table 1. US Stakeholders' views on genetic testing of children. Percentage agreeing that parents should be able to have their minor children tested

Condition	Genetics service providers (n = 1084)	Primary care physicians (n = 409)	Parents (n = 409)	Public (n = 988)
Familial hyper-cholesterolaemia[a]	81	89	93	84
Susceptibility to inherited cancers[b]	70	81	93	84
Alcoholism	48	76	N/A	N/A
Huntington's disease	27	66	N/A	N/A
Alzheimer disease[c]	25	58	62	53
Requests to test minors for adult-onset disorders[d]	38	20	N/A	N/A

[a] Additional description (high risk of heart disease) added to primary care physician survey. Patient and public surveys said a condition that is 'preventable (such as heart disease)'.
[b] Described as a condition that is 'treatable if found early (such as breast cancer)' on patient and public surveys.
[c] Public and patient surveys said 'a condition that is neither treatable nor preventable'.
[d] The only statistically significant difference among US specialists was in having had requests to test ($p < 0.0000$). 2% of obstetricians, 27% of paediatricians, and 31% of family practitioners reported such requests.

knowledge of Huntington's disease or Alzheimer's disease. 43% said the information was necessary to make decisions about medical management. Fewer mentioned parental autonomy (13%) or preparing for the future (9%) as reasons for testing. More (25%) mentioned protecting the child's autonomy as a reason not to test, and 13% said there was no need to know at this time. Few (3%) mentioned the possibility of stigmatization as a result of testing. A substantial minority (38%) had had requests to test children for adult-onset disorders. The most frequently reported condition for which parents sought testing was Huntington's disease.

In contrast to geneticists, the majority of both primary care physicians and parents thought parents should be able to have their minor children tested for both Huntington's and Alzheimer's. Most members of the US public thought parents should be able to have their children tested for conditions that were 'treatable if found early' (84%), or 'preventable' (81%). A slight majority (53%) approved of testing for conditions that were 'neither preventable nor treatable'. These results point to a dichotomy between geneticists and the rest of the US medical community, whose views closely paralleled those of their patients.

In a separate question, asked only in the US, Germany, Japan, and Hungary, 20% of US geneticists would urge a woman with a dominant breast cancer gene to have her 13-year-old daughter tested now; 71% of parents said it was moderately (22%), very (17%), or extremely (32%) likely that if they themselves had a breast cancer gene they would encourage their own 13-year-old daughters to be tested now. Some parents wrote in that they would impose radical lifestyle changes on a daughter who had a breast cancer gene, including a mother who, apparently believing that pregnancy prevented breast cancer, said that she would make her daughter get pregnant at age 16 and would keep her pregnant every year thereafter.

3.2 Testing children: international views

In English-speaking nations other than the US (Australia, Canada, South Africa, UK) the percentages of geneticists who would test for Huntington's, Alzheimer's or alcoholism were lower than in the US (*Table 2*). The low percentages who would test in the UK (2%, 2%, and 4%) are especially striking (*Table 3*). This reluctance to test may stem in part from greater experience with requests for testing in the UK than anywhere else in the world except Switzerland and the Netherlands. In the UK, 76% had had requests to test children for adult-onset disorders, as compared with 38% in the US and 49% in western Europe. Percentages in northern/western Europe of practitioners who would test are also lower than in the US (Table 2). In southern Europe, eastern Europe, the Near East, Asia and Latin America, however, majorities would test children for all conditions listed (*Tables 2 and 3*). Willingness to test in these regions appears to stem from cultural beliefs about the rights of parents and their authority over their children.

European data appear in *Table 3*. Denmark and Switzerland stand out from other northern/western European nations as having higher percentages who would test for Huntington's or Alzheimer's disease, though they were still in the minority. Russia, the Czech Republic, and Greece had the highest percentages of practitioners who would test for these disorders in southern/eastern Europe, while Italy had the lowest percentages. As in the US, fewer women than men geneticists would test for these conditions.

Table 2. Testing children: international perspectives (n = 2903 providers). Percentage of geneticists who thought parents should be able to have minor children tested

Area	Huntington's disease	Alzheimer's dementia	Alcoholism	Cancer genes	Familial hypercholesterolaemia	Had requests to test children for adult disorders
USA	27	25	48	70	81	38
Other English-speaking nations	9	9	19	62	69	65
Northern/western Europe	16	14	27	51	63	49
Southern Europe	66	62	66	80	89	34
Eastern Europe	73	69	77	85	91	16
Near East	55	54	75	88	93	45
Latin America	79	76	83	89	90	15
Asia	75	57	63	82	80	17
Total	38	34	49	70	78	36

3.3 Telling children about test results

If a test is carried out, whether and when to tell the child the results becomes a question. On the one hand, it may be difficult for parents to conceal the test results entirely. Their demeanour toward the child may convey some change in perception of the child's health. On the other hand, there is at present no way of making certain that the child will have access to the results, if desired, upon reaching adulthood.

Table 4 shows the ages at which providers and parents would tell a child the results of genetic tests. Geneticists in the UK have been included for comparison. Substantial minorities of geneticists and primary care physicians would wait until a child reached the age of legal majority (18 in the US) before telling the child about a test result for familial hypercholesterolaemia or a genetic cancer syndrome. These responses appear to be somewhat contradictory to the stated purpose of testing, which is to begin treatment or monitoring early. If not told, the child may wonder about the purpose of medication, diet or medical testing. Even fewer parents than providers would tell the child about test results for these two conditions before the age of legal majority.

The majority of geneticists would not tell the child about susceptibility to alcoholism before legal majority, perhaps because they considered genetic susceptibility stigmatizing (*Table 4*). Primary care physicians, however, would tell the child early, perhaps as part of an 'early intervention' programme to prevent alcoholism.

The majority of geneticists would not tell children the results of tests for Huntington's disease or Alzheimer's disease before legal majority (*Table 4*). Primary care physicians were almost evenly divided, with a small majority favouring waiting until the age of legal majority. Among parents, 31% would tell the child the results of an Alzheimer's test before majority, 55% would wait until majority, and 14% would never tell the child.

Table 3. Testing children: International perspectives, full data (*n* = 2903) providers. Percentage of geneticists who thought parents should be able to have minor children tested

Country (number of respondents)	Huntington's disease	Alzheimer's dementia (AD)	Alcoholism	Cancer genes	Familial hypercholesterol-aemia	Have had requests to test children for adult disorders	Would tell a minor result of test for AD
United States (total = 1084)	**27**	**25**	**48**	**70**	**81**	**38**	**18**
Other English-speaking nations (total = 269)	**9**	**9**	**19**	**62**	**69**	**65**	**16**
Australia (15)	13	13	7	64	73	93	0
Canada (136)	12	11	26	63	71	56	22
South Africa (16)	25	25	67	81	81	50	0
United Kingdom (102)	2	2	4	58	67	76	12
Northern/western Europe (total = 448)	**16**	**14**	**27**	**51**	**63**	**49**	**12**
Belgium (15)	15	15	50	62	92	57	17
Denmark (28)	33	33	33	56	67	37	14
Finland (22)	18	18	23	41	64	41	13
France (75)	21	19	37	72	85	51	9
Germany (255)	14	10	24	45	56	44	12
Netherlands (26)	11	11	18	36	57	84	13
Norway (9)	11	11	11	56	78	67	0
Sweden (12)	8	8	25	50	42	58	0
Switzerland (6)	40	20	33	80	67	83	20
Southern Europe (total = 96)	**66**	**62**	**66**	**80**	**89**	**34**	**1**
Greece (12)	90	70	70	80	90	40	0
Italy (22)	48	48	57	76	91	38	0
Portugal (11)	64	64	64	82	91	27	0
Spain (51)	70	66	70	82	88	33	2

Eastern Europe							
(total = 163)	**73**	**70**	**78**	**88**	**93**	**16**	**5**
Czech Republic (81)	73	70	78	86	95	30	5
Hungary (36)	58	56	58	75	81	17	0
Russia (46)	86	80	93	91	93	12	7
Near East							
(total = 45)	**55**	**54**	**75**	**88**	**93**	**45**	**18**
Israel (23)	41	38	59	82	91	77	15
Turkey (22)	67	63	90	94	94	18	20
Asia (total = 410)	**75**	**57**	**63**	**82**	**80**	**15**	**12**
China (252)	86	56	67	86	83	18	6
India (23)	74	70	56	82	87	14	9
Japan (113)	59	55	59	74	73	13	25
Thailand (22)	84	84	92	96	96	4	16
Latin America							
(total = 232)	**79**	**76**	**83**	**89**	**90**	**17**	**10**
Argentina (19)	69	69	85	85	85	23	0
Brazil (74)	69	66	80	88	90	8	15
Chile (16)	75	77	88	93	94	19	6
Colombia (15)	64	64	79	71	79	21	8
Cuba (14)	100	100	100	100	100	0	15
Mexico (64)	82	77	77	84	85	32	4
Peru (14)	100	100	93	100	100	7	0
Venezuela (16)	73	67	73	100	93	25	15

Table 4. Telling minors the results of genetic tests. Percentage agreeing that child should have option of knowing test results [a]

Condition	UK Geneticists (n = 102)	US Genetics service providers (n = 1084)	US primary care physicians (n = 499)	Parents (n = 409)
Familial hypercholesterolaemia[b,c]				
Before age of legal majority	56	58	73	54
At age of legal majority[d]	43	41	27	45
Susceptibility to inherited cancers[e]				
Before age of legal majority	47	46	56	47
At age of legal majority[d]	52	52	43	52
Alcoholism				
Before age of legal majority	20	41	68	–
At age of legal majority[d]	68	55	30	–
Never	12	4	2	–
Huntington's disease				
Before age of legal majority	12	20	46	–
At age of legal majority[d]	78	73	49	–
Never	10	7	5	–
Alzheimer's disease				
Before age of legal majority	12	18	44	31
At age of legal majority[d]	76	73	50	55
Never	12	9	6	14

[a] On patient questionnaire, 'when should parent tell the child the results?'
[b] Additional description '(high risk of heart disease)' on primary care physician survey, and as a condition that is 'treatable if found early (such as heart disease)' on patient survey.
[c] The only statistically significant difference among US specialists was on familial hypercholesterolaemia ($p = 0.01$). 80% of paediatricians and 63% of obstetricians would tell the child before legal majority.
[d] 18 years of age in US.
[e] Described as a condition that is 'treatable if found early (such as breast cancer)' on patient survey.

Responses of geneticists in the UK and US were generally similar, but significantly more geneticists in the US would tell minors about susceptibility to alcoholism or about Huntington's disease.

3.4 Rights of late adolescents

Should a 16-year-old be able to refuse genetic testing if the parents wish it? In

Table 5. A 16-year-old's right to refuse testing. Percentage who thought a 16-year-old at risk should be able to refuse testing if parents want her/him tested

Condition	US genetics services providers (n = 1084)	US primary care physicians (n = 499)
Preventable or treatable if diagnosed early[a]	22	28
No effective prevention or treatment	96	84
Late-onset	95	68

[a]Differences among US specialities were significant ($p < 0.003$) for treatable disorders. 41% of obstetricians and 22% of other specialists thought a 16-year-old should be able to refuse testing.

our survey, we chose this age because it is two years below the age of legal majority (18) in the US, but it is an age at which courts have allowed adolescents to make some health care decisions. Responses in *Table 5* show differences in professional culture. Paediatricians were more likely than other specialists to say that a 16-year-old should not be able to refuse testing for a treatable disorder. For these paediatricians, medical benefit outweighed the 'child's' autonomy. Among geneticists, women were more likely than men to respect the 16-year-old's autonomy.

4. Cultural and clinical contexts

4.1 Testing in the context of autonomy

Americans' overwhelming cultural belief in autonomy, described earlier, was reflected in survey responses, especially those of parents and primary care physicians (*Table 6*). Columns 1 and 2 compare responses of geneticists in the UK and US. Although there are many similarities, especially in regard to supporting patients' decisions even if the professional disagrees and also in regard to a woman's right to decide about abortion, there are also areas of significant difference that indicate greater attention to autonomy in the US. For example, 4% in the UK and 36% in the US thought patients were entitled to whatever services they could pay for out-of-pocket; 43% in the UK and 90% in the US thought professionals owed patients a referral for procedures that the professional morally opposed; 43% in the UK and 65% in the US thought they owed referrals out-of-state or out-of-country; 14% in the UK and 55% in the US thought they owed referrals for sex selection; 46% in the UK and 90% in the US thought parents had a right to know fetal sex.

Many US physicians appear to have forgotten that they can ethically refuse a service that provides no known medical benefit (Beauchamp and McCullough, 1984; Brett and McCullough, 1986). Instead, they feel that they have to honour all patient requests, or at least offer a referral. Fear of lawsuits did not emerge as an issue in the write-in comments. Although much of the US medical practice is driven by lawsuits, patients cannot sue for refusal of services that are not customary, including testing minors for adult-onset disorders or carrier status.

Table 6. Views on autonomy. Percentage agreeing with statements

	UK genetics professionals (n = 102)	US genetics professionals (n = 1084)	US primary care physicians (n = 499)	Parents (n = 409)
Rights to services				
Withholding any requested service is paternalistic	51	56	57	69
Prenatal diagnosis should be available to any woman who requests it	47	41	42	80
Patients are entitled to any service they request and can pay for out-of-pocket	4	36	26	59
Rights to referral				
When patients ask for a procedure that a provider is unwilling to perform for moral reasons, the provider owes patient a referral, if procedure is legal	93	90	82	86
If law forbids a procedure in state/country, should refer outside state/country	43	65	55	49
A provider who refuses sex selection should offer a referral	14	55	41	49
Rights to know/not know				
Parents should be told fetal sex if they ask	46	91	87	91
After taking a test patients should have right not to know the results	88	82	62	41
Rights to decide				
Genetic counsellors should support all patients' decisions	99	90	not asked	79
A woman's decision about abortion should be her own, without any intervention from anyone	69	85	65	59
Parents should have the right to choose the sex of their children	3	10	11	13

Belief in a 'right to referral' for such services may be an Achilles heel undermining attempts of professional bodies to bring an orderly, rational approach to the situation. Sometimes referral becomes an efficient way of getting insistent patients out of the office without having to engage in lengthy discussion. Our survey questions on sex selection indicate that professionals who had experience with such requests were more likely than others to choose the alternative of referral rather than acceding to or refusing patient requests. (Referral was not given as an alternative in the questions on testing children.)

In the US, parents who follow a chain of referrals may succeed in their efforts. In Europe, those who are able to travel to southern or eastern European countries will probably also succeed (*Table 3*).

Many US geneticists 'draw the line' at insurance payment, whether public or private, for services of which they disapprove, notably sex selection, or pre-natal diagnosis without medical indications (Wertz and Fletcher, 1998). They apparently believe that withholding payment may prevent a practice from becoming widespread, but would leave a 'safety valve' for insistent demands by allowing patients to pay out-of-pocket. Such line-drawing also helps to sustain professionals' self-image as upholding a moral framework. Actually, payment may be the least ethical place to draw lines, because it supports and fosters a two-class system of medical care.

Readers should pay special attention to Column 4 (parents' views) in *Table 6*. In general, parents were more autonomy-oriented than either group of providers. (The major exception, parents' denial of a 'right to know', resulted from their belief, expressed in written comments, that no one would take a test in the first place unless they wanted to know.) The parents' views may represent the wave of the future, particularly in the United States, as medicine becomes a set of big businesses that try to market themselves to consumers. What consumers want is every service they ask for, without limit. They believe that nothing should be withheld (this would be a 'denial of patients' rights') and that patients are 'entitled' to whatever service they request, as long as they can pay for it out-of-pocket.

4.2 Testing in the current context of clinical services

In the US, parents may approach testing laboratories directly (depending on the state). In a recent survey of testing children among 105 laboratories in the Helix research network (an informational network of laboratories, mostly university-based) that also provided clinical services, 45% reported that they had provided tests directly to consumers, without a physician intermediary (Wertz and Reilly, 1997). Of the 46% that had policies on testing asymptomatic children, most had only fragmentary or single-disease policies, such as refusal to test for Huntington's disease. Yet 22% of those who performed Huntington's tests had tested asymptomatic children under the age of 12. The laboratories surveyed were not commercial laboratories. There are no published studies of the private sector.

It is probable that most parents will not go directly to laboratories. As genetics becomes part of general medicine, primary care physicians will become the gatekeepers to genetic testing. Not only do they favour testing, perhaps uncritically, but their knowledge of genetic disorders and the impact of these disorders on people's lives may be woefully inadequate. For example, in our survey, only 58% knew that Huntington's disease was autosomal dominant and only 22% knew that familial hypercholesterolaemia was autosomal dominant; 19% thought that life expectancy with fragile X syndrome was less than 20 years and 35% thought that people with fragile X could not have children. Many thought that Turner's syndrome (71%) and neurofibromatosis (48%) involved mental retardation. Many thought that boys with cystic fibrosis (64%) or Klinefelter syndrome (45%) could have biological children. And 30% thought that trisomy 52 (a trick question, as humans have only 46 chromosomes) was pre-natally diagnosable.

5. Discussion

In the future, there will be more requests to test children. Since both parents and primary care physicians seem to agree on this issue, it appears that more children may be tested. If multiplex tests become available for newborn screening, knowledge-hungry parents may request comprehensive, voluntary testing in addition to the government-mandated screening, thus pushing the issue of 'testing children' back to the newborn period. In the absence of research showing demonstrable harm, or of a successful lawsuit from someone tested as a child, it may be difficult for physicians to counter requests for testing in the absence of medical benefit, though both public and private insurance may refuse to pay for it. Although such testing may become more widespread than professional bodies currently recommend, increased public education about genetics may reduce the possibility of stigmatization, and legislation about employment and insurance may prevent discrimination.

Primary care physicians' views appeared to be closer to the views of parents than did the views of geneticists in the US survey. This was especially true of general questions about patients' rights to services upon request. Sometimes geneticists' views seemed to be almost an aberration in medicine, born of a speciality that sees comparatively few patients, spends considerable time on each one, is research-oriented, and places extensive emphasis on patient education.

In discussing the issues, it may make sense to separate carrier testing from the presymptomatic testing done for Huntington's disease, breast cancer, or other adult-onset disorders, as the Genetic Interest Group (GIG) has done in arguing in favour of carrier testing (Genetic Interest Group, 1995). As knowledge about genetics increases, carrier stigmatization may decrease. Carrier testing could enable children to begin thinking about their reproductive futures early. Even if the major benefit is reduction in parental anxiety, the potential harms of carrier testing may be minor in comparison to those from presymptomatic testing, which GIG opposes.

Research on the psychosocial effects of childhood testing should be supported. Until data are available, it is difficult to predict the areas of greatest benefit or harm. Meanwhile, the statements by the various professional bodies offer thoughtful guidance.

The responses in *Table 4* suggest a need to rethink the question of when to tell children about test results. This will depend in part on the purpose of the test. If the test was done to encourage prevention or to start early treatment, the knowledge may help to enlist the child's cooperation. The benefits of knowledge should be weighed against the dangers of stigmatization, especially for conditions like alcoholism, where there is no proven method of prevention.

There is no set age at which children or adolescents become 'ready to know' the results of testing (Grisso, 1981; Grisso and Vierling, 1983; Leiken, 1982). Many adults are never ready to know. If testing has been done for carrier status and if the child wants to know (and can understand the basic medical facts), it may be reasonable to disclose the results, keeping in mind that the knowledge may affect family relationships or change the child's self-image, unless disclosure is accompanied by thorough education and counselling.

For untreatable, late-onset conditions it would be best to preserve the child's autonomy to wait until legal majority and then ask the young adult if she/he wishes to know. Disclosing results of presymptomatic testing may cause more harm than good, but there may be cases in which the child wishes to know, especially if there is an ill sibling. The child or adolescent should be emotionally prepared, reasonably competent, and should wish to know, independently of the wishes of the parents.

The type of disorder and the number of years between testing and possible onset will affect decisions about when to disclose results. For example, in the very rare families where Huntington's disease may appear at age 18 or 20, it could be helpful to tell children before the age of legal majority, provided that they really wanted to know and had thorough psychological assessment and counselling. For breast cancer, where the earliest age of onset is 25 and there is a 20% likelihood of never developing the disease, the harm of telling a 15-year-old girl may outweigh the benefit, even if she says she wants to know. However, if she has been tested and is not told, this means that testing will have been carried out solely for the benefit of the parents. In many cases, however, the child will have already picked up clues from the parents about the test result.

There is, however, no way of making certain that the child will have the opportunity to learn the results, if desired, upon reaching adulthood (Fanos and Johnson, 1995). If there is no clear opportunity for the adult to learn the results of her/his childhood testing, this would argue for waiting until adulthood to conduct testing, unless there is a direct medical benefit to the child.

The statements by major professional bodies have argued that decisions about testing should be made in the best interests of the child, not the parents. It may be important to recognize the limits to parental autonomy and the importance of relinquishing some individual benefits in favour of a greater good. Patient responsibility, not often stressed in English-speaking countries, deserves greater emphasis.

References

Ambuel, B. and Rappoport, J. (1992) Developmental trends in adolescents' psychological and legal competence to consent to abortion. *Law Hum. Behav.* 16: 129–154.

American Medical Association Council on Ethical and Judicial Affairs (1995) *Testing Children for Genetic Status.* Code of Medical Ethics, Report 66. American Medical Association, Chicago, IL..

American Society of Human Genetics Board of Directors and American College of Medical Genetics Board of Directors (1995) Points to consider: ethical, legal, and psychosocial implications of genetic testing in children and adolescents. *Am. J. Hum. Genet.* 57: 1233.

Beauchamp, T.L. and McCullough, L.B. (1984) *Medical Ethics: The Moral Responsibilities of Physicians.* Prentice-Hall, Englewood Cliffs, NJ.

Bennett, R. (1976) Allocation of child medical case decision-making authority: a suggested interest analysis. *Va. Law. Rev.* 62: 285–330.

Bloch, M., Adam, S., Wiggins, S., Huggins, M. and Hayden, M.R. (1992) Predictive testing for Huntington's disease in Canada: the experience of those receiving an increased risk. *Am. J. Med. Genet.* 42: 499–507.

Brett, A.S. and McCullough, L.B. (1986) When patients request specific interventions: defining the limits of the physician's obligation. *New Engl. J. Med.,* 315: 1347–1351.

Brock, D.W. (1989) Children's competence for health care decision-making. In: *Children and Health Care: Moral and Social Issues* (eds L.M. Kopelman and J.C. Moskop). Kluwer Academic Publishers, The Netherlands, pp. 181–212.

Buchanan, A.E. and Brock, D.W. (1989) *Deciding for Others: the Ethics of Surrogate Decision Making.* Cambridge University Press, Cambridge, pp. 215–266.

Clarke, A. and Flinter, F. (1996) The genetic testing of children: a clinical perspective. In: *The Troubled Helix: Social and Psychological Implications of the New Human Genetics* (eds T. Marteau, and M. Richards). Cambridge University Press, Cambridge, pp. 64–176.

Clow, C.L. and Scriver, C.R. (1977a) The adolescent copes with genetic screening: a study of Tay-Sachs screening among high-school students. *Prog. Clin. Biol. Res.* **18**: 381–393.

Clow, C.L. and Scriver, C.R. (1977b) Knowledge about and attitudes toward genetic screening among high-school students: the Tay–Sachs experience. *Pediatr* **59**: 86–96.

45 CFR 46.408, US Dept. of Health and Human Services. *45 Code of Federal Regulations 46, Subpart D – Additional Protection for Children Involved as Subjects in Research.* 48 Federal Register 9818.

Ewald, L.S. (1982) Medical decision making for children: an analysis of competing interests. *St. Louis Univ. Law J.* **25**: 689–733.

Fanos, J.H. (1997) *Sibling Loss.* Lawrence Erlbaum, Mahwah, NJ.

Fanos, J.H. and Johnson, J.P. (1995) Barriers to carrier testing for adult cystic fibrosis sibs: the importance of not knowing. *Am. J. Med. Genet.* **59**: 85–91.

Fanos, J.H. and Nickerson, B.G. (1991) Long-term effects of sibling death during adolescence. *J. Adolesc. Res.* **6**: 70–82.

Genetic Interest Group (1995) *GIG Response to the Clinical Genetics Society Report, 'The Genetic Testing of Children'.* GIG, London.

Green, M. and Solnit, A.J. (1964) Reactions to the threatened loss of a child: a vulnerable child syndrome. *Pediatr* **34**: 58–77.

Grisso, T. (1981) *Juveniles' Waiver of Rights: Legal and Psychological Competence.* Plenum Press, New York, NY.

Grisso, T. and Vierling, L. (1983) Minors' consent to treatment: a developmental perspective. *Prof. Psychol.* **9**: 412–427.

Harper, P.S. and Clarke, A. (1990) Should we test children for 'adult' genetic diseases? *Lancet* **335**: 1205–1206.

Holder, A.R. (1977) *Legal Issues in Pediatrics and Adolescent Medicine.* John Wiley and Sons, Inc., New York, NY.

Holder, A.R. (1988) Disclosure and consent problems in pediatrics. *Law Med. Health Care* **16**: 219–228.

Holder, A.R. (1989) Children and adolescents: their right to decide about their own health care. In: *Children and Health Care: Moral and Social Issues* (eds. L.M. Kopelman and J.C. Moskop). Kluwer, Dordrecht, pp. 161–172.

Institute of Medicine, Committee on Assessing Genetic Risks (1994) *Assessing Genetic Risks.* National Academy Press, Washington, DC, p. 276.

Lapham, E.V. *et al.* (1996) Genetic discrimination: perspectives of consumers. *Science* **274**: 621.

Leiken, S. (1982) Minors' assent or dissent in medical treatment. In: *President's Commission for the Study of Ethical Problems in Medicine and Biomedical and Behavioral Research, Making Health Care Decisions.* US Government Printing Office, Washington, DC, pp. 175–191.

National Commission for the Protection of Human Subjects of Biomedical and Behavioral Research (1977) *Report and Recommendations: Research Involving Children.* US Dept. of Health, Education, and Welfare, Washington, DC, DHEW publication OS 77-0004, **1**: 1213.

National Society of Genetic Counselors (1995) Prenatal and childhood testing for adult-onset disorders. *Perspect. in Genet. Counsel.* **17**: 5.

Page, A. *et al.* (1992) Attitudes of high school students toward carrier screening for cystic fibrosis. *Am. J. Hum. Genet.* **51** (suppl.), Abstract 17.

President's Commission for the Study of Ethical Problems in Medicine and Biomedical and Behavioral Research (1982) 'Making Health Care Decisions'. US Government Printing Office, Washington DC, USA, Vol. 1, pp. 60, 183.

Schliefer, J.T. (1980) *The Making of Tocqueville's 'Democracy in America.'* University of North Carolina Press, Chapel Hill, NC, pp. 124–125.

Science (editorial) (1996) ELSI's cystic fibrosis experiment. *Science* **274**: 489.

Science (editorial) (1997) Gene tests get tested. *Science* **275**: 782.

Shryock, R.H. (1962) *Medicine and Society in America: 1660–1860.* Cornell University Press, Ithaca, NY.

de Toqueville, A. (1951) *Democracy in America,* Vol. I. (ed P.J. Bradley) Alfred A. Knopf, New York, NY, p. 43, (original edition 1835).

Tyler, A., Morris, M., Lazarou, L., Meredith, L., Myring, J., and Harper, P. (1992) Presymptomatic testing for Huntington's disease in Wales 1987–1990. *Br. J. Psychiatry* **161**: 481–488.

Wadlington, W.J. (1973) Minors and health care: the age of consent. *Osgoode Hall Law J.* **11**: 115–125.

Wertz, D.C. (1997) Is there a 'women's ethic' in genetics?: a 37-nation survey of providers. *JAMWA* **52** (1): 33–38.

Wertz, D.C. and Fletcher, J.C. (1998) Ethical and social issues in prenatal sex selection: a survey of geneticists in 37 nations. *Soc. Sci. Med.* **46**: 255–273.

Wertz, D.C., Fanos, J.H and Reilly, P.R. (1994) Genetic testing for children and adolescents: who decides? *JAMA* **272** (11): 875–881.

Wertz, D.C. and Reilly, P.R. (1997) Laboratory policies and practices for the genetic testing of children: a survey of the Helix network. *Am. J. Hum. Genet.* **61**: 1163–1168

Appendix 1. The genetic testing of children: Report of a working party of the Clinical Genetics Society

1. Conclusions and recommendations

1   The predictive genetic testing of children is clearly appropriate where onset of the condition regularly occurs in childhood or there are useful medical interventions that can be offered (e.g. diet, medication, surveillance for complications).

2   In contrast, the working party believes that predictive testing for an adult-onset disorder should generally not be undertaken if the child is healthy and there are no medical interventions established as useful that can be offered in the event of a positive test result. We would generally advise against such testing, unless there are clear cut and unusual arguments in favour. This does not entail our recommending that families should avoid discussing the issues with younger children, but rather that formal genetic testing should generally wait until the 'children' request such tests for themselves, as autonomous adults. This respect for autonomy and confidentiality would entail the deferral of testing until the individual is either adult, or is able to appreciate not only the genetic facts of the matter but also the emotional and social consequences of the various possible test results.

     In circumstances when this type of testing is being contemplated, there should be full discussions both within the family and between parents and genetic health professionals (clinical geneticists or non-medical genetic counsellors); the more serious the disorder, the stronger the arguments in favour of testing would need to be.

3   For some disorders, there is insufficient evidence to know whether a diagnosis

The Genetic Testing of Children, A.J. Clarke (ed.).
© 1998 BIOS Scientific Publishers Ltd, Oxford.

in childhood is helpful in the medical management of the possibly (not yet) affected child. Research in these areas will be worthwhile and important. When such research is planned, however, it will be important to incorporate a social and psychological evaluation of the genetic testing, as well as a technical and more strictly medical evaluation, because the results of the psychosocial evaluation may be critical in future clinical judgements if the medical benefits remain uncertain or are shown to be minor. Furthermore, the psychosocial study of testing for these conditions, where the existence of possible medical benefits justifies the study of the testing, may throw light upon the likely psychosocial effects of testing for other disorders; hence such studies may be of more general applicability.

4   The situation with regard to testing children for their carrier status for recessive disorders and balanced, familial chromosomal rearrangements is more complex. In general, the working party would make a presumption against testing children to determine their carrier status, where this would be of purely reproductive significance to the child in the future.

4A Circumstances may arise, however, in which the genetic testing of children could be helpful in the provision of accurate information to other family members. Even in families with apparently balanced chromosomal translocations, however, we think that this occurs only occasionally. It is important that children in such families are not tested 'as a routine', but that each situation is considered on its merits so that children are tested only when the results will contribute to the counselling of other family members. Otherwise, if the results would only be of future reproductive concern to the child, then it is wiser to defer the testing until the child is able to understand the issues and requests testing in person.

4B Where such (carrier) testing is, or has been, taking place, it would be useful to institute prospective and retrospective psychosocial evaluations of the impact of the testing on the children and their families, so that future policy can be guided by evidence rather than conjecture and anecdote.

4C If the testing is not to be performed in childhood, then a certain obligation rests upon the health care system and the family together to ensure that testing is offered when the child is older. While testing in childhood may allow parents and physicians to feel that they have done their duty, this may still leave both parties with an obligation to ensure that the tested child is offered counselling (and possibly an updated genetic test too) when he or she 'comes of age'.

5   There are additional factors to be considered with a healthy but 'at risk' child referred for adoption, in so far as the results of the testing might influence decisions made on behalf of the child. However, it should not be assumed that genetic (predictive or carrier) testing will be required before a suitable placement can be achieved. In each case, we would advise discussion between the medical adviser to the adoption agency and a clinical geneticist. The important factors other than the possible laboratory test results need to be identified for future attention in advance of any test being performed.

6   Because some of these recommendations (1–5) are likely to diverge from the practice and beliefs of many within the medical profession, and even within

clinical genetics, it is important that further discussion and debate take place. We believe that the medical profession should work towards a consensus on these issues before such tests become more widely available through commercial laboratories, which may pay little respect to the goal of coupling laboratory testing with the provision of counselling and support as a package of genetic services. The ability of the working party to arrive at such a consensus, despite our holding different views initially, suggests that this may be a realistic goal. It will also be important to extend this work to achieve a broader consensus across the professions of nursing, social work and the law as well as medicine.

2. Background

1 Opportunities for genetic testing of children

Technical advances in recent years have greatly enhanced our ability to identify individuals carrying faulty genes. It has been possible for some years to identify those carrying adult polycystic kidney disease by ultrasound scanning before the onset of symptoms; the healthy carriers of some haemoglobin disorders can be identified by routine, automated haematological analysis. However, it is now possible to identify those apparently healthy persons likely to develop one of a large number of inherited disorders by molecular genetic methodologies. It is also possible to identify those who carry faulty genes or chromosomes that will cause them no health problem, but which may result in problems for their future children. Much of this new testing has become possible because the techniques of molecular genetics can be applied to a person of any age, and do not depend upon the at risk person manifesting early signs of the disorder. However, the principles involved are no different for molecular genetic testing than for more traditional diagnostic techniques.

2 Ethical concerns raised by genetic testing of children

One early concern addressed the ethics of testing children at risk of developing Huntington's Disease (HD) in later (adult) life (Craufurd & Harris, 1986). A consensus was reached by those involved in the predictive testing of those at risk of HD, who decided that children should not be tested (World Federation of Neurology, 1989; Bloch & Hayden, 1990). The knowledge of HD gene status can be burdensome for an adult, even when the result is low risk (Huggins et al., 1992), and only a minority of at-risk adults in fact choose to undergo such testing – about 10–15% (Bloch et al., 1992; Crawford et al., 1989; Tyler et al., 1992). The right of the child to decide in adult life whether or not to be tested would be removed if their genetic status was determined at the request of parents or others; such requests have been made on many occasions (Morris et al., 1988, 1989). To what extent is HD unique, and to what extent has the discussion about HD served to awaken us to similar issues raised by the testing of children for other disorders? (Harper & Clarke, 1990).

Similar ethical concerns do arise in the case of other adult-onset, neurodegenerative disorders such as Alzheimer's disease and prion dementia (Collinge et al., 1991).

We already know that discrimination against persons on the basis of their genetic constitution has been practised by insurance companies in Britain and in North America (Billings *et al.*, 1992) and has even been practised by employers against healthy workers who happen to carry a recessive disorder, such as sickle cell trait (Bowman, 1991). If children are to have predictive tests performed, they may lose the opportunity to obtain life insurance before clarifying their genetic status. Harper (1992, 1993a) has suggested a moratorium on the use of genetic testing by insurance companies to avoid generating new problems without an opportunity for reflection and for the development of a socially accepted consensus.

Two areas were regarded as potentially problematical. (1) Predictive testing of apparently healthy children (that is, presymptomatic testing) for late-onset disorders, usually of autosomal dominant inheritance, in which clinical manifestations are unlikely until well into adult life (and in which early treatment or surveillance for complications would not be helpful). (2) Testing healthy children to determine their carrier status for inherited disorders that would have no implications for their own health, but might affect the health of their future children (usually autosomal recessive or sex-linked inheritance, or balanced chromosomal rearrangements).

In both these areas, the child's future autonomy, their ability to decide for themselves whether or not to be tested, could be undermined. In addition, the confidentiality to which any adult being tested would be entitled, would have been breached when the results were disclosed to the child's parents.

The potential harms caused by childhood genetic testing might include damage to the child's self-esteem, distortion of the family's perceptions of the child, loss of future adult autonomy and confidentiality, discrimination against the child in education, employment or insurance, and adverse effects on the child's capacity to form future relationships. Such testing breaches the policy of providing counselling before or in parallel with the testing process, when viewed from the perspective of the child being tested; this could damage the professional attempts to provide genetic services of counselling and testing together as a package.

Although concerns about predictive and carrier testing in childhood have been voiced, there may be advantages to such testing, including the opportunity for the child to adjust to circumstances, the fostering of openness within the family, the resolution of parental uncertainty and (for carrier status tests) ensuring that testing has been offered to the whole family. These points will require consideration.

Clearly, considerations of the possible harm done by genetic testing do not apply to a large number of disorders, where it is to the advantage of the child that the condition be diagnosed at the earliest possible stage. This applies wherever therapy, or surveillance for possible complications of a disorder, can usefully be started in childhood. Also, where an affected individual is likely to manifest the disorder during childhood, these ethical difficulties of presymptomatic genetic testing in childhood do not arise.

This working party was constituted to address these concerns.

3. Working party aims

The working party was asked to: examine current attitudes and practices; focus attention on any difficulties raised by the genetic testing of children; and make appropriate recommendations about future practice. These objectives led us to examine two questions. To what extent does such testing take place now? Under what circumstances is such testing justified, and when might it be better deferred until the child is older?

Even before embarking upon this project, we acknowledged that there were likely to be differences of opinion and practice amongst members of the medical profession. We found such differences, which were reflected within the working party. On the assumption that it is desirable, is it possible for health professionals to work towards a consensus on these issues? Some apparent differences of approach are likely to have arisen for historical reasons rather than from any major philosophical schism: how can such differences be resolved, so as to avoid arbitrary inconsistencies of practice?

4. Current practice and attitudes in Britain: a questionnaire survey

We wrote to those molecular genetic and cytogenetic laboratories engaged in clinical diagnostic work requesting information about the number and nature of the tests carried out on children. We also asked whether some requests were declined on ethical grounds.

We carried out a questionnaire survey of approximately 3000 health professionals in Britain (as described in the Appendix to this article), seeking information about their practice and attitudes concerning the genetic testing of children. Detailed results are given in Appendix 1 of this article.

This survey has confirmed that there is widespread testing of children for genetic disorders, but most of it falls outside our remit: it either involves testing children for disorders that usually become manifest during childhood (or in which there is a significant risk of this happening), or it is otherwise clearly appropriate as part of good medical practice. By this, we mean that these tests are used in management decisions concerning therapy, or the surveillance for complications of genetic disorders.

Among geneticists and coworkers, the predominant view is that no requests for testing but those initiated by a child's parents or medical advisers should be entertained. This view is shared by paediatricians, except that they would be willing for adoption agencies to initiate such testing. Respondents consider genetic tests in childhood to be justified if a useful intervention can be made as a result (diet, treatment, surveillance for complications), or if medical management may be otherwise improved. There is also general agreement that predictive testing is best avoided amongst untreatable late onset disorders such as HD, prion protein and Alzheimer dementias, and other neurodegenerative disorders.

In contrast to the consensus on those topics, there was a lack of consensus concerning the wisdom of testing for other late-onset disorders for which no useful interventions would be available. Sixteen percent of respondents considered that predictive testing in childhood was generally undesirable. There was also some disagreement about whether or not such tests were justified even where some clinical intervention is possible. Predictive testing, it was thought by some, might also be helpful in clarifying the natural history of some conditions.

There was a wide range of views on the question of testing children for gene carrier state where this would have no health repercussions for the child, as with the carrier state for recessive disorders and chromosomal rearrangements. It is certainly common practice amongst paediatricians to test the healthy sibs of affected individuals to see if they may be carriers of, for example, cystic fibrosis or of a balanced chromosomal translocation.

There were also differences of attitude concerning the right of parents to arrange genetic testing for their children. Geneticists and coworkers were less likely to view parental wishes alone as sufficient to justify testing than were paediatricians and others. Only a minority of respondents had a firm age limit below which they did not test children's genetic status, and these age limits varied widely.

Finally, it was clear from the questionnaire replies that geneticists and coworkers are more wary than paediatricians or haematologists of using phrases that could be regarded as paternalistic, eugenicist or directive. Thus, geneticists and coworkers did not support the notion of genetic testing in childhood to increase 'responsible' attitudes to reproduction, whereas other groups did support such sentiments. These differences may arise from different understandings of the meaning or value implications of the word 'responsible', or may indicate that the genetics community has become sensitized to such language, and no longer regards it as acceptable terminology. Alternatively, those involved in clinical genetics may have decided that 'responsible' reproductive behaviour should be encouraged, but that genetic testing in childhood does not, or in practice is unlikely to, result in such 'responsible' adult reproductive behaviour. We cannot distinguish these possible explanations from the replies we received.

5. Current practice and attitudes in Britain: a prospective study

Because the retrospective gathering of data is difficult, and is likely to be unreliable, we initiated a prospective survey of genetic testing in children (under 16 years) for 12 months from 1/1/92. Seventeen molecular genetic laboratories supplied data, although it was conceded that these were incomplete even when considerable pains had been taken to gather information on all cases. We also circulated clinical geneticists, Clinical Genetics Society (CGS) members and consultant paediatricians. These data therefore only provide an indication of the pattern of childhood testing in Britain, not an accurate measure of current activity, and the numbers listed must be a considerable underestimate. No attempt was made to enquire into the tests carried out by biochemistry or haematology laboratories, and the generation of carrier status results by newborn screening programmes (e.g. for certain haemglobinopathies) was not studied.

We have listed in *Table 5* in Appendix 2 of this article the numbers of tests reported to us as either performed or deferred in 1992, grouped under the major categories of test. We were informed about a total of 165 tests performed, and 37 tests deferred. Much the most frequent test performed was that for cystic fibrosis carrier status. Forty-two tests (one quarter) were predictive, the rest being carrier status tests. There was no evidence that the tests on children were usually performed in later childhood; the age distribution of children tested appeared to be random, and was not concentrated in the early teenage years. Two exceptions to the apparently random distribution of ages were the tests for fragile-X normal transmitting males (the youngest reported was at age 4; before 4 years, the test would also identify children likely to manifest learning problems, and would therefore be diagnostic) and for familial adenomatous polyposis (FAP) coli (where 9 of the children were tested in the age range 9–11 years, presumably as an aid to management, and the remaining 13 children were tested in the age range 0–6 years).

6. Attitudes of family and patient support groups

A letter and a brief set of questions were sent to 108 separate family support groups affiliated to the Genetic Interest Group (GIG). We hoped to arouse interest in these issues, to elicit accounts of the experiences of families in which tests had been performed in childhood, and to discover the attitudes of those affected by genetic disorders and of members of their families. We received 78 replies; some were from whole groups or local branches of a group, and others were from individuals within groups. The purpose of this enquiry was not to ballot the members of these specific disease-oriented groups, but to obtain information about the range of opinions held within these groups by individuals whose lives have been touched by the various genetic conditions.

It is possible to group the replies into a few very broad categories. Forty-one replies favoured a policy of not performing predictive or carrier genetic tests in childhood.

Another group of 14 replies indicated a willingness to carry out carrier tests in childhood but not predictive tests, although some reservations were made by a few respondents, such as the child having the right to decide at an appropriate age.

The third group of 19 replies indicated a willingness for predictive tests to be performed in childhood. Interestingly, this group was about evenly divided as to whether or not tests for carrier status should also be performed on children.

Four replies were unclassifiable.

Specific comments from respondents are listed in Appendix 3 of this article.

7. Legal considerations

Recent developments in child law, most importantly the Gillick decision in 1985 (*Gillick* v *W. Norfolk & Wisbech HA*, 1985) and the Children Act 1989, have considerably altered the legal status of children. Whereas it was previously possible to regard children as little more than the property of their parents, this is no longer the case. Instead, parents have responsibility for the care of their children, rather than rights over them, and are expected to act in their best interests. Parental decisions about testing for genetic disorders should therefore be made according to whether the child will benefit, not in order to relieve the anxieties of the parents. If necessary, in cases of dispute, a court can be called upon to determine whether a particular child should be tested, although this would be rare in practice. The courts have suggested that children should be given the chance to take decisions on medical care for themselves if it is possible to wait until they are able to do so without risking their health (Re D, 1976).

The law ensures that decisions about testing are taken jointly by parents and professionals (Re J, 1990). In effect, parents have a veto because testing cannot be carried out without their consent, although a general consent to diagnostic testing would be sufficient without a full explanation of the conditions being tested for. The consent of any one person with parental responsibility will suffice, unless a court order specifically deals with the matter. Fathers who have never been married to the mother will not normally have parental responsibility. However, in England and Wales, they may obtain it by applying for a court order giving it to them, or making a 'parental responsibility agreement' with the mother and having it registered with the court.

Where older children are concerned, they will be able to consent on their own behalf if they have sufficient understanding of the issue to make a choice (Gillick v W. Norfolk & Wisbech HA, 1985) [Age of Legal Capacity (Scotland) Act 1991, s. 2 (4)]. By statute, it is presumed that children of sixteen have this capacity [Family Law Reform Act 1969, s. 8.; Age of Legal Capacity (Scotland) Act, s. 1], but if a child below that age can understand the test being proposed and its significance then they too will be able to consent. Even if a child is mature enough to be able to give a consent, recent English cases have established that parents can still authorize treatment, even against the child's wishes (Re R, 1991; Re W, 1992; Montgomery, 1993). However, even if it is legal to do so, it may well be considered unethical to test for genetic status against the wishes of a child with a good understanding of the issues. Where children do not have the intellectual capacity to agree to treatment, parental consent must be obtained. Such consent may be given in respect of children until they reach the age of 18.

Although children cannot be tested without consent, either their own or that of their parents, health professionals cannot be forced to offer care that they consider inappropriate (Re J, 1992). Thus in cases where it would be contrary to the child's interests to be tested, it is legitimate to refuse to do so even if the parents request it. Nor are professionals obliged to reveal information on the genetic status of a child when they believe it would harm the well-being of the child. In principle, parents are entitled to see their children's records under the Access to Health Records Act 1990 (applying to manual records created after 1 November 1991) and the Data Protection Act 1984 (for computerized records). However, health professionals are entitled to refuse access where they believe that revealing the information would cause serious harm to the physical or mental health of the child or others.

The general principles of malpractice law also require health professionals to give due consideration to the welfare of the child patient. They will be negligent if they fail to practise in a manner accepted as proper by a responsible body of professional opinion in their specialty (Bolam v Friern HMC, 1957). Thus, legal standards reflect professional ones. As long as practitioners conform to responsible medical practices, they will be safe from legal action. This means that as professional ethics develop, legal requirements will also change to reflect the greater understanding. Given the general consensus that it is unwise to test children for HD, to do so might well be regarded as negligent except in very particular circumstances. In other areas, however, both testing and refusing to test would probably be acceptable providing the child's position was considered. However, an over-rigid policy might be found negligent because it was not addressed to the particular child's needs and would therefore be unacceptable to professional opinion. Witholding testing could only leave a doctor open to litigation if it appeared to be negligent on the above test.

It should be emphasized at this point that most potential disagreements amongst family members, or between family members and health professionals, about the desirability of genetic testing for individual children can be avoided by sensitive counselling. This discussion of the legal issues should not be taken to imply that recourse to the law is likely to be required on a regular basis.

8. Issues relating to adoption

The issues relating to genetic disease and genetic testing in the context of adoption are complex, and will not be considered here in detail. They are discussed more extensively elsewhere (Payne, 1993; Turnpenny et al., 1993). However, we will raise the question as to whether there are particular considerations that might justify the genetic testing of a child being considered for adoption, restricting our attention (as elsewhere in this report) to tests of (unaffected) carrier status, and to predictive tests for adult-onset disorders.

Prospective adoptive parents may have a keen interest in the genetic status of a child that they are considering for adoption. Like most people (if given the choice) adopters will usually want healthy children, and the adoption agency is under an obligation to gather information about the child's circumstances, including the health of the child and of members of the family. All relevant available information is then given to prospective adopters so that they can make as informed a decision as possible when deciding whether to accept a child. Good practice indicates that all available information should be passed on.

Adoptive parents are mostly from the 10% of involuntarily infertile couples, but some will have made a conscious decision to limit their family because of a known history of an inheritable disorder. Their personal experiences are likely to affect their views on genetic conditions in a prospective adoptive child.

Tests of carrier status

It is our view that there will usually be no particular justification for testing the child any earlier than would be the case with a child still in the birth family. Adoptive parents have the same legitimate interests in the health of their children and grandchildren as do biological parents. However, there are specific difficulties that may arise from carrier testing in the context of adoption.

First, the adoptive family may be informed about a family history of possible relevance to the adopted child's future reproductive plans, but may fail to remember this or to pass on the information to the child at an appropriate age.

Secondly, the adoptive parents may focus on the possible carrier state of their child and accord it more emotional significance than is warranted.

In these situations, the lack of familiarity wih the condition in question may complicate the task of discussing the issues with the growing child, leading to over- or under-emphasis on the possibility of 'genetic risk'. Difficulties of communication may also arise when a genetic disease in the birth family comes to light after the adopted child has been settled in the new family. Similar problems of communication may arise in reverse if the adopted child develops a genetic disease after settling in the adoptive family.

Openness between adoptive parents and child, and the maintenance of links between the adoption agency and both the adoptive and the birth families, may help to minimize these problems. However, the questions of principle relating to carrier testing in childhood are no different in the context of adoption from the issues encountered in other families, even if the pathways of communication are more vulnerable to rupture. They do not amount to a reason for ensuring that the tests are carried out pre-placement.

Predictive testing

For adult-onset conditions where predictive genetic testing is available, and where the adoptive child is at risk, the question arises as to whether or not the fact of the child's adoption (prospective or established) is a justification for carrying out a predictive test which might not otherwise be performed until the adopted child became an adult and chose to undergo testing.

Arguments for testing include those in *Table 1*, and the specific point that appropriate carers may more easily be found for the child (if pre-placement).

Arguments against testing also include the list in *Table 3*, and some specific points related to adoption – that the diagnosis will label the child and affect the (already difficult) process of identity development, that it is irrelevant to the needs of the child for acceptance as he/she is, and that it may cause more problems by excluding paternity.

The arguments will have to be weighed in each case, but their force will not differ greatly from the standard case of a child in the original birth family, unless it proves difficult to find suitable prospective adoptive parents for a child at risk of a late-onset genetic disorder because of the uncertainty surrounding the child's possible genetic status. Then, either the decision to put the child forward for adoption, or the decision about genetic testing, will need to be reconsidered. In practice, this situation may arise only infrequently, but will call for a careful consideration of the child's overall best interests when it does so.

In general, it would seem best, wherever possible, to find adopters who can accept the child as a whole, and subsequently participate in any testing that is appropriate for the child as a confirmed member of their family.

Finally, it should be noted that the status of a child at risk of developing a genetic disorder (even in adult life) has some parallels with the status of a child with special needs. These include financial implications for the local authority, as a child who is at 'substantial risk of developing a serious disorder, known at the time of adoption' may require an adoption allowance, payable until the age of 18.

Summary

Inherited disorders have a wide range of significance in adoption because of the multiplicity of viewpoints that exist between the members of the adoption triangle (adopted person, birth family, adoptive family) and the Adoption Agency. When inherited disorders arise in the context of adoption placement decisions, they may be highly significant and have an important bearing on the outcome of an adoption.

The management of genetic disorders in adoption should commence with determining the attitudes and expectations of adopters with regard to present or future disability. Specific issues such as predictive and carrier testing are the same as for any similarly situated individual at genetic risk, with the additional consideration of the effect of the decision (and, if a test is performed, the possible test results) on the child's placement. The legal situation regarding consent to testing in the adoption process has not been clarified judicially.

9. Discussion

The two central questions to be answered, or at least considered, are:

What is good clinical practice with regard to predictive or carrier testing in childhood?

and

Who has the responsibility or the right to arrange predictive testing or carrier testing in childhood: the families, the individuals (the children as future adults), adoption agencies or doctors?

Unfortunately, there is a dearth of firm evidence on which to base policies. Various blends of ethics, anecdote and prejudice are likely to shape people's attitudes, including our own. While there is real concern about the abrogation of a child's future autonomy as an adult, and about the loss of confidentiality entailed in childhood testing, we do not know whether harm actually results from such testing (but see below). There are concerns that the parents' attitudes to the child may change, and that unfavourable parental expectations about a child's future mental, physical and emotional development could be self-fulfilling. These factors could result in low self-esteem in the child, but there is little evidence that genetic tests produce such damage in practice. This may be because such testing does not cause harm, or because such testing in the past has been limited in scope, and the resulting harm has not come to light because evidence of it has not been sought. We have had to arrive at our responses to these questions in the absence of adequate empirical data, relying upon our various experiences and upon our concern to maintain the generally accepted principles of medical ethics.

Predictive testing (where there is no direct health benefit to the child).
'What is good practice with regard to predictive testing in childhood?'

The working party has arrived at a consensus view that there should be a general presumption against such testing, although we recognize that circumstances may arise when testing could be appropriate. We now set out the reasoning that has led us to this common understanding.

There is the precedent of the decision made by those performing predictive testing for HD, a consensus reached on the basis of the supposed likely effects of such testing and not from the experience of harmful effects (the consensus was reached before such testing of children was ever undertaken). In families with HD, individuals may be singled out for no rational reason as being destined to develop the disorder. Such family 'preselection' of children as being likely to develop HD is recognized as having potentially profound adverse effects on the emotional development of the unfortunately labelled child (Kessler, 1988). These effects could be even more severe if medical evidence supported, and perhaps directed, the collusive family dynamic that has selected one child as the 'victim'.

Adults undergoing predictive testing for HD have extensive counselling, in which they are required to confront the effects that the test is likely to have on their whole life. Insurance (of life and health), home ownership, career, marriage and reproductive plans may all be deeply affected by the results of the testing. Only some 10–15% of at-risk adults choose to accept such testing (Craufurd *et al.*, 1989; Tyler *et al.*, 1992). It may be difficult for parents to appreciate the reluctance to undergo testing that is felt by many adults at risk of HD, and which their at-risk child may also come to experience. A sensitive exploration of the issues will often help parents to appreciate the limitations of their perspective, and how their feelings may differ from those of their child in the future; this may lead them to reconsider their request for testing. However, where the family has already 'preselected' the child, as being affected or unaffected, they may continue to press for testing. In these circumstances, it could be particularly inappropriate to perform the test.

There is little evidence as to ill effects resulting from predictive testing in childhood for Huntington's disease or for other disorders, although the fundamental principles involved are much the same for any late-onset condition, especially for other neurodegenerative disorders. Only limited evidence is available concerning the Swedish newborn screening programme for alpha-1-antitrypsin deficiency, which was discontinued because of severe adverse psychosocial consequences (Thelin *et al.*, 1985). However, it is clear that predictive testing in childhood for late-onset disorders, even in the context of a high-risk family situation, can raise as many problems and as much anxiety as is generated by continuing anxiety about the child's genetic status, and the knowledge that a child will develop such a disorder may – at least in some family contexts – cause worse problems than continued uncertainty. At present, we have no means of identifying those families that would be helped by having uncertainty resolved by genetic tests and distinguishing them from other families in which the results of testing would be harmful. Indeed, there is not even any agreed means of deciding whether or not a family might have been helped or harmed by such an intervention, and what timescale should be considered in coming to a judgement on this.

Given the dearth of evidence, it is the ethical consequences of childhood testing (loss of adult autonomy and confidentiality, and the possibility of causing harm to the developing child), together with the limited empirical evidence that is available, that has led us to advocate this cautious policy, erring towards a presumption of non-maleficence (*'primum non nocere* – the first goal is to cause no harm').

Furthermore, there are now good legal reasons for professionals to make decisions in this area primarily on the basis of the long-term best interests of the child; failure to do so could lead to professionals later being sued for acting against the interests of the child, even if this was at the request of the parents.

Despite our judgement that it is wise to avoid predictive tests in childhood, we recognize that experience and information may accumulate over time, and the apparent disadvantages of predictive testing for HD in childhood may not apply in practice to all other late-onset disorders – or even to Huntington's disease in every situation and for all time. The situation will certainly change when effective interventions become available for these conditions.

If the child stands to gain some health benefit from predictive testing, then it is obviously good practice to make the testing available. However, predictive testing carried out to satisfy the curiosity of a medical practitioner (except in the context of certain types of research, see below), or to relieve parental anxiety, would not seem to us in general to be appropriate. There may be circumstances where such testing is justified, but we expect these circumstances to be unusual. In our judgement, the possible emotional harm to the child, the abrogation of their autonomy and the breaching of their confidentiality, will generally outweigh the possible benefits that we can see.

An additional caution about the use of predictive tests in childhood must be noted: gene testing is not a substitute for good clinical assessment. This issue has been raised previously in the context of HD, where the occurrence of juvenile-onset cases is infrequent but well recognized. If an at-risk child develops clinical features compatible with the early stages of juvenile HD, then some alternative explanation for the suspicious clinical features may still be more likely to be correct, because the early signs of HD can be gradual and non-specific. In such circumstances, performing a predictive genetic test (not 'presymptomatic' in the usual sense of the word) may be considered, but it is unlikely to be helpful (Bloch & Hayden, 1990; Craufurd et al., 1990). Showing that a child has a mutation in the HD gene does not demonstrate that the clinical features are caused by HD, and may discourage the further diagnostic efforts needed to arrive at the correct explanation. If the child's symptoms are caused by another disorder, then an unfavourable, predictive test has been carried out on a minor without their consent.

Similar scenarios can arise with myotonic dystrophy and other mendelian disorders. Identifying children as being at increased risk of developing one of the common, multifactorial diseases (cardiovascular disease, hypertension, diabetes mellitus, cancers, ...) has the same drawbacks as predictive testing for late-onset, monogenic diseases unless there is some demonstrably effective therapy, or unless the testing is carried out as part of a therapeutic trial in which the allocation of treatments depends upon identifying the subjects with increased risk at the outset.

Carrier testing

The arguments here are less clear, and no consensus exists amongst health professionals. The working party has again arrived at a consensus view that carrier tests in childhood should generally be deferred. The arguments for and against testing can be (and indeed already often are) discussed openly in genetic counselling sessions with the parents of children who might be carriers of a genetic disorder. In most cases, this will rapidly lead to a consensus that testing can be deferred without any penalty, to allow the child to eventually participate in the decision as an autonomous person. We present the contrary arguments and the reasons for our view in the next six sections.

(i)   It may be better for a child to know about their carrier status in childhood: this may then be accepted as a simple matter of fact, without the emotional problems that could arise from a later disclosure.

This argument would only be valid if the one alternative to testing in childhood were to be silence on the issue within the family. We would like to emphasise that a decision NOT to test a child for carrier status is NOT intended to discourage a family from talking about the possible carrier status of the child. In practice, talk about such matters may actually be easier without firm knowledge; it is not difficult to explain to a child that the decision as to when and whether to test is being left for them to decide once they are older. This explicit granting of control to the child may actually be helpful in enhancing their self-esteem, and in allowing them to come to terms with possibly unfavourable results once adult.

We recognize that 50% of 'possible' carrier children would be shown NOT to be carriers by testing in childhood, and that life might be simplified for them by childhood testing. However, it is our judgement that the possible benefits experienced by these children must be balanced against the possible harm done by identifying carriers in childhood. Such harm might be especially likely in a family where some children are shown to be carriers, and others are shown not to be. There may also be families in which being identified in childhood as NOT a carrier could be emotionally disadvantageous.

(ii) Testing their child may decrease the level of anxiety of the parents simply by giving them this knowledge

BUT it may well increase their anxiety if the child is shown to be a carrier, and the testing may need to be repeated in any case, as techniques and methods improve.

(iii) There is no direct evidence to indicate that carrier testing in childhood is harmful

NOR is there good evidence of benefit, and some evidence has accumulated suggesting that knowledge of carrier status can have adverse effects, which might be magnified if the carrier (or the non-carrier) is identified involuntarily when young, and is labelled in this way by the whole family unit. We have arrived at our judgement on the basis of the currently available information, and recognise that further research is needed to examine these issues.

Problems may arise with carrier testing because of stigmatization or the fear of stigmatization, even in an area where the carrier frequency is high (Stamatoyannopoulos, 1974). Adult carriers of Tay-Sachs have been shown to view their future health with less optimism than others (Marteau *et al.*, 1992); it is not clear how this would affect those carriers identified in childhood. An earlier study has shown that 19% of Tay-Sachs carriers were worried about their carrier status at follow-up some years later (Zeesman *et al.*, 1991), and 10% of sickle cell and thalassaemia carriers in Rochester, NY are unable to state that being a carrier will not damage their health (Loader *et al.*, 1984). We do not know whether children identified as carriers will have the same degree of misunderstanding, or less, or more. The wider issues of social stigma have also caused concern even in the recent past, particularly in relation to sickle cell carrier screening in the USA (Bowman, 1991; Kennen & Schmidt, 1978).

Some evidence is emerging of the distress caused by carrier testing for cystic fibrosis in the healthy adult sibs of affected individuals (Fanos & Johnson, 1993). The limited experience with population screening for cystic fibrosis in Britain indicates that many adults, even some of those related to known carriers (Trembath *et al.*, 1991), decline the offer of carrier screening tests. Uptake is influenced predominantly by how the test is offered (Bekker *et al.*, 1993; Watson *et al.*, 1991), so public enthusiasm for the tests may be substantially less than the enthusiasm of the health or health research professionals, and the future adult enthusiasm of the present child cannot be taken for granted.

(iv) Carrier testing in childhood avoids urgent carrier testing in young teenage pregnancies (often concealed until late pregnancy)

Sexual activity, however, is not a reason for genetic testing, and it should not be presupposed that such teenagers will all want carrier testing and possible pre-natal diagnosis, calling for a decision about a possible termination of pregnancy (in fact, could some teenage pregnancies in at-risk families be concealed so as to avoid such tests?). A number of alternative practical approaches can be adopted. First, the question of their possible genetic risk can be put to the adolescent within the family setting or by their family doctor, and referral for genetic counselling can be arranged if and when they choose. To push such counselling or testing upon an adolescent (or anyone) is unacceptable, and it is likely to generate more problems than it 'solves'. Second, adolescents who may be carriers of a genetic disorder, like other adolescents, should be given proper information about human reproduction and contraception.

(v) Testing in childhood ensures that the testing is at least performed, and may absolve the doctor of his responsibility in this regard

However, imparting the information to the family does not ensure that the correct information will be given to the child in an appropriate fashion and at an appropriate age. If parents in practice find it difficult to impart the facts of sexual intercourse and human reproduction to their children, will they find it easier to discuss genetic disease? Anecdotes reported to us suggest not.

When carrier tests are carried out in childhood, we believe that there is still a responsibility on the medical profession to ensure that the future adult is offered genetic counselling at an appropriate age. This is both to ensure that the adults have the opportunity to discuss the issues relevant to themselves, and because the facts given to the parents, in connection with the particular disease prognosis or diagnostic testing, may no longer apply because of advances in medical knowledge.

Carrying out carrier tests in childhood, then, does not absolve the doctor of responsibility for ensuring that the future adult is offered genetic counselling at an appropriate age. While the initial scenario may work well in some families, it must be remembered that many families do not conform to the stereotype nuclear family structure. In a social context with parental fluidity and serial step-parenting, and because of the emotional responses of guilt, blame, reponsibility and denial in relation to their genetic disease in many families, it is not possible to rely upon the effectiveness and sensitivity of intergenerational communication. In addition, to set up testing of children in this way runs counter to a core principle of clinical genetic practice, by separating the testing from the counselling by a gap of years. Finally, the carrier testing methods may change, so that further analysis may be appropriate in any case once the child is adult and beginning to plan a family.

(vi) If testing is not performed in childhood, how may a doctor fulfill his responsibility to offer testing to all relevant individuals in the family?

There are several possible ways of ensuring that a child is offered testing at an appropriate age. These include:

(1) a clear transfer of responsibility from doctor to family, so that the onus of offering genetic counselling and testing to the child (once adolescent or adult) lies with them. This may be the only reasonable option if the family is planning to emigrate.

(2) a clear transfer of responsibility to the child's family practitioner, who can prompt a genetics referral when the child is old enough to be directly interested in the issues.

(3) continuing involvement with the family in the form of an active genetic register, with regular, long-term follow-up to keep in touch with family members even when they move area (Burn et al., 1989; Harper et al., 1982; Norman et al., 1989; Read et al., 1986).

In general, we would favour the third option, of involvement on an active follow-up register. However, it is clear that this is the most expensive option, because the resources required to establish and maintain such registers are considerable. We must also emphasize that testing in childhood does not in itself remove the physician's responsibility to ensure that the child is offered counselling when older.

Summary

The above arguments tend towards the view that carrier testing for genetic disease in childhood may not be the most appropriate approach, at least if practical alternatives can be devised and established. Inadvertent carrier testing (from research or from pre-natal diagnoses that indicate an unaffected, but carrier fetus) should be avoided wherever possible. Research is needed to ascertain the psychosocial consequences, the adverse effects or the benefits, of carrier testing in childhood.

Unsought information in research and clinical practice

It is possible for unsought information to be generated about a family in the course of research, for which consent has not been obtained, and which is unlikely to be of use in the research. This is particularly likely to occur when blood samples have been collected on all available family members as part of a preliminary gene localization (linkage) study, when these samples are stored in a research unit, and when tests are carried out on the stored research samples some years later once a specific test has been developed. This amounts to genetic testing without counselling or consent, and is likely to arise when samples have been collected indiscriminately from all available family members, and when the samples are tested wholesale in a manner for which consent was not obtained from the family (because the test was then not available). This can generate difficult practical and ethical problems, but these can often be avoided by careful advance planning, attention to detail in research consent procedures, and a clear separation of research work from service results and clinical records. This area has been discussed by Harper (1993b), and guidelines have been proposed. A specific effort on the part of clinical and laboratory staff will often be required to ensure that information is not generated on family menbers which has not been requested by them; once such unsought information is available, it may be difficult to prevent its being passed to the family, perhaps inadvertently and possibly resulting in serious family distress.

Predictive testing carried out as part of a research study into the consequences of early diagnosis may well be legitimate, but attention should be paid to the guidelines of the Medical Research Council Working Party on Research on Children (Medical Research Council, 1991) and the British Paediatric Association's Guidelines for the Ethical Conduct of Medical Research Involving Children (British Paediatric Association, 1992).

Unsought information may also result from the genetic testing of children in standard clinical practice, and we report one scenario that illustrates some of the problems that can result. In a family in which some individuals carried a balanced chromosomal rearrangement, the children were tested cytogenetically when very young, and at least one child was found to carry the rearrangement in balanced form. The family subsequently attended for counselling because they did not know how to give their teenage daughter her (positive) test results; it was decided to offer her a test once she expressed interest, as if she had never been tested in early childhood, rather than tell her that she had in fact been tested before and that her parents had known her test result for years.

Who has the right or responsibility to arrange for the genetic testing of children?

Parents. There is no clear consensus on this issue among British professionals. One legal view from North America would hold that parents are likely to have the right to the full disclosure of any genetic information about the child to which they want access (Pelias, 1991), although this has not yet been tested in the courts. It has also been suggested that geneticists may have an obligation to carry out such testing, and it has been emphasized that each child's case must be considered individually; no general policy would be defensible (Sharpe, 1993). The legal situation in Britain is different, and to our minds more satisfactory, in that any decisions must be made on the basis of the best interests of the child, although again no blanket policy would be defensible. Given that the genetic test results will have no medical management implications for the child, and that the uptake of carrier and late-onset predictive tests by adults in Britain is low, the arguments in favour of parental rights to demand this information are weak, although the parents may have the right to demand alternative means of ensuring that testing and counselling are offered when the child is older.

Adoption agencies. Adoption agencies are recognized by paediatricians as having a legitimate interest in the genetic status of prospective adoptive children. This view is reasonable, in so far as the result of such testing will influence the likelihood of a child being accepted by suitable adoptive parents. However, it may be better for a child to be adopted by parents whose willingness to adopt is not dependent upon the results of predictive or carrier genetic tests.

Medical practitioners. Where testing is to the possible medical advantage of the child, it is clearly the duty of medical practitioners to ensure that it is carried out. Where testing may be of interest to the future adult, for health reasons or to permit informed reproductive decision-making, the offer of counselling (and possible testing) should be made once the individual is mature or in early adult life; this may require the establishment of an active genetic register.

With respect to carrier testing, or to predictive testing for a late-onset disorder, the medical profession need not assume any obligation to initiate such tests. When such testing of children is requested by others, however, doctors have a right to refuse to carry it out if they consider that it may cause harm to the child or is not in the child's best interests.

Acknowledgements

We thank all those who responded to our various questionnaires and letters, professionals and lay societies and support groups. We thank Miss Jean Dunscombe for her hard work and organizational ability in ensuring the dispatch of so many documents, and for co-ordinating the work of the working party. We thank the professional bodies who distributed questionnaires to their members in their own mailings. We also thank Michael and Sue Aldred for their assistance in designing the questionnaires, the data preparation staff of the University of Wales College of Medicine for imprinting data onto floppy disks, and Iain Fenton for his assistance in extracting this information once again. We would also like to thank Professor Martin Bobrow for his diligent reading of several drafts of the report and his very constructive comments upon them.

We are grateful to the Marie Stopes Research Fund, administered by the Galton Institute, for its financial support of the questionnaire study incorporated into this report.

Members of the working party

Dr Angus Clarke (chair)
Senior Lecturer in Medical Genetics
Department of Medical Genetics
University of Wales College of Medicine, Heath Park, Cardiff

Dr David Fielding (BPA representative)
Consultant Paediatrician
Department of Paediatrics
Countess of Chester Hospital, Chester

Mrs Lauren Kerzin-Storrar
Principal Genetic Associate
Department of Medical Genetics
St Mary's Hospital, Hathersage Rd, Manchester

Dr Helen Middleton-Price
Principal Scientist
Department of Genetics, Institute of Child Health
30, Guildford St, London WC1

Mr Jonathan Montgomery
Lecturer in Law
Faculty of Law
University of Southampton
Highfield, Southampton

Dr Heather Payne
Senior Clinical Medical Officer, Fostering and Adoption
Clinical Directorate of Community Child Health
Lansdowne Hospital, Cardiff

Dr Emily Simonoff
Senior Clinical Scientist
MRC Child Psychiatry Unit
Institute of Psychiatry
De Crespigny Park
Denmark Hill, London

Miss Audrey Tyler, MBE
Honorary Research Officer
Department of Medical Genetics
University of Wales College of Medicine, Heath Park, Cardiff

References

Bekker, H., Modell, M., Denniss, G., Silver, A., Mathew, C., Bobrow, M. and Marteau, T. (1993) Uptake of cystic fibrosis testing in primary care: supply push or demand pull? *Br. Med. J.* 306: 1584–1586.

Billings, P.R., Kohn, M.A., de Cuevas, M., Beckwith, J., Alper, J.S. and Natowicz, M.R. (1992) Discrimination as a consequence of genetic testing. *Am. J. Hum. Genet.* 30: 476–482.

Bloch, M. and Hayden, M.R. (1990) Opinion: Predictive testing for Huntington disease in childhood: challenges and implications. *Am. J. Hum. Genet.* 46: 1–4.

Bloch, M., Adam, S., Wiggins, S., Huggins, M. and Hayden, M.R. (1992) Predictive testing for Huntington disease in Canada: The experience of those receiving an increased risk. *Am. J. Med. Genet.* 42: 499–507.

Bolam v Friern HMC [1957] 2 All ER 118 24.

Bowman, J.E. (1991) Invited editorial: Prenatal screening for haemoglobinopathies. *Am. J. Hum. Genet.* 48: 433–438.

British Paediatric Association (1992) *Guidelines for the Ethical Conduct of Medical Research Involving Children*. BPA Ethics Advisory Committee.

Burn, J., Church, W., Chapman, P.D., Gunn, A., Delhanty, J. and Roberts, D.F. (1989) A regional register for familial adenomatous polyposis: congenital hypertrophy of the retinal pigment epithelium as a means of carrier detection. *J. Med. Genet.* 26: 207.

Collinge, J., Poulter, M., Davis, M.B., Baraitser, M., Owen, F., Crow, T.J. and Harding, A.E. (1991) Presymptomatic detection or exclusion of prion protein gene defects in families with inherited prion diseases. *Am. J. Hum. Genet.* 49: 13561–1354.**Q1**

Craufurd, D. and Harris, R. (1986) Ethics of predictive testing for Huntington's chorea: the need for more information. *Br. Med. J.* 293: 249–251.

Craufurd, D., Dodge, A., Kerzin-Storrar, L. and Harris, R. (1989) Uptake of presymptomatic testing for Huntington's disease. *Lancet* ii: 603–605.

Craufurd, D., Donnai, D., Kerzin-Storrar, L. and Osborn, M. (1990) Testing of children for 'adult' genetic diseases. *Lancet* 335: 1406.

Fanos, J.H. and Johnson, J.P. (1993) Barriers to carrier testing for CF siblings. *American Society of Human Genetics Meeting*, New Orleans, October 1993. Session 22, Abstract 51.

Gillick v W. Norfolk & Wisbech HA [1985] 3 All ER 402.

Harper, P.S. (1992) Genetic testing and insurance. *J. Roy. Coll. Phys. Lond.* 26: 184–187.

Harper, P.S. (1993a) Insurance and genetic testing. *Lancet* 341: 224–227.

Harper, P.S. (1993b) Research samples from families with genetic diseases: a proposed code of conduct. *Br. Med. J.* 306: 1391–1394.

Harper, P.S. and Clarke, A. (1990) Should we test children for 'adult' genetic diseases? *Lancet* 335: 1205–1206.

Harper, P.S., Tyler, A., Smith, S., Jones, P., Newcombe, R.G. and McBroom, V. (1982) A genetic register for Huntington's Chorea in South Wales. *J. Med. Genet.* 19: 241–245.

Huggins, M., Bloch, M., Wiggins, S., Adam, S., Suchowersky, O., Trew, M., Klimek, M.L., Greenberg, C.R., Eleff, M., Thompson, L.P., Knight, J., MacLeod, P., Girard, K., Theilmann, J., Hedrick, A. and Hayden, M.R. (1992) Predictive testing for Huntington Disease in Canada: Adverse effects and unexpected results in those receiving a decreased risk. *Am. J. Med. Genet.* 42: 508–515.

Kenen and Schmidt **Q3** (1978) Stigmatisation of carrier status: social implications of heterozygote screening programmes. *Am. J. Pub. Health* 68: 1116–1120.

Kessler, S. (1988) Invited essay on the psychological aspects of genetic counselling. V. Preselection: a family coping strategy in Huntington Disease. *Am. J. Med. Genet.* 31: 617–621.

Loader, S., Sutera, C.J., Segelman, S.G., Kozyra, A., Rowley, P.T. (1991) Prenatal hemoglobinopathy screening. IV Follow-up of women at risk for a child with a clinically significant hemoglobinopathy. *Am. J. Hum. Genet.* 49: 1292–1299.

Marteau, T.M., van Duijn, M. and Ellis, I. (1992) Effects of genetic screening on perceptions of health: a pilot study. *J. Med. Genet.* 29: 24–26.

Medical Research Council (1991) *The ethical conduct of research on children* (Report of working party). MRC, London.

Montgomery, J. (1993) Consent to health care for children. *Journal of Child Law* 5: 117–124.

Morris, M., Tyler, A. and Harper, P.S. (1988) Adoption and genetic prediction for Huntington's disease. *Lancet* ii: 1069–1070.

Morris, M., Tyler, A., Lazarou, L., Meredith, L. and Harper, P.S. (1989) Problems in genetic prediction for Huntington's disease. *Lancet* ii: 601–603.

Norman, A.M., Rogers, C., Sibert, J. and Harper, P.S. (1989) Duchenne's muscular dystrophy in Wales: a 15 year study, 1971 to 1986. *J. Med. Genet.* **26**: **Q1**

Payne, H. (1993) Genetic disease and adoption. In preparation. **Q2**

Pelias, M.Z. (1991) Duty to disclose in medical genetics: a legal perspective. *Am. J. Med. Genet.* **39**: 347–354.

Read, A.P., Kerzin-Storrar, L., Mountford, R.C., Elles, R.G. and Harris, R. (1986) A register-based system for gene tracking in Duchenne's muscular dystrophy. *J. Med. Genet.* **24**: 84–87.

Re D [1976] Fam. 185.

Re J [1990] 3 All ER 930.

Re J [1992] *The Times*, 12 June.

Re R [1991] 4 All ER 177.

Re W [1992] 4 All ER 627.

Sharpe, N.F. (1993) Presymptomatic testing for Huntington Disease: is there a duty to test those under the age of eighteen years? *Am. J. Med. Genet.* **46**: 250–253.

Stamatoyannopoulos, G. (1974) Problems of screening and counselling in the haemoglobinopathies. In: *Birth Defects: Proceedings of the Fourth International Conference* (eds A.G. Moultsky and F.J.B. Ebling). Excerpta Medica, Amsterdam.

Thelin, T., McNeil, T.F., Aspegren-Jansson, E. and Sveger, T. (1985) Psychological consequences of neonatal screening for alpha-1-antitrypsin deficiency. *Acta Paediatr. Scand.* **74**: 787–793.

Trembath, R.C., Green, J., McMahon, C., Malcolm, S. and Pembrey, M.E. (1991) Carrier testing for CFTR gene mutations in relatives identified through a cystic fibrosis prenatal testing service. Response rates and determinants of uptake. *Clinical Genetics Society*, Belfast.

Turnpenny, P.D., Simpson, S.A. and McWhinnie, A.M. (1993) Adoption, genetic disease and DNA. *Arch. Dis. Child.* **69**: 411–413.

Tyler, A., Morris, M., Lazarou, L., Meredith, L., Myring, J. and Harper, P.S. (1992) Presymptomatic testing for Huntington's Disease in Wales 1987–1990. *Br. J. Psychiatr.* **161**: 481–489.

Watson, E.K., Mayall, E., Chapple, J., Dalziel, M., Harrington, K., Williams, C. and Williamson, R. (1991) Screening for carriers of cystic fibrosis through primary health care services. *Br. Med. J.* **303**: 504–507.

World Federation of Neurology (1989/90) Research Committee Research Group Ethical issues policy statement on Huntington's disease molecular genetics predictive test. *J. Neurol. Sci.* (1989) **94**: 327–332; and *J. Med. Genet.* (1990) **27**: 34–38.

Zeesman, S., Clow, C.L., Cartier, L., Scriver, C.R. (1984) A private view of heterozygosity: eight-year follow-up study on carriers of the Tay-Sachs gene detected by high school screening in Montreal. *Am. J. Med. Genet.* **18**: 769–778.

Appendix 1: results of the questionnaire study

Molecular genetics laboratories

We sent questionnaires to 35 of the laboratories listed in the Clinical Molecular Genetics Society handbook, targeting those that we thought most likely to be involved in diagnostic work and including all Regional NHS laboratories. We received replies from 16 laboratories, of which half did not carry out testing that would fall within our definition as being potentially problematic.

Some laboratories carry out the genetic testing of young children to determine their disease status, and generate carrier status information incidentally. This information is not then usually witheld from the family. Several laboratories carry out predictive testing for disorders such as familial adenomatous polyposis, myotonic dystrophy and adult polycystic kidney disease. Testing is also carried out for some X-linked disorders that can affect females (OCT deficiency, fragile-X mental retardation). In these conditions there is a distinction to be drawn between testing for a child who may be clinically affected and one in whom evidence of carrier status is being sought. Six of the laboratories regularly test young children specifically to determine their carrier status for recessive conditions such as cystic fibrosis or sex-linked muscular dystrophy (Duchenne or Becker). Pre-symptomatic diagnosis of Leber's optic atrophy has been carried out in a young girl. Two centres specifically stated that it was their policy not to carry out any such predictive or carrier tests on children.

Cytogenetics laboratories

Ten of the 23 regional or sub-regional cytogenetics laboratories to which questionnaires were sent, provided us with detailed information about their testing of children, with particular emphasis on the determination of their carrier status for familial chromosomal rearrangements. Several other laboratories responded with general statements and observations. The picture that emerges is that most laboratories examine samples from several children each year in order to determine whether or not they carry a familial chromosomal rearrangement. However, several reasons were given to account for some of these tests that do not necessarily apply in all situations. First, some tests are carried out at birth to confirm a prenatal test result. Second, the results of the test may be used to assist with the interpretation of test results in other family members. However, there is probably still a substantial number of children tested where these considerations do not apply, and where the questions raised by our working party do need to be addressed.

It was put to us that cytogenetics laboratories generally accept samples referred from paediatricians, geneticists or other clinicians in good faith, without questioning the ethical or clinical judgements involved; it is primarily with the clinicians that these issues need to be explored. The same argument could be employed by molecular laboratory staff. Whether or not this argument is an adequate response to these issues may deserve wider discussion.

Clinicians and genetic coworkers (nurses, counsellors, etc.)

The questionnaire was circulated to members of several professional groups, including members of the Clinical Genetics Society (550, including some paediatricians and genetic coworkers), members of the Genetic Nurses and Social Workers Association (73), consultant members of the British Paediatric Association (990), members of the British Paediatric Cardiac Association (100), members of the British Society for Haematology (780), the Society for Endocrinology (1000), the British Paediatric Neurology Association (20 non-BPA members), Association of British Neurologists (500), the British Association of Paediatric Surgeons (87), general surgeons with an interest in familial adenomatous polyposis (44) and to selected ophthalmologists (11). Because of the importance of obtaining responses from a representative group of paediatricians, and because of the low response rate from paediatricians as a whole, a second copy of the questionnaire was distributed to the non-respondents among a randomly selected group of 10% of consultant paediatricians *Table 2.*

Table 2. Clinical questionnaire

Specialties of responding clinicians	Responses	Questionnaires sent
Consultant clinical geneticists	32 (58%)	55
Genetic coworkers	17 (23%)	73
Consultant paediatricians (consultant paediatrician members of British Paediatric Association)	337 (34%)	990
Other members of CGS	14 (3%)	470
'10%' group of paediatricians (of all consultant members of British Paediatric Association)	50 (42%)	118
'FAP' surgeons	14 (32%)	44
Paediatric surgeons	14 (16%)	87
Ophthalmologists	7 (64%)	11
Haematologists (45 adult, 6 paediatric)	51	
Other physicians	30	

The proportion of respondents for each professional group is not precise because of some overlap of memberships between different groups (e.g. CGS and the Genetic Nurses and Social Workers Association; CGS and the British Paediatric Association).

The low response rate from 'other' CGS members appears to have resulted from the reluctance of non-consultant clinical geneticists to reply to a questionnaire that required them to list 'their' cases in which childhood testing for genetic disorders was proposed; such cases, it was felt, would be best reported by the appropriate consultant so as to avoid a duplication of reports.

Predictive Testing. Only 184 of the 512 respondents (36%) had received requests to carry out predictive testing of 902 children. In addition, some respondents were involved in newborn screening programmes for haemoglobinopathies and other disorders. Many of the tests discussed were potentially of health benefit to the child (e.g. hyperlipidaemias, medium chain acyl-CoA dehydrogenase deficiency, family cancer syndromes, haemoglobinopathies, cystic fibrosis, alpha-1-antitrypsin deficiency) or were for disorders likely to become manifest in childhood (adrenoleukodystrophy, Becker or Duchenne muscular dystrophy). Requests for predictive testing were declined or deferred by a number of individuals, principally for neurodegenerative disorders (HD, Charcot-Marie-Tooth disease, myotonic dystrophy) and also for retinitis pigmentosa (RP), autosomal dominant polycystic kidney disease (APKD) and familial adenomatous polyposis coli (FAP). Some requests for the predictive testing of children for myotonic dystrophy, RP, APKD, and hypertrophic obstructive cardiomyopathy (HOCM) were accepted and carried out.

Only a few of these requests originated from outside the nuclear family, typically from adoption agencies (pre-placement).

Carrier Testing. A greater proportion of respondents (41%) received requests to determine the carrier status of healthy children for inherited disorders. As in predictive testing, most of these professionals were clinical geneticists, paediatricians or haematologists. The bulk of such testing involved familial chromosomal rearrangements, Duchenne and Becker dystrophies, cystic fibrosis and the haemoglobinopathies; almost all of these tests were carried out, but some respondents who discouraged such tests did not report the number of tests requested. Carrier status tests for X-linked visual disorders, coagulopathies and fragile-X mental retardation are also carried out regularly.

A number of requests for carrier testing were declined or deferred, e.g. for CF, DMD, chromosome translocations etc., although only a minority of respondents (8%) had a firm age limit below which they did not carry out such testing, and the age limit varied from 1 to 18 years. Of the 276 respondents who had received no requests, 28 discourage such tests in childhood, 123 refer all such requests elsewhere, and 130 had never been asked. Only a few requests were received from outwith the family. The numbers of children on whom testing was performed or requested was reported as 1706 children annually. Although this is probably only a proportion of the tests carried out, it is clear that a substantial number of children are being tested for their carrier status.

Attitudes

Predictive tests. Predictive tests in childhood were thought to be justified under some circumstances by 66% of respondents on the grounds of a health benefit to the subject, while 16% of respondents thought that such testing was not justified. Those who considered such testing to be helpful gave many examples of disorders where this could be so. These disorders included the hyperlipidaemias, FAP, Wilm's tumour, retinoblastoma and Marfan syndrome. Some respondents stated that they would only carry out such testing if they had available a useful intervention (e.g. treatment or surveillance for complications). A minority of respondents wished to offer testing for disorders where there are no clearly established useful interventions, including HD, HOCM, Leber's optic atrophy and myotonic dystrophy.

Attitudes to predictive genetic testing in childhood were explored further in very general terms. Respondents were asked to indicate the strength of their agreement or disagreement with a set of statements about genetic testing (*Table 1*). The seven points given as possible advantages of predictive testing in childhood were all regarded positively by a large majority of respondents, although not so many supported the notion that the child's attitude to reproduction as an adult would be 'more responsible' as a result of testing in childhood. On this question, most geneticists and all fieldworkers expressed disagreement, whereas a majority of paediatricians and others supported this judgement. A majority thought that testing in childhood would also permit more accurate genetic counselling of other family members.

Right to request genetic testing. The next question, concerning who has the right to request the genetic testing of a child, revealed marked differences in attitude between the members of different professional groups (see *Table 3*). Because the response rate to the questionnaire was poor (only 58% of consultant geneticists and 34% of consultant paediatricians), it is not clear how representative the

Table 1. Perceived advantages/disadvantages of predictive testing in childhood

Possible advantages of predictive testing in childhood

1 Relieves anxiety about possible early signs of the disorder.
2 Family uncertainty about the future is reduced.
3 More accurate genetic counselling becomes possible.
4 The child's attitude towards reproduction in adulthood will be more responsible.
5 Children who might benefit from genetic counselling in the future will be identified.
6 Practical planning for education and career, housing and family finances becomes possible.
7 Parental expectations of the child's behaviour become altered.

Possible disadvantages of predictive testing in childhood

1 Removes the child's right to decide whether or not to be tested in adulthood.
2 Parental expectations of the child's future reproductive behaviour become altered.
3 Damages the child's sense of self-esteem.
4 Generates unwarranted anxiety about possible early signs, before any genuine manifestation of the disorder.
5 Leads to future difficulties in obtaining life insurance.
6[a] Rarely leads to clarification of the genetic status of other family members.

[a] This is not so much a disadvantage as a lack of advantage, but was included as a disadvantage for the purposes of the questionnaire.

respondents are of their professional groups. However, the attitudes of the group of 10% of consultant paediatricians who were given a second mailing if they had failed to respond to the first, were very similar to those of the other responding paediatricians. It is therefore reasonable to interpret the questionnaire results as if the respondents' views are representative of their professional groups'.

A majority of genetics coworkers stated that not even parents should have the right to request the genetic testing of their child. While clinical geneticists were divided on this issue, they thought that this right should apply to testing for certain disorders only. In contrast, a majority of respondents overall (364 = 71%), and of paediatricians and haematologists in particular, thought that parents should have the right to request the genetic testing of their child. This was confirmed by the responses to a separate question (*Table 3ii*), suggesting that the 'right to request' was generally interpreted as a right to have the test performed. A very substantial majority of all groups thought that social services (unless having parental responsibility), legal authorities, the extended family and the education authorities should not have this right. Some generalizations concerning differences between professional groups can be made. Most haematologists who responded thought that medical practitioners but not adoption agencies should have this right. Paediatricians were divided on the question of doctors' rights and of adoption agencies' rights, but were more in favour of adoption agencies' rights than were other groups; clinical geneticists and fieldworkers would restrict the rights of both these groups of professionals.

Potential disadvantages of testing in childhood. Turning to the potential disadvantages of carrying out tests in childhood, a majority also agreed with the statements that such tests limited the child's future autonomy, could lead to difficulties obtaining life insurance, and could have an adverse effect upon parental expectations of the child's future reproductive behaviour (see *Table 1*). Additional potential disadvantages that were put forward included future difficulty in obtaining training and employment, and the disturbance of the parent–child relationship leading to impaired 'bonding' and the possible rejection of the child by the parent.

Table 3. Differences in attitudes between professional groups

(i) 'Do you agree that the following have the right to request that genetic testing (predictive or carrier status) be performed on a child – even if the result will have no direct health benefit for the child?'

	AFFIRMATIVE REPLIES (% of respondents in each professional group)			
	Parents	Extended Family	Adoption Agencies	Medical Practitioner
Geneticists (n=46)	57%[a]	6%	28%	21%
Coworkers (n=17)	29%	0%	12%	12%
Paediatricians (n=337)	75%	10%	55%	42%
'10%' Paeds. (n=50)	72%	16%	48%	40%
Haematologists (n=51)	76%	14%	37%	61%
ALL (n=512)	71%	11%	47%	41%

[a] Of the 27 geneticists (57%) who thought that parents should be able to arrange for genetic testing of their child(ren), 18 (39% of all the geneticists) would permit this for certain disorders only.

(ii) 'It is up to the family to decide whether or not their child(ren) should be tested for genetic disorders (for predictive or carrier tests)'.

	AGREE
Geneticists	46%
Coworkers	35%
Paediatricians	79%
'10% Paeds'	78%
Haematologists	84%
ALL	74%

Despite the lack of firm evidence on which to base their views, and the broad range of opinion found within each professional group, we did find significant differences of opinion between professional groups. Paediatricians were evenly divided as to whether or not the child's sense of self-esteem could be damaged by testing, whereas a small majority of geneticists and a large majority of fieldworkers thought that this could be an important disadvantage of such testing. A majority of all respondents also considered that a diagnosis in childhood could result in excessive anxiety about possible early signs of the disorder, just as a majority of respondents also considered that it could lead to the resolution of such anxiety.

The questions we had identified as issues to be considered were recognized as potential issues, and additional distinct issues were not identified. There was some evidence that respondents were more confident of recognizing potential advantages of testing than they were of recognizing the potential disadvantages (fewer responses to the statements of potential advantages of testing indicated neutrality, and substantially more responses indicated a strength of feeling).

One set of respondents thought that children might adjust to unwelcome genetic information more readily in middle childhood than in adolescence. Others expounded the opposite view, that such a process could be harmful or misleading. A large majority of all groups thought that children should be consulted on these questions once they are sufficiently mature.

There were several reasons for considering the process as potentially harmful. The disorder may be variable, so that the previous experience in the family merely causes confusion. Children may be singled out to their own detriment or that of their sibs. Healthy children may be 'medicalized' when they are still very impressionable. The test results may be given an excessive emotional weight in some families, generating disproportionate levels of anxiety. Some respondents thought that, by performing the test only at the specific request of the (now adult) individuals, one is allowing them a degree of control that may be helpful in their coming to terms with unwelcome information.

Supplementary questionnaire

Some months after the first questionnaire described above, a second questionnaire was distributed to members of the Clinical Genetics Society and to consultant members of the British Paediatric Association. This asked whether or not the respondent would be willing to arrange genetic testing of a healthy 5-year-old child for a number of different disorders (predictive tests for some dominant and sex-linked disorders; carrier-status tests for some autosomal recessive and sex-linked disorders and for balanced chromosomal translocations). It was hoped that differences in the acceptability of testing for different disorders might allow inferences to be drawn about the factors that influence professional attitudes.

Interpretation of supplementary questionnaire results. Responses were received from 260 paediatricians, 49 geneticists, 7 coworkers, and 8 haematologists – 324 in all (*Table 4*). The same trends appear in these data as in the longer, original questionnaire: paediatricians were generally much more willing for genetic tests to be performed in childhood than either geneticists or coworkers. Fewer firm opinions were expressed concerning the more uncommon disorders. There was little to suggest that the availability of an early treatment or surveillance for complications for a disorder made any difference to the readiness to test for it.

Respondents fell into two broad groups, that in which most tests were approved, and that in which most would be refused. The most important factor influencing responses seems to have been the respondents' attitudes to such

Table 4. Responses to supplementary questionnaire.

'Would you test a 5-year-old child whose parents wanted to know the genetic status for the following disorder(s)?'

Disorders	Geneticists (n=49)		Paediatricians (n=260)	
	Yes	No	Yes	No
Predictive Testing				
Adult polycystic kidney disease	27	17	198	29
Becker muscular dystrophy	30	15	189	38
Charcot-Marie-Tooth	11	28	154	46
Facio-scapulo-humeral muscular dystrophy	8	30	156	50
Huntington's Disease	2	47	100	89
Hyperlipidaemias	40	7	223	16
Hypertrophic obstructive cardiomyopathy	37	8	201	13
Leber's optic atrophy	20	18	152	40
Marfan syndrome	43	4	199	24
Multiple endocrine neoplasia	32	9	160	22
Myotonic dystrophy	19	26	182	42
Neurofibromatosis I	30	13	207	29
Polyposis coli	28	16	214	14
Prion-protein dementia	0	42	66	48
Retinitis pigmentosa	20	22	177	35
Spinal muscular atrophy	16	23	163	45
von Hippel-Lindau	36	8	143	22
Unaffected Carrier Status Testing				
Alpha-1-antitrypsin deficiency	19	27	132	92
Cystic fibrosis	22	27	170	74
Haematological disease	22	28	155	74
Spinal muscular atrophy	17	30	130	98
Tay-Sachs	20	30	142	88
Adrenoleukodystrophy	12	32	132	74
Becker muscular dystrophy	11	33	136	82
Choroideraemia	11	32	75	81
Duchenne muscular dystrophy	14	32	163	75
G6PD deficiency	13	32	139	90
Haemophilia (A and B)	13	33	159	74
Immune defects	12	31	139	79
Norrie's disease	12	30	81	77
Retinitis pigmentosa	11	32	130	78
(Balanced) chromosomal translocation	27	20	146	83

testing in general, and the nature of the various disorders seems to have had relatively little influence. Most respondents answered similarly for all the diseases for which they stated a view (although the proportion of questions answered with a 'Don't know' did vary between individuals). Only with Huntington's disease and prion dementia was there a suggestion of a widespread unwillingness to carry out predictive testing on a young child. Even with these conditions, it was striking that many paediatricians expressed a readiness to carry out presymptomatic predictive testing. The few paediatric haematologists surveyed expressed uncertainty about testing for most disorders, and were divided in their willingness to test children to determine their carrier status for haematological disorders. The further analysis of these responses, and the exploration of the reasons for this pattern of results, will need to be studied as a future piece of research, and is beyond the scope of this report. ˙

Appendix 2

Table 5. Prospective study of testing in molecular genetics laboratories, 1992

	Genetic Tests	Performed (Age Range)	Deferred
Alpha-1-antitrypsin deficiency (carrier test)	1	9 years	0
BCR	13	infancy–14 years	2
BMD predictive	4	1–12 years	0
CF carrier	68	infancy–15 years	9
DMD carrier	5	8–15 years	4
Fabry's carrier	1	3 years	0
FAP	22	infancy–11 years	0
FSH dystrophy	1	10 years	0
Fragile-X carrier	10	4–14 years	1
Fragile-X NTM	5	4–14 years	0
Haemoglobinopathy carrier	10	1–14 years	5
Haemophilia A carrier	8	3–14 years	0
HMSN I	1	birth	1
HOCM predictive	0		1 (age 4)
Huntington's Disease	0		2 (ages 5,14)
Kennedy's disease (predictive)	0		1 (infant)
Myotonic dystrophy	13	infancy–14 years	7
Intermediate SMA	1	3 months	0
SMA 3	1	10 years	0

Carrier and predictive tests for adrenoleukodystrophy, Wilson's disease and a few other conditions were deferred for a variety of reasons, sometimes including technical as well as ethical and counselling factors.

Abbreviations: BCR, balanced chromosomal rearrangement; BMD, Becker muscular dystrophy; CF, Cystic fibrosis; DMD, Duchenne muscular dystrophy; FAP, familial adenomatous polyposis coli; FSH, facio-scapulo-humeral; HMSN, hereditary motor-sensory neuropathy; NTM, normal transmitting male; SMA, spinal muscular atrophy.

Appendix 3: Attitudes of family and patient support groups

Comments on the replies from the group which favoured a policy of not performing predictive or carrier tests in childhood included remarks such as 'at least wait until child able to give informed consent', and 'once of an age to be sexually active'. This group included all six replies from adults with spinal muscular atrophy. Those opposed to the principle of predictive or carrier testing in childhood generally favoured the right of health professionals to make decisions about whether or not to proceed with tests requested by parents. Those in favour of carrying out such tests generally opposed this type of professional control, regarding it as excessively paternalistic.

Remarks written in general opposition to the principle of genetic testing in childhood included:

'testing will destroy the innocence of childhood.'

'good counselling – yes; testing – no.'

'if there is some preventive measure ... tests should be done, but otherwise no.'

'I don't think it should be done just for the parents' peace of mind ...'

'we feel it is not the parents' right, but the "child's" right in adulthood. Doctors should be able to refuse the tests in order to protect the child.'

'I wanted my son tested to see if he was a carrier of PKU ... but after long discussions my husband and I decided it would serve no purpose other than labelling him ... Other more serious disorders would need a lot more thought but I think the same decision should be made.'

'It is important, I think, for a child to grow up knowing that there is a chance that he/she is a carrier for something – but not until she/he is at the older age when they are thinking of partners and producing offspring should it be necessary to do any testing ...'

'predictive testing would be just another example of projected unethical genetic cleansing.'

'if as a result of trying to wipe out the disease, there were fewer remaining, then society may become less tolerant.'

Remarks written by those in general support of childhood testing included:

'A parent's job is hard enough. If there is the possibility that their child will become seriously ill at whatever age, then parents should have the opportunity to either allay their fears and the stress on the family, or alternatively to make provision for the support – moral, emotional, financial – of their child. A parent should have the right to ask for testing but (should) not be 'pressured' in any way. The decision to test or not should be with the parents.'

'Doctors should have no say in the matter.'

'I think it (predictive testing) is essential, and that persons with a potential disorder in later life should be made aware so that they can plan for their futures, and be prepared (for) what to expect.'

'This issue (carrier testing in childhood) is simpler (than predictive testing). I feel that testing of children in this case would be not only right, but vital. An informed decision on reproduction could be made.'

There were two further comments on the field of childhood testing that provide interesting perspectives for readers of this report:

'Decisions on carrier testing should be left to the child when they are old enough ... GPs should ensure that such children fully understand the risks before the decision to start a family. Support and counselling are needed and cannot be left to parents.'

'As with all matters relevant to a child, the fundamental criterion has to be what is in the best interests of the child. This may be a decision capable of being made by the parents or doctors, but not necessarily either.'

Appendix 2: EUROSCREEN survey of attitudes to the genetic testing of children among members of the European Society of Human Genetics

One of the components of the EUROSCREEN project – funded by the European Union to examine a range of bioethical issues – was a study of attitudes towards the genetic testing of children among geneticists across Europe. This was intended both to explore attitudes towards the genetic testing of children among genetics professionals in Europe and at the same time to raise the level of professional awareness of the topic as a potentially difficult area. The questionnaire was devised collectively by members of the EUROSCREEN subgroup (see below) with assistance from Dr. Priscilla Alderson, Miss Audrey Tyler and Ms. Lucy Brindle. The questionnaires were distributed by Mairi Levitt (of EURO-SCREEN) and this compilation of the data was prepared by Angus Clarke and Iain Fenton. The overall EUROSCREEN project was coordinated by Prof. Ruth Chadwick of the Centre for Applied Ethics at the University of Central Lancashire, Preston, UK.

Out of concern to keep the questionnaire as brief as possible, little information was requested about the respondents apart from their country of residence. The willingness of the geneticist to perform genetic testing on a child was determined in each of ten distinct scenarios. 'Willingness to test' was scored on a scale from 1 (very unwilling to test) through 3 (neutral) to 5 (very willing to test).

The questionnaire scenarios are now reproduced together with the overall

The Genetic Testing of Children, A.J. Clarke (ed.).
© 1998 BIOS Scientific Publishers Ltd, Oxford.

(pooled) results. Following this, differences between respondents from different countries are examined.

In each scenario, the parents of the child have requested the test in order to resolve their sense of uncertainty about the child's genetic status. The children to be tested are healthy.

Response Categories (for each scenario):

1 Very unwilling, or would refuse to test.
2 Unwilling, reluctant.
3 Neutral.
4 Willing.
5 Very willing, enthusiastic.

The number of respondents choosing each option is indicated.

SCENARIO 1: Carrier status for cystic fibrosis

A 5-year-old child who has an affected brother
Willingness to test

	1	2	3	4	5
Responses	31%	25%	19%	17%	6%

SCENARIO 2: Carrier status for cystic fibrosis

A 12-year-old child who has an affected brother
(a) If the child **wants** to be tested
Willingness to test

	1	2	3	4	5
Responses	15%	19%	19%	33%	14%

(b) If the child is **neutral**
Willingness to test

	1	2	3	4	5
Responses	29%	27%	26%	14%	3%

(c) If the child **does not want** to be tested
Willingness to test

	1	2	3	4	5
Responses	77%	11%	7%	4%	0%

SCENARIO 3: Gene status for fragile-X (carrier)

A developmentally normal 6-year-old child whose mother is an obligate carrier of fragile-X syndrome (the daughter of a normal transmitting male)

(a) If the child is a BOY
Willingness to test

	1	2	3	4	5
Responses	28%	29%	15%	23%	5%

(b) If the child is a GIRL
Willingness to test

	1	2	3	4	5
Responses	32%	26%	15%	21%	5%

SCENARIO 4: Gene status for fragile-X (predictive and carrier)

A 1-year-old boy whose mother is an obligate carrier of fragile-X syndrome (the daughter of a normal transmitting male)
Willingness to test

	1	2	3	4	5
Responses	12%	14%	15%	44%	13%

SCENARIO 5: Predictive testing for familial breast cancer

In a known BrCa1 family, would you be willing to test a 6-year-old girl?
Willingness to test

	1	2	3	4	5
Responses	57%	23%	7%	9%	3%

SCENARIO 6: Predictive testing for Huntington's disease

Request from the parents of a 2-year-old child, whose mother or father carries a known mutation in the HD gene; the child is therefore at 50% risk of HD
Willingness to test

	1	2	3	4	5
Responses	75%	11%	4%	6%	3%

SCENARIO 7: Predictive testing for Huntington's disease

Request from the parents of a 12-year-old child whose mother or father carries a known mutation in the HD gene
(a) If the child **wants** to be tested
Willingness to test

	1	2	3	4	5
Responses	49%	26%	4%	12%	5%

(b) If the child is **neutral**
Willingness to test

	1	2	3	4	5
Responses	70%	15%	7%	3%	2%

(c) If the child **does not want** to be tested
Willingness to test

	1	2	3	4	5
Responses	85%	9%	2%	1%	1%

SCENARIO 8: Predictive testing for Huntington's disease

Request from the mother of a 6-year-old girl, whose father is at 50% risk of HD but does not want to be tested (the girl is therefore at 25% risk)
Willingness to test

	1	2	3	4	5
Responses	71%	19%	3%	3%	2%

SCENARIO 9: Pre-natal testing for Huntington's disease

The parents in scenario 8 are separated, but the mother is pregnant and wishes for the fetus to be tested before she decides whether or not to continue with the pregnancy (the father remains unwilling to be tested, not wanting his risk-status to be clarified)
Willingness to test

	1	2	3	4	5
Responses	11%	12%	21%	42%	9%

SCENARIO 10: Carrier testing for cystic fibrosis

How willing would you be to offer population screening to identify carriers of CF (a) to schoolchildren of 10–14 years?
Willingness to test

	1	2	3	4	5
Responses	35%	31%	16%	12%	4%

(b) as part of the newborn screening programme?
Willingness to test

	1	2	3	4	5
Responses	47%	21%	12%	13%	5%

Interpretation of Overall Results

These results indicate a reluctance among the 189 respondents to undertake predictive genetic tests on young children, with more reluctance in relation to Huntington's disease than to breast cancer susceptibility (BRCA1). 162/189 (86%) were unwilling to test a 2-year-old and 161/189 (85%) a 12-year-old for Huntington's disease and 150/189 (79%) were unwilling to test a 6-year-old girl for BRCA1 mutations. If the 12-year-old child at risk of HD wanted to be tested then 142 would still be unwilling to comply, and if the child did not want to be tested then 177/189 (94%) would be unwilling to perform the test.

There was less reluctance to undertake carrier testing for cystic fibrosis either within families or in the general population, particularly in the context of a CF family when the child was old enough to express their wish to be tested. While 106/189 (56%) were unwilling to test a 5-year-old for cystic fibrosis carrier status and 105/189 (56%) were unwilling to test a 12-year-old, this fell to 63 (33%) if the (12-year-old) child wanted testing and rose to 167 (88%) if the child did not want testing. Rather more geneticists were unwilling to carry out population screening for CF-carrier status in children than were willing to offer testing within a family context – 125 (66%) were unwilling to offer CF screening to school children of 10–14 years and 128 (68%) were unwilling to offer carrier screening for newborn infants.

While only 48 respondents (25%) were unwilling to test, the 1-year-old son of a female carrier of the fragile-X syndrome, 106 (56%) were unwilling to test a developmentally normal 6-year-old boy and 111 (59%) were unwilling to test a developmentally normal girl of the same age. Once the child is old enough that the test has become essentially a carrier test – rather than a possible diagnostic test on the child – then fewer geneticists would be willing to perform it.

Most respondents would clearly allow their decisions to be modified by the stated preferences of a 12-year-old child.

The Respondents

We gathered information about the country of residence of the respondents and the nature of their employment – in clinical or laboratory work. The country of origin of the respondents is given in *Table 1*. It is clear that the response rate was poor – 20% overall – and that it varied enormously between countries; the respondents could not

Table 1. Rates of response to questionnaire by respondents in different countries (1995)

COUNTRY	Number of Respondents	Number of ESHG Members
Austria	4	12
Belgium	10	46
Britain	25	118
Denmark	5	37
Finland	12	35
France	12	85
Germany	30	60
Greece	5	32
Ireland	2	10
Israel	3	19
Italy	12	78
Netherlands	10	47
Norway	2	18
Portugal	1	23
Spain	5	58
Sweden	8	18
Switzerland	4	25
Turkey	4	40
Yugoslavia (former)	1	7
USA	10	67
USSR (former)	2	8
Others	21	89
TOTAL	189 (=20% of)	932

be treated as representative of their colleagues. We therefore decided not to search for statistically significant differences in responses between geneticists from different countries because any such findings would simply be uninterpretable.

Because there were more responses from Britain and Germany than from other countries, however, we did examine the pattern of responses from those countries. A 'willingness-to-test-profile' was derived for each category of respondent by taking the number of respondents from each category who chose each score in the range 1–5 and dividing this by the number of respondents. There are 16 questions and subquestions, so there are 16 points that can be distributed across the spectrum of 'willingness to test' from 1–5. The profile records the way in which these scores are distributed. We found somewhat less willingness to test children in Britain than in the overall set of responses, and markedly less willingness in Germany – in particular, the German respondents showed much less enthusiasm for testing. This pattern is summarized in *Table 2*.

Table 2. Patterns of 'willingness to test' across all questions in the questionnaire (maximum total score of 16 to be distributed across the 5 possible answers)

		Mean score		
		Overall	Britain	Germany
Least willing to test	1	7.23	7.64	9.53
	2	3.16	3.64	2.97
Neutral	3	1.90	1.64	1.70
	4	2.57	1.80	1.73
Most willing to test	5	0.80	0.72	0.03

In contrast, there were respondents from Turkey, Japan and the former USSR, as well as some individuals from Britain, Denmark, Greece, Italy and Spain, who declared themselves to be very willing to carry out a range of tests – for Huntington's disease as well as CF-carrier status. From the pattern of responses given by these individuals, it did not seem that comprehension of the English language was a problem. Much more consensus to defer tests in the absence of a direct medical benefit was evident in Scandinavia, The Netherlands, Germany and Austria.

From additional written remarks provided by some respondents and other informants, it seems that a consensus had been worked out within Institutes and Departments but not at the national level. Within Europe, where a consensus had been reached, it was usually to defer testing.

Comments from Respondents

A range of comments provided by respondents is now provided to indicate the diversity of views they expressed:

'I see only the benefits of childhood genetic testing'.

'Each case needs to be discussed individually'.

'Children accept genetic information better before puberty'.

'It's best to test for lots of conditions – then there's no problem if everyone's a carrier for something'.

'Beware stigmatization from early testing – avoid testing before results would be useful'.

'It would be paradoxical not to test children when pre-natal testing could have generated the same information – and might do so for a brother or sister'.

'Testing is OK even for Huntington's disease because a child can arrange his worldview around their genetic condition'.

'Predictive genetic testing in childhood is indicated only if there are likely to be clinical symptoms or problems early in life and if they can be alleviated or prevented by early knowledge of the child's genetic status, e.g. familial hypercholesterolaemia. Demands from parents for the testing of their children are responded to by careful and thorough genetic counselling about the reasons not to test.'

'Most 12-year-olds would need counselling rather than testing.' (In relation to scenario 7.)

'It seems paternalistic if a geneticist refuses the parents' wish to test their other children when such a diagnosis as CF has been made in one of their children. It may help the parents to oversee the consequences of the condition in the family and to have the feeling that they dealt with it carefully and responsibly.'

'After learning that her son's developmental delay was due to fragile-X, a mother had immediately wanted to know the situation of her healthy 2-year-old daughter. The test was carried out by the son's paediatric neurologist. The daughter had a full mutation. The mother said that she had told "everybody" about her daughter's situation and was going to tell her future kindergarten as well as school about it. She said she felt it was important to alleviate misconceptions about carriership of fragile-X and that her daughter's case would serve as "a missionary". I couldn't help silently wondering whether the daughter would have chosen this "career" had she been given the opportunity to choose'.

'A woman was an obligate carrier of the fragile-X syndrome because her father was a "normal transmitting male". Her son was tested and was normal but her 4-year-old daughter was shown to carry a full mutation. As soon as the result was known, the paediatric neurologist had informed the kindergarten about the possible special needs of the child, and had arranged rehabilitation and follow-up "just to make sure". Now, instead of a healthy daughter, the woman and her partner suddenly had an abnormal daughter'.

'Quite a few parents (nearly) panic when they learn that the cause of their child's disability is a genetic condition. They wish something to be done about it and one thing to do is to find out who else in the family has the gene. Experience has taught me that the more panic or anxiety there is, the fewer irreversible things should be done; a gene test is rarely, if ever, a solution that will help the family. After proper consideration from all perspectives, a gene test might sometimes be appropriate to solve some specific problem, but it is not a magic tool that will make all things better'.

Knowledge about the carrier status of one's child gives enormous power over the child's future life. Some parents, however, though hesitant, are seduced to use it; some feel it is their duty to use it, to guide the child to make the 'right' choices. How much power and control would these same parents have wanted their own parents to have over their lives (had it been a possibility)?

Acknowledgements

We are grateful to all those members of the European Society of Human Genetics who responded to this questionnaire in 1995, especially to those who sent additional comments and letters. We are grateful to the Commission of the European Communities for funding the Euroscreen project. This work forms part of the Euroscreen project and has benefited from discussions with members of the subgroup on the genetic testing of children – Drs. Kris Dierikx, Rogeer Hoedemaekers, Darren Shickle, Traute Schroeder-Kurth and Kirsten Rasmussen and myself. We are also grateful to many respondents for their considered comments on issues raised in the questionnaire; we are especially grateful for the substantial contributions provided by Dr. Ilkka Kaitila, Dr. Maarit Peippo and Prof. M. F. Niermeijer.

Index